THE GRIP OF CULTURE

ISBN 978-1-8380747-4-6

Set in Minion and Trajan.
Cover images: Midjourney.
Printed by KDP.
Published by The Global Warming Policy Foundation.
www.thegwpf.com.

THE GRIP OF CULTURE

Andy A. West

THE SOCIAL PSYCHOLOGY OF
CLIMATE CHANGE CATASTROPHISM

GWPF

Contents

APPENDICES

LIST OF FIGURES

List of Tables

Preface

Back in 2006, I had no strong interest in global warming. I was aware of it, and more so as the years passed, but was not motivated to inform myself more deeply; I had no reason to question the narrative of climate change. However, as 2007 dawned I received a gift: a DVD of Al Gore's film *An Inconvenient Truth*. A few weeks later I finally got around to watching it, and even before the end my interest was soaring. Whatever the status of the science it attempted to communicate, the film also featured something I'd been trying to understand for many years: not climate change, but the power of cultural narrative.

I've had a keen amateur interest in evolutionary studies for over forty years. This started with biological evolution, but eventually I concentrated on the products of another evolutionary system, namely 'cultural entities' (or, to use more familiar terms, religions and ideologies).

The study of cultural entities does not have a field of its own. In practice not even the one called 'cultural evolution' proved fruitful.* As a result, I've pursued material from social psychology, evolutionary psychology, anthropology, memetics,† and even neuro-science, including both popular and academic works. Eventually, I consciously articulated a goal to myself: to develop a model of how cultural entities work, which would also explain their interaction with rationality, as expressed either individually, or at scale through institutions such as democracy, the law, and science.

Considering my interest, it was rather embarrassing that before seeing Gore's film I'd completely missed a strong cultural element

* Cultural evolution, a cross-disciplinary field, examines how cultures evolve, but my chief interest became how they operate: how they may come to dominate a society and how they interact with societal institutions and other cultures.

† The field had its heyday from the 1980s, but was overegged, and later suffered a backlash. However, although the terminology developed became stigmatised, research in the field continues but tends to use different terminology.

to the climate change discussion. How far did its influence extend? How long had it been growing? I was by then quite familiar with the power of cultures; given enough latitude they can overwhelm reality in the public consciousness, and their grip can subtly extend into organisations of all kinds before anyone notices their fundamental irrationality.

In an attempt to answer such questions, I started gathering information about the climate domain, but, rather than the underlying science, I focused on the social psychology – narratives and framings, attitudes, and emotively triggered behaviours. However, one needs to know the furniture of a domain to study it, so I also became familiar with all the main areas of scientific debate in both the public and scientific spaces, the dynamics of which are very different. While this took a long time, my degree in physics at least enabled me to progress the task without constantly being stumped by the concepts or language employed.

Avoiding bias in this kind of investigation is best done without an investment in any of the competing narratives. Consequently, this book assumes neither a climate-change sceptical or an 'orthodox' position. Beyond a certain threshold, cultural behaviours aren't dependent on what is objectively correct; hence there's a real sense in which 'what is right' doesn't matter very much when attempting to characterise a potential cultural entity.* And making as few assumptions as possible about what might be right, is helpful.

Five years later I called a halt, and started to set out my thoughts about culture in the climate domain. However, what I had was a patchwork of knowledge and circumstantial evidence and assumptive glue, in which minor cultural characteristics were over-emphasised, and fundamentals were under-explored and not properly connected. The material wasn't incorrect, but it certainly couldn't function as an explanatory platform, or even the main route to one. Nevertheless, everything I had pointed to the existence of a full-blown cul-

* However, it is useful to confirm that narratives determined to be cultural do in practice clash with anything from what we think is the mostly likely range covering where the truth sits, as future history would define such truth. I use mainstream climate science as the gold standard for such comparisons, represented by the IPCC process (the AR5/AR6 technical chapters). However, it turns out that the dominant public narratives about climate change contradict not only the mainstream position, but sceptical science too.

tural entity based on 'climate catastrophism'. I had to figure out how I could clearly demonstrate its existence.

Prior to 2007, a significant stumbling block to my pursuit of the nature of cultural entities was that I didn't know any of the actual historic examples inside out. As noted above, one needs to know the furniture of a domain to study it in detail. I am far from being an expert in world religions or, say, the secular cultures of Communism and Fascism. Although these all share some basic features, a detailed narrative analysis, or the assessment of subtle cultural attitudes across (often historic) populations, requires extensive knowledge of the target cultures. The task of cleanly isolating the generic 'rules' of cultural entities is therefore hard; it is possible I might never have learned enough to reach my goal.

But if climate catastrophism was indeed a new cultural kid on the block, this difficulty disappears. Most of its narratives and features and expansion would be recorded on the Internet, for anyone to find and, being 'young', it would not yet have accumulated too much obscuring historical baggage. In addition, public attitudes to climate change are polled constantly, and the results are usually freely available, allowing for statistical assessments of huge numbers of people and national comparisons right across the globe. In short, it should be straightforward to test my hypothesis that a cultural entity was at work in the climate domain.

Having rebuilt my approach appropriately, I delved much more deeply into public attitudes on climate change, and so came across the excellent work of social psychologist Dan Kahan. And not just his papers, but his *Cultural Cognition* blog where his theories were discussed and challenged and iteratively advanced. This was an invaluable experience, which really moved me forward.

Kahan's work stresses the critical nature of cultural identity in explaining attitudes to climate change and other socially conflicted topics. His findings align extremely well with my own hypothesis about cultural entities. Cutting a long story short, it was also during those years of engagement at *Cultural Cognition* that I began to realise that the situation in the US is unique. American politics is so polarised that there are essentially two extra cultural entities in the mix, namely the Rep/Con and Dem/Lib political 'tribes'.* This, incidentally, is why the US gets a chapter to itself in this book. Retrospectively, I also realised that while this political polarisation makes US cultural attitudes more obvious – and more readily measurable – it also tends to mask the other factors that are in play, and it is these that are most important everywhere else.

From 2014, I began to put aspects of my work into the climate blogosphere, in order to see how well they might fly and to invite challenges. Regarding the latter, I was not disappointed! It was only after some years of developing my ideas in response to this that I began to turn my mind to measurement. How could I demonstrate with hard social data that a cultural entity was dominating the climate domain?

In truth, at the beginning of this trail I had not dreamed that measurements of the nature included in this book would be possible within my resources. I am no statistician, and I don't have an academic department to implement studies and surveys for me. However, it was at Kahan's *Cultural Cognition* blog that I started to see that the measurement problem might be far easier than I had originally anticipated. Harvesting data on cultural cognition amounts to asking people in different ways what they think. As noted above, there are reams of public surveys on climate change attitudes that do this from practically every conceivable angle across most nations in the world (albeit with greater coverage for Western nations). If these surveys couldn't indicate the presence of a cultural entity, it would be a major blow to my hypothesis.

I started the measurement trail at the end of 2019 and had some initial results in a bloggable form by April 2020. The clarity of the results and the simple explanatory model that emerged from this work truly surprised me. I've seen a lot of social psychology studies over the years, and clarity is not the typical term that springs to mind even where these studies are pretty useful. Unless anyone can think of a better explanation, the measurements (now greatly expanded for this book) do indeed robustly confirm a culture of climate cata-

* 'Rep/Con' and 'Dem/Lib' refer, respectively to the Republican/Conservative and Democrat/Liberal political tribes.

strophism across global publics.

It had long been at the back of my mind that, one day, I would have to pull all the functional descriptions and measurements into a single work. However, when an offer came from the GWPF to do exactly this, it was only after starting the task that it became clear my material needed a huge amount of additional work in order to make it readable, understandable and integrated. So here I take the opportunity to express immense gratitude to my editor Andrew Montford, for his insights and guidance, for a great deal of heavy-lifting he had to do on my behalf, and for the patience of a saint when dealing with my idiosyncrasies.

This book does not yet entirely achieve my goal of setting out a generic model of cultural entities. But I believe it is a big step forward, and a bigger step still in describing the specific characteristics of the culture of climate catastrophism.

Andy A. West

Chapter 1

Introduction

1.1 Scope

This book is about the social psychology associated with climate change, which can be characterised and measured across national publics without reference to the physical climate system, its future state, or how it responds to human emissions of greenhouse gases. This is the case because the social psychology has emergent characteristics of its own, which are unaffected by mainstream views on the science of the ocean-atmosphere system (as represented by the working-group technical papers of the UN Intergovernmental Panel on Climate Change – the IPCC), or indeed the arguments of the small minority of sceptical scientists. This is unsurprising when one notes that publics have, in general, no relevant scientific knowledge.

However, the public's lack of knowledge about climate science does not mean that their attitudes on the issue are capricious or bland or unfathomable. In fact, they are highly systemic and frequently strongly expressed. Most importantly, they are predictable across nations. This means in turn that the attitudes of national publics must have a systemic motivation. As we shall see, this turns out to be cultural. And it is not only cultural in the sense of, say, tribal political or religious influence determining attitudes, but in the sense that the climate-change domain supports an independent culture *of its own*.

This book does not address the conduct of climate science, or the differences of opinion among its practitioners. The attitudes of national publics to climate change are largely unaffected by the scientific arguments. However, differences of opinion are mentioned in passing, and the pronouncements of some individual scientists who propagate the (cultural) narrative of climate catastrophe are discussed. But there is one factor related to climate science that is emphasised throughout, namely that claims of certain near-term global climate catastrophe in the absence of dramatic action are *not* supported by mainstream science (see Section 5.2.2).

Nevertheless, a narrative of climate catastrophe still dominates public discourse, a fact that indicates a potent culture is operating throughout society. This book characterises that culture – the culture of climate catastrophism, or simply 'climate catastrophism' – details some of the mechanisms by which it works, and explains its roots in human evolutionary history. At the book's centre is a set of measurements that demonstrate its existence, and which in turn can accurately predict, across nations, real-world phenomena such as the level of deployment of renewable energy or the prevalence of climate activism.

The culture described here is not some unique phenomenon spawned by the challenging issue of climate change. In fact, strong cultures of this kind are universal, and reflect group behaviours ingrained in humanity by our evolutionary trajectory. Unsurprisingly, therefore, they all share the same fundamental characteristics, honed via evolution in a multitude of older (religious) cultures.[1]

Although cultural belief is normal and ubiquitous, cultures can have significant downsides. A new culture brings great risk to society, because it can potentially undermine democracy, the law, and even science itself.

Though much in this book may seem unintuitive or even surprising, the changes that climate catastrophism is engendering in society – from activist antics to wind-farm subsidies, from the posturing of presidents and prime ministers to the sacrifices of millions of individuals who just want to help 'save the planet' – are not the result of a hoax or conspiracy or delusion or greed or mental flaws or mendacity or nefarious intent. But they do not arise from rational processes either. Instead, they are ultimately a function of emotive reactions deep within our subconscious. Such reactions are typically felt both passionately and *honestly*. They are also completely and utterly normal.

1.2 Book layout

Much of the material in this book first appeared as guest-posts at *Climate Etc*,[2] the excellent blog of Professor Judith Curry. *Climate Etc* covers physical climate change, climate policy, the operation of science – especially under conditions of deep uncertainty – as well as some of the related social psychology. Professor Curry actively fosters input from a wide range of perspectives, both within and beyond the climate domain. I've hugely appreciated the opportunity to be a guest voice at *Climate Etc* and I thank Judith not only for that, but also her encouragement, without which this book would likely never have been written.

Chapter 2 introduces the idea of a cultural entity, and considers, by means of a thought experiment, everything that one might expect to result within society when such an entity is in play. In Chapter 3, I step back to fundamentals. Where do cultures of this kind come from? Why are they so successful? And what generically causes their observable features? It's impossible to explore in depth all the many features that indeed we do expect, so the following chapters cover those that are most important in understanding cultural entities, or fulfil a highly visible role in the climate domain.

Chapter 4 explores how cultural entities affect children, most importantly through co-opting them as prophets and proselytisers. It compares the role of child prophet in the climate domain with the same role in a purely religious culture. This chapter also examines the widely acknowledged psychological pressures that climate catastrophism places upon children.

The main vehicle of a culture, its 'DNA' so to speak, is a diverse population of emotive memes linked to a common existential theme. In the climate domain, this role is fulfilled by the 'Catastrophe Narrative', which is explored comprehensively in Chapter 5. These narratives trigger specific behaviours in adherents.

A common feature of all strong cultures is the demonisation of critical voices and out-groupers. In the modern era, and in particular in the climate domain, demonisation has been delivered, with considerable success, using the idea of 'denialism'. This is covered in Chapter 6.

When looking at the reactions of publics to cultural entities, it is important to understand that active disbelief – emotive or instinctive rejection – is just as important as belief – emotive cultural commitment. Moreover, belief and disbelief are not merely mirror images of each other; they are, to a large extent, independent. Chapter 7 provides a comprehensive picture of the origins and effects of instinctive cultural disbelief, or 'innate scepticism', in particular with respect to conflicted science topics.

Chapters 8–10 cover a range of hard-data measurements of climate catastrophist culture across more than 60 countries, with explanations of why cultural features set out in earlier chapters produce the observed patterns. I also show that the same patterns occur in a different domain – one that is inarguably cultural. However, America's very highly polarised public leads to more complex patterns, which are covered separately, in Chapter 11.

Armed with these measurements, we can make real-world predictions! For instance, the deployment of renewable energy and the prevalence of climate activism across nations are both predictable from cultural attitudes, and do not result from the climate or climate exposure of countries, or any technical or even rational policy. These aspects are covered in Chapters 12 and 13.

Chapter 14 briefly returns to the list of cultural characteristics first shown in Chapter 2, and re-examines them in light of the knowledge from all the other chapters, before looking in more detail at cultural impacts on morality and the law. Chapter 15 compares some features of historical cultural entities, particularly millennarian ones, with climate catastrophism, before moving on to some concluding thoughts. One of these I will mention here: given the importance of climate catastrophist culture in determining public attitudes and real-world impacts, such as climate policy, it is critical to understand why it pervades our society and how it works. As we shall see, the last thing anyone – from the highly concerned to the sceptical – should want is a culture dominating the issue of climate change.

1.3 Companion files

A series of online companion files provide supporting information to this book. Each is introduced in the first chapter or section where it is relevant. The companion files can be downloaded from the book's website at www.thegwpf.org/culture.

The files are as follows:

- The CN-Archive is an archive of Catastrophe Narrative quotes referenced in Chapter 5.
- The Excel-Ref provides all the charts and source links related to climate-change attitudes, policy and activism.
- The Excel-Ref1 provides charts and source links for the 'religion only' chart in Chapter 9.
- The online appendix provides auxiliary information that isn't critical to the print version of the book.

Chapter 2

A POTENT NEW CULTURAL ENTITY IN OUR SOCIETY

2.1 A cultural entity

There are many definitions of 'culture' or 'a culture' used in academia, and still more in common usage. My own definition does not mean 'a night at the opera' or 'a traditional activity', say, but a 'cultural entity': a social movement or group bound by strong beliefs. Religions are common examples. I could have used the term 'ideology' rather than 'culture', but this tends to be used primarily for political movements (so only secular, not religious ones).

The terms stretch across boundaries too, so, for instance, one can refer to a 'religious ideology', or a 'secular religion'. As we shall see, the latter expression is appropriate for the culture within the climate-change domain, and indeed it is intuitively understood in general debate. However, because its use can lead to over-literal comparisons with the surface features of religion, I only use it sparingly. Instead, I mainly use 'a cultural entity', a term that encompasses both religions and ideologies, reflecting the fact that they have the same underlying mechanisms. As shorthand, 'a culture' is easier.

2.2 Climate culture intuitively perceived

I've long since lost count of the parallels drawn between the climate-change movement and religion or religious-like practices.[3] Those who make such claims are mostly sceptics, and they base their claims on gut feelings, rather than social analysis or data. They often draw the worst possible conclusions, with claims of hoax or conspiracy. Nevertheless, they correctly discern that a cultural phenomenon of some kind has emerged in the climate-change domain.

Similar parallels are, however, drawn by the 'orthodox' side too. Some describe climate change as a 'transformative culture', or equivalent; they too discern a cultural phenomenon. However, typically stressing this as a positive feature, the orthodox side fails to appreciate that, as we shall see, cultures of this kind do not so much communicate a 'truth', as manufacture it. A few perceive that a strong cultural angle does represent a problem of some kind. For instance, climate scientist Hans von Storch (who is certainly not a sceptic in this area) and cultural scientist Werner Krauss observed in their 2013 book *The Climate Trap* (emphasis mine):[4]

> The climate scientist [von Storch] had the suspicion that climate science was dragging around a 'cultural rucksack' that was influencing the interpretation of the data. The cultural scientist [Krauss], with regards to the appearances by some climate scientists in the media and the roles they were readily assigned, was reminded of *weather-wizards and shamans of foreign cultures*...Without really being aware of it, *climate scientists had taken over the role of prophets*: they predicted the imminent end-of-the-world if society did not fundamentally change soon, reduced its emissions, and behaved more sustainably with the environment. The problem was not only the message, but also that they were often completely way in over their heads [in their] role as mediator between nature and society.

In the blurb for his book *Apocalypse Never*,[5] which among other environmental solutions advocates for nuclear power to reduce emissions, environmentalist Michael Shellenberger says (emphasis mine):

> What's really behind the rise of apocalyptic environmentalism? There are powerful financial interests. There are desires for status and power. But most of all there is a desire among supposedly secular people for transcendence. This spiritual impulse can be natural and healthy. But in preaching fear without love, and guilt without redemption, *the new religion* is failing to satisfy our deepest psychological and existential needs.

As noted above, these assumptions of the existence of a strong culture – of a secular religion – in the climate domain were based on surface observation and gut feel rather than data. However, Chapters 8–13 provide measurements from hard social data showing that they are correct.

2.3 A thought experiment

Despite the significant and persistent commentary about 'religious' behaviour, there doesn't appear to be formal recognition that the dominant narrative – namely that global climate catastrophe is certain within decades if we fail to take drastic action – is indeed sponsored by a culture, and *not* by mainstream science; the IPCC's Assessment Reports do not support such a claim at all. On the contrary, those who study cultural behaviour in our society – sociologists, social-psychologists and others – appear *en bloc* to have bought the claim that this Catastrophe Narrative is simply an output of the 'hard' physical sciences. Why would they even attempt to apply their knowledge in this domain? Unless of course to try and explain 'deniers' (see Chapter 6).

What might sociologists and social psychologists think if they were not blinded by 'the' science? It turns out that by simply knowing a cultural entity was in play, they could conclude a great deal. Consider the following thought experiment. Professor Crusoe, an expert on bio-cultural evolution, cultural evolution and social psychology, is returning from a field trip in the Pacific. He is shipwrecked and ends up stranded on an isolated island for 35 years, until finally he is rescued by a passing ship. On the voyage back to civilisation, the sailors tell him that the whole world is now hugely worried about man-made climate change and has been spending trillions on the issue. Professor Crusoe of course knows nothing of climate change, but before any other detail gets discussed, one nerdy sailor happens to add that he'd read a new exposition in *Cultural Cognition*[6] magazine, which said that the climate-change narrative and the movement it inspires have been shown to be products of an 'emergent culture'. For someone with Professor Crusoe's expertise, this knowledge is enough for him to draw an extraordinary number of conclusions about the effect of the climate-change phenomenon on society. These can be seen in Box 1.

Box 1. Professor Crusoe's list

1. In large sectors of the public, there will be a common understanding of an existential threat and how society should respond to it (a cultural consensus).
2. The consensus will promote statements that are presented as all-explaining and/or indisputable.
3. The consensus will be actively policed, via status control – those criticising the consensus will be sidelined or downgraded – and emotive pressure – using fear and guilt to suppress dissent.
4. There will be uncritical acceptance, and possibly even adoration, of authority figures – and perhaps of 'prophets' too – who promote the cultural narrative.
5. Key information that makes the cultural entity and its narrative vulnerable to attack will be restricted to elite guardians.
6. Cultural adherents will place groups in society into an approval hierarchy, ranging from perpetrators of cultural misdeeds at one extreme, to their victims at the other. For a global culture, the groups could be large, incorporating whole ethnicities or nations or major demographic divisions.
7. Alongside the narrative of existential risk, there will also be the prospect of salvation, and rebirth and renewal, to be achieved through cultural conformance. This vision of hope will be propagated by the prophets in (4).
8. The visions of existential threat and salvation set out in the narrative will create unrealistic anxieties, fears, guilt, hopes and inspiration across society. This will cause immense bias – and a corresponding loss of objectivity – towards the culture in all areas of endeavour connected with it.
9. Double standards will prevail. Transgressions – legal, moral or ethical – of adherents will be overlooked, while those of critics will be harshly punished. This will create a feeling of threat, a shifting moral landscape and possibly large changes to the law (if the culture has been active long enough).
10. The narrative will be used to justify all sorts of social and infrastructural changes that benefit the culture, independent of their

true usefulness to mankind, and even if some are damaging.

11. There will be organisations that are particularly convinced of the narrative and, as a result, are particularly emotive in their cultural advocacy. They will police the internal ranks of the culture, and act as missionaries and recruiting sergeants (think Jesuits).
12. Many cultural adherents, especially those in the above advocacy groups, will self-identify with the culture, and will therefore be especially instinctive and emotional in their support, at the expense of reason.
13. False claims of conspiracy will be made against the culture (cultures emerge from subconscious processes).
14. Dissenters from cultural orthodoxy will be demonised, and possibly persecuted if the culture has penetrated society far enough.
15. Nevertheless, unless the culture completely dominates elites, expert opinion in the cultural domain will be highly polarised.
16. A large proportion of the public, possibly a majority, will remain unconvinced of the cultural narrative.
17. The culture will attempt to form coalitions with other cultures, both religious and secular.
18. Whole national leaderships and rafts of lesser authorities will have bought into the culture.
19. Huge resources will pour into activities that benefit and promote the culture, yet which do not help deliver the salvation it promises; they may even hinder it.
20. It will be claimed that there is no doubt about the existential threat. Nevertheless, the cultural narrative will slowly evolve.
21. There will be icons – visual reminders of the narrative. The evolution in (20) means that some will be de-emphasised or set aside for new ones, having lost their usefulness for some reason.
22. There will probably be positive elements to the culture (cultures are by no means all bad).
23. The societal effects are being caused by the cultural narrative, not any real-world phenomena.
24. Etcetera.

Not all experts in the field would be happy with his list; there is, for instance, still strong (even bitter) resistance in some quarters of social science to reductionism – the idea that complex phenomena can be broken down into simpler processes – which is inherent in Professor Crusoe's approach.[7] Many would also differ over its appropriate application. Some would be happy characterising dead and possibly primitive cultures in this way, but would be apprehensive about doing so for living and more sophisticated ones.

Others – for example, psychologist and meme proponent Susan Blackmore[8] or public academic Richard Dawkins[9] – would be quite comfortable characterising living religions this way. That said, they'd likely both be horrified by the possibility that this characterisation was applicable to the social aspects of climate change. However, such concerns mostly reflect the biases of the individuals more than the validity of the characteristics; most social scientists today appear to mistakenly believe that the narrative of certain global catastrophe is not cultural, but an output of the 'hard' sciences.

Professor Crusoe is, because of his long absence, free from the bias about the climate domain, seeing it with fresh eyes and an open mind. With his expertise, his insights would also go deeper than a list of outward characteristics; he would know that fundamental drives from our deepest evolutionary history are at play in the background, and moreover that these would be objectively and directly detectable.

Despite being able to scribble his list down in a few minutes, he would of course know nothing about precise details; the new culture of climate catastrophism arose while he was marooned. So, for example, he wouldn't know the particulars about consensus policing in the climate domain (Point 3), for instance the so-called '28Gate' affair,[10] in which a meeting mostly attended by advocates agreed to de-emphasise sceptic views in the BBC's output. Nor about Climategate,[11] the 2009 email leak from the University of East Anglia, which revealed attempts by climate scientists to restrict data to their privileged circle (Point 5). Similarly, he would not know that the authority figures dominating the climate domain (Point 4) include the UN leadership, with the list of revered prophets including US politician Al Gore, NASA climate scientist James Hansen and, more recently, the child environmental activist Greta Thunberg.

He would know nothing of the cultural pressure for a crash 'Net Zero' program, which fulfils the salvation narrative in Point 7, nothing about WWF and Greenpeace (or more latterly Extinction Rebellion), who fulfil the Jesuit role set out in Point 11, and nothing about the sceptics who claim 'hoax' and 'conspiracy', just as predicted in Point 13. He has never heard of 'denialists' or 'deniers', an emotive stigma label attached to legitimate questioners (Point 14), or the Democrat/climate-culture coalition in the US (Point 17), the Paris climate-change agreement (Point 18), or the billions poured into renewable energy (Point 19). Likewise, he is oblivious that the once prominent icon, the 'Hockey Stick',[12] was de-emphasised because it wasn't sufficiently decoupled from reality and so became disadvantageous to the culture (Point 21).[13] He just knows that the kinds of things on his list happen in strong cultures.

We will return to Professor Crusoe's list later. As we will see, the circumstantial evidence, the theory, and the measurements confirm all of his conjectures. Despite his long absence and a complete lack of knowledge about the climate-change issue, he can correctly deduce a whole host of features and behaviours from a single fact about the climate domain, namely that it is dominated by a cultural entity.

2.4 Use what we know

For about 150 years, we've been learning how cultures work and evolve. Great progress has been made on a wide range of topics, from the linked evolution of culture alongside our genes, to insights on social-thinking[14] – how we think in groups – and much more. Nevertheless, a great deal of mystery remains. For instance, what are the fundamental differences between cognitive processes that drive group and individual behaviours? This question is still being explored from different directions by researchers in fields including anthropology, biocultural evolution, psychology, and neuroscience.

Public behaviours regarding climate change have obvious cultural characteristics. Knowing this, we can apply the 150 years of accumulated knowledge about how cultures work to better understand the climate domain. The kind of questions we want to answer include:

- How did the cultural narrative of climate-catastrophe become so dominant?
- Why are those who question the cultural narrative demonised rather than treated rationally? By what means does this happen?
- Across the world, who are the people actually supporting the narrative, and why? Likewise, who resists it, and why?
- How could the culture of climate catastrophism have such an impact in a world where, in theory, the law, democracy and science are supposed to run society?
- How can the culture have directed such enormous spending to projects that will not only fail to solve the purported problem, but may well cause harm along the way?

To answer these questions fully, we must begin by stepping back to the fundamentals.

Chapter 3

CULTURAL ENTITIES: DEEP ROOTS AND KEY FEATURES

3.1 The 'purpose' of cultural entities

Cultural entities arose in our deep evolutionary past, through a process known as 'cultural group selection'.[15] Their 'purpose' is to enable the behaviours that lead to group cohesion and altruistic cooperation,[16] which is hugely beneficial to groups of humans seeking to survive in an environment that in part consists of other groups exhibiting the same behaviour.[17] The most altruistically cooperative groups adapt better to their environment, hence they are selected,[18,19] as are genes[20] that support both more and stronger cultural behaviours (the process is termed 'gene-culture co-evolution'[21]). As this has been happening for countless generations, these particular genes now hugely influence our brain architecture. In other words, the cultural behaviours that support altruistic cooperation in humans are deep and instinctive, built into the very way that our minds work.

We can split 'altruistic cooperation' into its two parts: altruism and cooperation. Although the terms overlap, and are sometimes used synonymously, strictly speaking altruism is more about 'selflessly helping others', while cooperation means 'working together for a shared goal'. The latter implies shared values and outlooks that allow cooperative projects to proceed; it's no good having selfless motivation but no idea how to work together or what to work on. In this sense, gene-culture co-evolution results in both cooperation and altruism.

Altruism does not always mean individuals making serious sacrifices for their cultural group, but it can certainly run to this.

Throughout history, millions have suffered or died defending their religious beliefs. In recent times, millions volunteered to fight in both World Wars, defending their cultural (and in this case geographically located) group against enemies doing exactly the same. Even today, the existence of suicide bombers shows there are still people willing to make the ultimate sacrifice for the perceived benefit of their faith.

3.2 The features of cultural entities

Altruism and cooperation rest in practice on an array of subsidiary behaviours and features (Box 2). We can propose these as the elements of a cultural engine that shapes the detailed social characteristics in Professor Crusoe's list in Chapter 2. Some of the elements in Box 2 are well accepted and some less so.

We should be able to detect the effects of the cultural engine in mass public attitudes to climate change, patterns of activism and even in public policy outcomes. In later chapters, this expectation is confirmed through measurements of these factors across international publics, supported by complementary observations (for instance cultural narrative from public authorities, and cultural behaviours in, along with impacts upon, children).

The emergent nature of these features makes them challenging to describe in isolation; they all lean on each other. However, characterising cultural entities in this manner is still a useful way to proceed, even though it means that the explanation of each separate feature may have to reference several of the others.

In the next few sections, the list in Box 2 is expanded into detailed descriptions. This level of detail is both enlightening and very useful, but the rest of the book can in principle be grasped without it, so long as the features are taken at face value. As a result, readers can, if they choose, skip past these descriptions to Section 3.3, which expands on the idea that cultural groups also exhibit behaviours that are more than just the sum of the features listed in Box 2, and which may therefore seem counterintuitive.

3.2.1 Cultural coherence: reliable identification and signalling

For cultural groups to function, it is critical that group members are able to reliably identify each other. This allows them to access group benefits (including the altruistic acts of others), and also means that cheats or interlopers from competing groups can be identified and dealt with.

Reliability of identification is achieved by the repeated issuing of cultural signals, through a variety of methods. For instance, tattoos are a recurrent feature of primitive cultures (and indeed many subcultures in the modern world), because until recently they couldn't be erased, ensuring this badge of membership was effectively permanent. Other indicators come in the shape of clothing or wearable icons (many Christians still wear a cross). While these can be imitated by infiltrators, they would normally be deployed alongside culturally distinctive behaviours, such as dietary choice, prayer rituals, or frequent repetition of approved parts of the cultural narrative (which may be complex if referring to a text such as the Bible or the Koran). Interlopers would soon be spotted by the orthodox, because it would be hard for them to imitate all of these signals at once and get their complex interrelationships right.*

3.2.1.1 Cultural signals aren't always what they seem

While some of the cultural signals noted above are obviously badges of membership, others, especially those conveyed by narrative, may not on the surface appear to be about membership at all. For instance, messages delivered in support of, say, 'the fight against capitalism', or 'the fight against certain global climate catastrophe', don't admit to membership of the Communist Party or of the climate catastrophist culture (the latter is rarely recognised as a culture anyhow). However, where the messaging stems from emotive commitments to the corresponding cultures, they are indeed membership proclamations *and nothing else*. They are *not* meant to be taken at face value.

So, for example, when people who know nothing of energy technologies call for a crash replacement of fossil fuels with renewable energy, they are really just saying 'I'm in the club of climate culture', thus garnering them a welcome from other club members. Some renewable energy actually ends up being implemented as a consequence, but a culture doesn't require that its adherents live as though

* See also Endnote 24 on signalling via ritual.

Box 2. Features of cultural entities

1. *Cultural signalling*: Crucially, altruistic acts aren't forthcoming for out-groupers. Hence it is hugely important that in-group members can be reliably identified. The signalling of cultural identity is covered in Section 3.2.1. As Section 3.2.1.1 clarifies, this signalling doesn't always *seem* to be about identity.

2. *Cultural commitment*: If people could flit in and out of groups at will, freeloaders could avoid altruistic costs by leaving the group, or maybe obtain benefits by faking the signals of group membership in the first place. Hence, as covered by Section 3.2.2, group membership is costly and a deep emotive commitment, one that bypasses rationality and is subject to constant reinforcement. Also, cultures not only produce emotive commitment, but, for some people, emotive rejection instead. See Section 3.2.2.1.

3. *Cultural narrative*: Genes do not carry cultural information, only biological information that builds individuals who can exhibit cultural behaviours. Cultural information is passed on via symbols and rituals, and crucially, as Section 3.2.3 explains, by a cultural narrative (written or spoken), which we can think of as the 'DNA' of a cultural entity.

- Cultural narratives push emotive hot buttons in us. See Section 3.2.3.1.
- Cultural narratives have to be false in order to work; Section 3.2.3.2 explains why.
- Cultural narratives consist of a spread or 'population' of narrative variants, which, as explained in Section 3.2.3.4, is crucial to the operation of a cultural entity.

4. *Cultural identity*: The identity signalled in 1 above is established at the same time the emotive commitment in 2 is forged in individuals, by instilling the cultural information noted in 3. For modern cultures, most of the information comes from cultural narrative.

- Children are especially primed for absorbing a cultural identity, making them ideal hosts for the continuance of a parental/local culture, yet also highly attractive targets for an invading culture. See Section 3.2.4.1.

5. *Cultural policing*: Section 3.2.5 introduces cultural policing, behaviour exhibited to some extent by all group adherents that maintains group cohesion against freeloaders, internal group schisms (cultural heresies) and also competitive cultural groups. Particular cultural organisations or individuals may specialise in policing activity.

- Given that the cultural narrative is the main vehicle of a cultural entity, much policing activity comes down to narrative policing, detailed in Section 3.2.5.2.
- Extreme policing may include out-group demonisation and the pressure of fear. As noted by Section 3.2.5.3, those identified as out-groupers will be shunned, in some cases demonised.

6. *Cultural consensus*: Strong group cohesion, along with a policed cultural narrative, essentially makes a powerful group consensus on the issues promoted by the cultural narrative. This cultural consensus, explored by Section 3.2.6, is the source of the shared values and outlooks for the cultural group.

- As Section 3.2.6.1 makes clear, a cultural consensus is very different to a rational one!

the demands are real. Indeed, on the short timescales proposed, the energy-system transformation envisaged is a fantasy.

The irrationality of signalling fantasy goals stems ultimately from the fact that cultural commitments are emotive. As discussed in the next section, they bypass our rationality. And as we will see in Chapter 12, the pattern of renewables deployment across nations can be shown to be cultural, and therefore a case in point.

3.2.2 Cultural commitment: emotive conviction

Cultural groups evolved for an advantage that comes at the communal level, but adherents also typically derive individual benefits from their membership, namely security and a share of the group's success. However, as noted above, they may also be required to sacrifice their personal interests for the sake of the group; some may even have to make the ultimate sacrifice. For this system to work, commitment to the group must be both strong and lasting. Ideally, loyalty will hold in practically every circumstance, and must also be automatic, so that the group can respond to rapidly changing events. To achieve all this, the commitment of group members is founded in deep emotive mechanisms that bypass rationality.

Forging such a commitment is accomplished by various means that act in concert. These may include the constant repetition of cultural behaviours that over time become ingrained habits – for instance, praying, fingering a talisman, making a sign or speaking cultural slogans – or rituals and ceremonies,[22,23] perhaps enhanced by rhythms or music or chanting or trance, or, in more primitive societies, with elements such as a sacrifice or a drug/drink-induced state. Other approaches include the use of initiation rites, witnessed by the community elite, or the granting of privileges, such as the approval of shortcuts to authority (think of the Masons) or control of the right to breed (the religious sanction of marriage). And throughout their lives, adherents are subject to a flood of emotive cultural narrative,* which, among other tricks, invokes deeper instincts, such as the need to care for children or the fear of the unknown. Bear in mind that these activities and the emotive reactions they prompt are the result of a very long evolutionary process, so are very closely intertwined.

* Cultural narrative is explored at Section 3.2.3.

Once achieved, the deep emotive commitment to the culture acts to some extent as a self-latching mechanism: resulting cultural biases make rational arguments or competitive cultural narratives much less attractive. Nevertheless, they remain an issue for cultural entities; adherents could potentially be persuaded back out of their cultural conviction. However, cultures have evolved means to mitigate this possibility and maintain commitment – see Section 3.2.4.1 on the role of children and Section 3.2.5.1 on cultural policing.

3.2.2.1 Emotive rejection and cultural balance

Cultural disbelief of an alien culture is usually a function of cultural belief in a local or native culture. That is to say, believers in a culture will instinctively reject a set of competing values promoted by outgroupers. However, a culture can also be instinctively rejected by its own adherents, if it asks too much of them, becomes too decadent, or betrays its original cultural values. In other words, as well as keeping out the competition, cultural disbelief can also act as a balance, preventing adherents from becoming slaves to their own culture, a development that would be unlikely to benefit the group.

In most discussions of cultural features that I've seen, cultural disbelief seems to be underplayed, but it is just as important as cultural belief. A way to think of cultural disbelief is as an instinctive or 'innate' scepticism. In this book, I normally use the latter term, because it reflects the manner in which disbelief is typically expressed, as sceptical statements.

Chapter 7 explores innate scepticism in detail. As we shall see, it is not merely the opposite – a mirror image, so to speak – of cultural belief.

3.2.3 Cultural narrative; the 'DNA' of a cultural entity

Cultural narratives aren't consciously scripted as part of some deliberate plan that foresees eventual cultural dominance. They arise through a process that depends on subconscious behaviours. They are in essence emotive stories, collections of compatible memes that prosper by selection; those that better engage our emotions will more successfully propagate. The emotions they engage are in turn primed, via long gene-culture co-evolution, to be triggered by the kind of stories found in cultural narratives. In short, cultural narratives are emergent.

A useful way to think about cultural narratives is that they are the 'DNA' of a cultural entity.* While actual DNA encodes our propensity to cultural behaviours, cultural narratives encode the particular culture (religious or secular) within which we express those behaviours. And, as with germ cells in biology, which pass gene information to the next generation, the core of a cultural narrative is protected, policed (see Section 3.2.5.2), in order to stop the essence of the culture from becoming too disrupted as it propagates. This role makes cultural narratives the most fundamental property of a culture: its vehicle or carrier.

The main religious faiths provide familiar examples of cultural narratives. Their core replicative material is protected from overly fast change by being written down in hallowed works, such as the Bible and the Koran. However, there are also clouds of related narratives around these works: hymns and popular songs, sermons and pious sayings, and religious books and pamphlets. This material is disseminated by an array of sub-faiths, ranging (in the case of Christianity) from gnostic to 'born again' to ultra-traditional, through to a variety of 'tamed' groups that accommodate many modern realities, but which all still cite the Bible as their authority and guide.

3.2.3.1 Cultural narratives and emotive hot buttons

Cultural behaviour has evolved over a very long time, with genetic and cultural elements intertwined. As a result, the emotive narrative assemblage that 'carries' the culture, and the subconscious responses to it, work hand-in-glove. The narrative touches emotive 'hot buttons' in those who are regularly exposed to it: anxiety, fear, guilt, inspiration, hope and so on.[24] Successful narratives push these buttons in combination, and also apply both stick and carrot – threats of doom tightly coupled with hope of salvation. Successfully pushed buttons will result in cultural belief and further propagation of the narrative.

* This is a helpful metaphor, not a literal comparison. DNA works in a very different way to a cultural narrative, and is subject to very different constraints. Indeed, for the purposes of this book we don't even care how cultural heritability works. We just need to recognise that the advent of culture in humans has led to group behaviours that are intertwined with cultural narratives.

The narratives are somewhat arbitrary; the surface details don't matter so long as the buttons are pressed. Only the underlying emotive content matters; the narrative doesn't even have to make sense.

3.2.3.2 Fairy-tales: cultural narratives are necessarily false

However, cultural narratives are actually worse than arbitrary, because they are necessarily divorced from sense.[25,26] This is because any cultural narrative based upon reality or logic is far too easily challenged by out-groupers or those within the culture who may tend towards the sceptical. Any narrative that is even 'too close' to reality essentially leaves open a route through which it might eventually be fact-checked. If the narrative were to fail such a check, its authority would be undermined. And even if it passed, the high emotions that sustain the culture would soon fade; truths tend to become mundane.

Even if the relevant facts aren't yet determinable, a cultural narrative that lends itself to logical analysis will still fail. Different experiences within any large group will lead to different *reasoned* opinions on the narrative, as adherents argue the logic. But the existence of differing opinions subverts the only purpose of a cultural narrative, which is to promote cohesion of the group. Also, the very act of reasoning tends to negate the emotive engagement that cultures rely upon to maintain deep commitment among adherents.

None of these problems occur if the core theme of the narrative is a blatant falsehood; in essence, a fairy tale. A fairy tale works precisely because it is so divorced from logic or reality, which means it is distanced from reasoned challenge too; it convinces purely *emotively*. Reason is cornered into saying only 'this is nonsense', an argument that is easy to defeat via emotion (especially in children, who have less knowledge and also less experience of rational reasoning). The fairy-tale nature of cultural narratives is also consistent with the emotive selection that leads to their emergence. From the perspective of a cultural entity, the bigger the lie the better.

So cultural narratives are *necessarily false*. However, there is a cost, in that falsehoods will lead to emotive rejection by some listeners; not because it's a lie *per se*, but because the narrative is subconsciously detected to be cultural, as noted in Section 3.2.2.1.

3.2.3.3 *Cultural narratives claim authority*

For credibility, cultural narratives must claim to be backed by a significant authority. Later religious narratives, for instance, claim the backing of an omnipotent (yet conveniently invisible) deity. Some cultural narratives have claimed backing by science, or cult political figures (often justified further by a re-write of history), or 'inalienable' features such as 'our blood' or 'this land of our forefathers', or 'the divine right of kings'. For maximum impact, they will also attempt to establish an existential importance and claim the moral high ground.

3.2.3.4 *Narrative variant populations*

Cultural narratives are structured as an overarching theme or 'core narrative', to which many sub-narrative variants are linked.*[27] This structure is important to how they operate. Cultures and their themes compete with one another, but there is also some internal competition between the sub-narratives of a single culture. Policing activities (Section 3.2.5.2) limit it – the culture's objective is group cohesion – but if these efforts fail, a heresy is born.

The population of narrative variants covers a large range of emotive strengths and story angles. It is essentially a set of co-evolving memes that targets all of our emotive hot-buttons. The more emotive variants are preferentially propagated and so rise to prominence, which is how the main theme first attracts attention. However, once a cultural group is established, it isn't as simple as 'the most emotive takes all'. Emotive variants are polarising: although they are attractive to some people, they also invoke cultural disbelief – innate scepticism – in others. Too much disbelief will cause a backlash that harms the group. In contrast, less emotive variants invoke less disbelief, so less backlash, and will likely help to recruit a larger number of adherents, although less ardent ones.

Some less emotive variants even incorporate a limited amount of realism, which is to say that they are not pure fantasy. Such variants buffer the main fairy-tale theme from reality, while retaining the link to the core theme. In this way, they help to disguise its true nature

* Narrative variants form a population of memes. Social psychology has recognised the existence of such populations, but has not yet fitted them into the broader framework of research in the field. Interested readers can find further details in Note 27.

and its blatant falsehoods. These variants are especially useful when the society supporting the culture is challenged by the intrusion of reality, for example when, say, resource shortages or pandemic or war threaten to focus minds on reality and not the narrative, thus threatening to break the culture's grip. Narrative evolution based upon a rich population of variants enables cultures to adapt in the face of such challenges.

Thus the population of narrative variants is in a kind of dynamic equilibrium, with group success balanced against the success of individual memes. A narrative population that is optimum for the culture's short- and long-term survival balances the strongest emotive attraction for new recruits against the least emotive backlash (as per Section 3.2.2.1), and other optimisations too, will prosper.

It is worth re-emphasising that although individual sub-narratives are crafted by individuals, the narrative population as a whole is emergent, which is to say it is selected subconsciously by people who come into contact with it. There is no conscious plot to construct the narrative variant population (the evolution of which may span millennia, as is the case for the mainstream religions). A few people might attempt to use the culture to their own ends, but these efforts will inevitably fail because the narrative will quickly evolve out of control.

The charts in Chapters 8–10 reveal a range of public attitudes in response to survey questions about climate change that have different emotive strengths and different reality constraints; these questions are essentially probing the range of attitudes that the narrative population of climate catastrophism evokes.

3.2.4 Cultural identity and the role of children

The loyalty of core adherents must be deep, automatic and unquestioning if a culture is to function. For that to happen, its details must become part of their personal identity, so that they serve it instinctively, emotively, and in an unreasoning way.

3.2.4.1 Programming children

The details of the culture that adherents must absorb encompasses a complex set of symbols, narratives and behaviours. They must also develop a thorough understanding of what the culture considers moral – morality being a key part of cultural identity; a shortcut to what is approved cultural behaviour.

Such information is not passed on via biology; genes encode only very generic cultural behaviours. Instead, it is instilled, ideally in childhood. As well as having no preconceptions, children seem to have a natural affinity for cultures (Chapter 4 delves further into this hypothesis). They also tend to be shielded by families and the immediate community to reduce the chance of outside cultures subverting them.* Once they reach adulthood, their minds are far less malleable, and so they are more likely to remain in the culture of their birth.

In some ways, cultural conversion can be perceived – in adults at least – as an 'infection' that overwhelms the natural defences of those who are susceptible. However, for children a culture is simply 'received', and as such their attachment is far more enduring. The young are therefore a source of vigorous adherents, who will likely remain loyal for the long-term. In addition, their supposed innocence makes them particularly valuable, because it makes them more believable and therefore more effective, both as proselytisers and, more rarely, as prophets. (This is why children more commonly fill the latter role than adults do.[28])

3.2.4.2 Subverting children

Since children are so valuable to cultures, an upstart alien one will try to capture the young by insinuating itself between them and the native culture. This is easiest where education has been devolved to enterprises that are separate from the parents, and potentially even the local culture, such as a modern school system. Teachers may have few or no links to the local culture. Infiltrating a school system is therefore far easier than subverting the shared culture of all the parents and the local community. A culture that can successfully subvert the children of its rivals gains a substantial body of committed proselytisers and potentially a child prophet or two.

3.2.5 Cultural control

Cultural group cohesion isn't just delivered by persuasive narratives

* At least historically; many modern liberal democracies have effectively ceased to do this.

and appropriate signalling; group members are incentivised to keep up their contributions, to shun heresies and above all to remain loyal. This is done through 'policing' activities.

3.2.5.1 Cultural policing

Some cultural adherents – those who identify especially strongly with the cultural narrative – make it their business to fulfil a policing role: emphasising the group boundaries (what signals or behaviours or expressions will mark people as definitively in-group, or out-group), fending off attacks from competing cultures, exhorting the less fervent to up their game, suppressing internal scepticism, opposing heresies, and doing whatever else is necessary to prevent the cultural consensus from weakening. Praise, reward, guilt, emotional blackmail, denigration, demonisation (see Section 3.2.5.3) and other techniques may be deployed as policing tools, although demonisation is in principle only used internally if waverers cannot be returned to the fold by gentler means. These techniques all heighten emotion, which is in the culture's interest because it prevents any drift towards rationality.

While entire organisations within the culture can take on a policing role (for example the Spanish Inquisition), it's also a communal activity that ordinary individual adherents may choose to participate in. And while the above descriptions sound dramatic, in practice much policing activity is mundane. Cultures encourage every adherent to contribute to keeping every other adherent blinkered (to any concepts or realities that the culture considers unacceptable), and this is a kind of low-level policing activity.

It's important to remind ourselves that all cultural behaviours are ultimately due to subconscious bias. While some individuals knowingly engage in policing activities, most think only that they are 'doing the right thing' (in order to mitigate some existential threat), that they are morally justified, and that time will prove them to have been on the right side of history. Whether that proves to be the case or not depends upon whether the culture concerned becomes successful enough to have a large hand in writing that history. If it doesn't, mistaken belief is not usually considered an excuse for committing dire acts.

3.2.5.2 Narrative policing

A lot of policing activity is centred on the core narrative, which is the essence of a cultural entity. This involves protecting it from challenge, reinforcing what views are acceptable, what discourse is allowed and what falls beyond the pale. This is not a straightforward matter. For instance, policing activity must always emphasise the main emotive theme but simultaneously tolerate the less emotive sub-narratives, which, as noted in Section 3.2.3.4, serve a purpose too.

For an established culture, this policing may be formal. For instance, in the case of religion, blasphemy laws may be part of everyday life. More often it is informal: the subjecting of those who challenge the narrative to tirades, outrage, emotive pressure, withdrawn services or, if the non-conformers still refuse to fall into line, demonisation (see Section 3.2.5.3) and perhaps social 'cancellation'. Such treatment is meted out to those, for instance, who legitimately challenge the narrative of certain global climate catastrophe, and those who legitimately challenge extreme trans-rights narratives by merely affirming the scientific fact that 'a woman is an adult human female'. From even the first days of the Third Reich, those who were brave enough to challenge anti-Semitic cultural narratives were treated in the same way; it was only later on that the policing of such dissident views was formalised in law.

3.2.5.3 Out-group demonisation and the pressure of fear

The term 'demonisation' is derived from religious practice. While modern secular cultures don't call out non-conformers as demons or agents of the devil, they can nevertheless make baseless and very derogatory accusations. For instance, calling all conservatives 'nasty', or all left-wingers 'loony', or implying that anyone from an opposing culture is nefarious, is hardly uncommon. However, subtler forms of demonisation exist too, for instance through use of terms such as 'anti-science' or 'denier' (the latter explored in Chapter 6).

The fear of being demonised keeps waverers from expressing scepticism. It's rather like the instinctive fear of catching a disease. Indeed, people subject to demonisation are perceived as 'cultural lepers' who should be shunned, along with their contagious views.

3.2.6 Cohesion + cultural narrative = cultural consensus

When a cultural narrative is successful, the group whom it emotively convinces will become highly cohesive, adopting accepted behaviours, and encouraging and reinforcing them in others. In short, they 'sing from the same hymn-sheet'. This agreement about what to say and how to act is referred to as a 'cultural consensus'; it shapes the attitudes of the whole group, although individuals may be affected in different ways, as though they are singing different parts of the same work, like the sections of a choir.* Many consider the terms 'culture' and 'social consensus' to be effectively synonymous. For instance, anthropologists are able to determine the nature and range of an unfamiliar culture by mapping where a consensus is upheld, and how strongly.[29]

In fact, a powerful consensus may even set the agenda for those who oppose it. Moreover, that agenda will usually be cultural. Because most resistance is instinctive (see Chapter 7), it will be a function of cultural mechanics, and so irrational. For instance, the excesses of sixteenth century Roman Catholicism didn't lead to a rational, atheist response and a rejection of the cultural, but instead to the competing religion of Protestantism. In similar fashion, much of the resistance to climate catastrophism stems from emotive beliefs, such as the idea that there is a conscious plan of elites to achieve population control; in other words that there is some kind of conspiracy at work. These ideas are propagated by opposing cultures, such as the Rep/Con tribe in the USA.

3.2.6.1 Cultural consensuses are not rational...

Because cultural narratives are fairy tales (Section 3.2.3.2), consensuses based upon them do not in any way reflect an agreement about reality; they are emergent artifacts of subconscious group behaviours. A cultural consensus should therefore never be confused with a rational one, such as that formed by a jury in a criminal trial, or by scientists considering which one of competing hypotheses best fits the evidence. However, cultural consensuses can pose as scientific or rational consensuses. This helps them present their core narratives as all-explaining and/or indisputable.

* See Section 9.7.1.1.

3.2.6.2 ...but they are net advantageous...

Cultural consensuses represent a huge advantage to groups. The shared values enable cooperative working on projects sanctioned by the culture's elite. A familiar example is the great pyramids of Egypt. At first glance these vast structures might seem like a burden upon society, but they are now known to have been produced by willing volunteers who were honoured when returning to their home villages. They stimulated deep social integration and spread a wide range of skills through Egyptian society: from writing, finance, administration, and logistics to engineering. For example, the logistical expertise later enabled the Pharoahs to maintain a healthy, trained, fully equipped and well-supplied army of 15–20,000 men, with full medical and engineering backup, in the field, hundreds of miles from home. There are many other examples of cultures achieving such feats, not all of them are reflective of purely religious values. For instance, the large-scale irrigation works created by the first Sumerian cities, such as those at Ur, had a direct practical purpose too, although they were invested with religious significance in order to deliver the extraordinary communal effort required.[30]

3.2.6.3 ...despite downsides

Cultural entities can still have downsides. For instance, cultural policing can be too zealous, employing bullying and producing excessive fear, suppressing innovation and independent inquiry, and harming individuals and society as a whole. Cultures will suppress some scandals, where this is in their interest, meaning that society will continue to be harmed; they will magnify others, where this suits them, leading to witch-hunts for supposed perpetrators. Cultures can fixate on policies that serve their narratives but are nevertheless damaging to society, or at least decrease the preparedness and increase the risks associated with real-world challenges, such as famine or pandemics or natural disasters.

3.3 Culture as a self-contained 'entity'

All of the features described in Section 3.2 are mutually self-reinforcing, having evolved as a single system that delivers evolutionary advantage. To realise this advantage, the adherents in a highly cohesive group are subconsciously steered to shared interests, so there's a meaningful sense in which cultural groups can be viewed as bounded 'entities' in their own right. In other words, cultural groups have characteristics that are only expressed at the group (not individual) level. Similarly, they appear to have an 'agenda' of their own; it doesn't belong to any of the group's adherents as individuals.

For an example of a group-level characteristic, the evolution of the group, what path it takes next and how it responds to challenges, is not a function of any individual view. It stems from the entire population of cultural narrative variants (Section 3.2.3.4), which have their existence in the speech and writings of numerous individuals. As this population is maintained by evolutionary processes (as these apply to the narrative variants, not to the humans who propagate them), no individual in the group determines, or even knows, how the group will evolve.

Biology furnishes a useful analogy, which, so long as it isn't taken too literally, can help illuminate the agenda of a cultural entity.[31] Viruses and prions can infect large populations of animals, say, or indeed humans. The status of viruses as living entities is marginal – they can't replicate on their own, for instance. Prions are even more basic. However, they have both evolved complex strategies to optimise their spread and defeat the elaborate immune systems of the victim (or 'host') population, including the spawning of many variants and homing-in on specific host vulnerabilities. It is accepted that those strategies fulfil the 'agenda' of the virus or prion attackers, which is purely to survive. Cultural entities, as expressed in their many narrative variants, are similarly not alive – they are neither sentient or agential – but have the same agenda of survival.

Cultural entities must have conferred net advantages on their human hosts (their adherents), or we wouldn't have evolved to support them. The relationship is therefore normally more akin to symbiosis than to infection by a virus. However, the evolutionary interests of a culture might not always coincide with those of its adherents. If its survival is at stake, it may lead its followers into behaviours that are highly damaging to them: a religious war can be seen as two cultures sacrificing their stocks of adherents in an attempt to avoid destruction.

To put it another way, the culture has an independent existence, and its own 'agenda', distinct from those of its individual adherents. This is why I refer to cultural 'entities'.

3.3.1 Different perspectives of human behaviours

The idea that cultures have an agenda of their own, and one that is potentially harmful too, is unnerving, and some might think that it leaves no space for individual intellect and free will. However, this is not the case. In practice, few people, if any, will be complete slaves to a culture. Though membership influences behaviour, there is plenty of scope for personal intellect too, especially as cultures work in narrow domains, and may therefore have little impact outside them. We can think of a cultural entity as a distributed app that runs on brain architecture; it occupies only some of any particular adherent's thoughts, and biases only some of their behaviours, and to different degrees for different individuals. People may also belong to several cultures at once (and reject others),* and their behaviours will vary in each.

Rational institutions – democracy, the law, and science – are not slaves to cultural agendas either. They undoubtely constrain cultural entities, although, as we will see in Chapter 14, cultures constantly attempt to undermine them too.

In summary, cultural entities, individual intellect and free will, and rational institutions are all completely compatible with each other; the existence of one does not preclude the existence of any of the others.

3.3.2 Cultural entities: good or bad?

Throughout our evolution as *Homo sapiens sapiens,* and possibly before, cultural entities have been a huge net benefit. This is one reason we're so susceptible to them. Among other things, they allow common action to be achieved in the face of the unknown (and, in an age before science, practically everything about how the world

* Membership of more than one culture typically means these will be allied, or at least largely non-overlapping in scope. People would experience severe internal contradictions if they were competing: cultures evolved to be exclusive.

worked was unknown). Cultures can manufacture consensuses as strong as any achieved by science, not through rationality and evidence, but instead through emotion. Because they're based on fairy-tales they can have downsides, but these have historically been greatly outweighed by the advantages.

Today the situation is less clear. Some people imply that cultural groups are still net beneficial. For instance, biologist David Sloane Wilson[32] advocates for mankind to live in deliberately designed ('pro-social') groups, as justification pointing to the fruits of evolutionary group selection:* altruism, co-operation rather than conflict ('doves not hawks'), and transparency between group members.

However, the same evolutionary selection process has bequeathed us a more problematic heritage too: noble cause corruption, biases that bypass our rationality (not usually a good thing!), cultural censorship, out-group demonisation, cultural conflict, and ideological fixations that, when expressed as irrational policies, can lead to human and environmental damage on a civilisational scale.

Moreover, the upsides of cultures are inseparable from the downsides – that has always been the case – because they are part of the same emergent system. While the upsides are the result of cooperation, people are not doing so rationally,† but because they are 'forced' to do so by the culture.‡ It is such forcing that brings about the downsides. The cultural forcing mechanism has proven to be highly successful in evolutionary terms, and it hasn't gone away. But society today is more complex, and the risk of harm is much greater. In particular, it will undermine the rational institutions (democracy, the law, and science). And nowadays, there are many more competing cultures, which may cause all sorts of difficulties, for instance clashing with one another or hijacking the education system or a field of science. Worse, each has the potential to cause far more damage if they do gain power and influence; when our most potent technology

* Which includes cultural group selection.

† The system has roots that predate rationality in any case.

‡ This doesn't mean physical force, and indeed, as already noted, people sometimes volunteer even for onerous cultural contributions. However, they are subconsciously motivated to do so by cultural mechanisms, including emotional manipulation (see Sections 3.2.2 and 3.2.3.1), and policing (see Section 3.2.5.1), which includes the threat of demonisation (see Section 3.2.5.3).

was fire, and the most powerful weapon a spear, the possibility of doing extreme harm was far less than today, when advanced technologies abound.

So, while cultural groups still deliver benefits, their potential harms are now much greater. It is therefore hard to say whether they remain net beneficial. However, in some ways the question is moot. The propensity to form cultural groups is embedded in our DNA,* so in practice we cannot get rid of them even if we want to. Moreover, without them, we would lack such vital qualities as team-spirit and national identity, such important tendencies as the religious inspiration to do good deeds, and the impetus from political movements that enables society to address difficult issues. Even some passion for science is somewhat cultural – seen as a noble enterprise pursued by a respected fraternity. Without it, society would be the poorer.

So, if we can't live without cultures, we must learn to keep them tamed. If we fail to do so, there is a risk of extraordinary net harms to humanity, notwithstanding some ongoing benefits. This is clear from historical examples, for instance 1930s Fascism - the culture that arose from a potent mix of anti-Semitism, National Socialism, and the science of eugenics.

Climate catastrophism is not abhorrent in this manner, but may well be net negative, as a result of the costs and restrictions it imposes on society, and the environmental damage caused by the policies it spawns (bio-fuels, renewable deployments, rare-earth mineral needs and so on), all of which have little effect on the physical climate system in any case (Chapter 15).

Moreover, the harms will spread far outside of the climate domain. Because cultural entities are 'blind', their narrative variants can take hold in entirely unrelated areas, helping them in a march towards a universal grip on society.† The large number of things that are (often bizarrely) claimed to be caused by climate change is a manifestation of this process.‡

* In complex ways that would defy any gene-editing we can currently foresee.

† Cultures will try to impact more and more aspects of society until they are ubiquitous. If successful, they will inevitably interact with other, older cultures.

‡ My favourites include an increase in UFO sightings in the UK, the bringing down of Air France flight 447, and the extinction of redheads.

3.3.3 Cultural entity brands

Throughout this chapter I've used religion to demonstrate the features of cultural entities, because this is the 'brand' of culture with which we're most familiar. In this context, spiritual practices such as the worship of ancestors or animal spirits can be counted as primitive 'religions'; they were expressions of the first cultural entities.

It's worth noting that there is a spectrum of opinion about the evolutionary origins of religion, between two extremes: by-product theory, in which its emergence was incidental to some other evolutionary advantage, and adaptionism, in which it was advantageous in its own right.[33] As noted in Section 3.1, (gene-cultural) groups are a selective advantage in themselves, and with religion being a cultural reflection and reinforcer of group identity, the interpretation in this work falls into the adaptionist category. In other words, religions of any type[34] are (or at least were, historically) an advantageous adaptation in their own right (although, on shorter timescales, religious groups will be distinguished purely by culture rather than by gene-culture). So, in this respect my theoretical flag is firmly planted: religions are not merely a by-product of some other evolutionary advantage.

Even the modern main faiths are millennia old, and have roots going back millennia more. They arose in less complex times, when humanity's store of knowledge was far smaller, and have competed in the same social-psychological domain (simple creation myths, omnipotent deities, life-after-death and so on) for an extremely long time. While in principle they're all different cultural entities, sub-brands as it were, and will act as such in some ways, this shared origin and evolution has resulted in more about them being common than is different. So, for instance, if their most basic values are interrogated (this is done in Section 9.6), they all respond as though they are part of the same cultural entity. And also, fortunately for the measurements conducted in this book, they all react in exactly the same manner to climate catastrophism.

However, religions aren't the only cultural entities, and as long as cultural narratives press the right hot-buttons (Section 3.2.3.1), their subject matter is, in principle, arbitrary. The weakening of the grip of religion, especially in the West, has left a kind of 'cultural vacuum' that various secular cultural entities have arisen to fill. The more extreme of these ideologies have become almost as familiar as religions, and more notorious; for instance, Soviet communism and, as noted in Section 3.3.2, National Socialism.

Two more recent examples can be found in the shape of extreme trans-rights culture, and so-called 'anti-racism' culture (based on Critical Race Theory). Among many fantastical claims, the fairy-tale of the former violates basic biology, its adherents preaching that the term 'women' does not equate to an 'adult human female', while the fairy-tale of the latter violates basic history to claim that everything about Western societies, whose majorities happen to consist of people with pale skins, is racist and bad, whereas everything about societies whose majorities have darker skins is noble and good (unless they're too Western, that is, in which case they are 'white in principle', so also racist and bad). The concurrent claim that individuals alive today are personally culpable for their society's historic misdeeds is similarly bizarre.

This is all 'situation normal' for cultural entities; the best way to think of their narratives is not from the perspective of the adherents who push them, but from the perspective of the cultural entity itself; they serve its agenda.

3.4 Key cultural features we will explore

Expounding on how all the generic features set out in this chapter manifest in the culture of climate catastrophism would take up a whole shelf of books and a lifetime of further research. I therefore explore in depth just four features that are key to understanding climate catastrophism, before moving on to measuring the culture's presence across nations.

For some features, a generic treatment, applicable to all cultures, is possible. For others, such as the cultural narrative, which is unique to each culture, only a more specific approach makes sense. The details of the next four chapters, and the approach taken in each, are as follows:

- Chapter 4 (mixed): the cultural role of children as prophets and mass proselytisers; climate catastrophism's psychological abuse of children.*
- Chapter 5 (specific): the cultural narrative of imminent global climate catastrophe, and its population of meme variants.
- Chapter 6 (largely generic): the misframed concept of 'denialism', which allows modern secular cultures (especially climate catastrophism) to demonise dissenters en masse, but without being perceived as doing so.
- Chapter 7 (largely generic): innate scepticism; the instinctive bulk reaction in publics against cultural invasion. This is *not* rational scepticism!

* Related: the cultural response of children as expressed in the School Strike for Climate movement, is measured across nations in Chapter 13.

Chapter 4

CHILD PROPHETS AND PROSELYTISERS

Section 3.2.4 introduced two roles of children in strong cultures. Firstly, they can act as high-profile prophets, their perceived innocence and protected social status meaning they (and hence also their message) can't be attacked. Secondly, because they are primed to pick up cultural templates, they are more easily recruited than adults, and so are a source of new adherents and low-level proselytisers. This chapter examines both roles within the culture of climate catastrophism. Because children are sometimes motivated by hard realities to act in ways that seem superficially very similar, my approach focuses on distinguishing the cultural roles from such cases.

4.1 Serious scenarios for children: reality or culture?

4.1.1 Children protesting

When do children band together to try and make a communal voice of protest heard by society? It typically happens in two different scenarios. The first is when they (and possibly adults too) have been genuinely wronged. The second is when they have been frightened by a fairy-tale cultural narrative, incorrectly thinking that there is some real and present threat or harm (Section 4.4.2). In both cases some action is sought from adults, in order to remove or mitigate the claimed problem.

4.1.2 Children in charge

When do society's leaders respond to children's protests, accepting major policy demands made by a child representative? Again, there are two scenarios to consider. The first is where there is indeed a gross wrong that demands redress, whether the harm has cultural roots or not. If children are actually suffering – psychologically or physically – and whatever the cause, this is essentially a hard-reality issue of present harm. The second scenario is where the children are seeking policy remedies for some culturally instilled fear. In these circumstances, societal leaders may well be 'on board' with the culture already; successful ones usually have strong instincts to avoid the stigma that comes from refusing a 'wronged' (according to the accepted cultural narrative) child. If this is the case, they are primed to accept the children's demands and are emotively disabled from refusing them. Cultural bias blinds them to downsides, however serious, and to costs, no matter how large.

4.1.3 Frightening our children

When do we find it institutionally acceptable to frighten children? While our first thought is perhaps that this should never happen, on further consideration there are once again two scenarios where it is considered morally acceptable. The first is where children must be taught a respect of dangerous hard realities that are beyond adult control. An example is the programme of gas-mask training for children that took place during World War II.[35] Because adults were not guaranteed to be available to help, the training involved a certain amount of fear, in the hope that children would then keep themselves safe autonomously. The second scenario occurs when a culture has defined the moral landscape (or alternatively, when an up-and-coming culture is attempting so to do), and sanctions fearmongering. An example is scaring children about sin or Hell or the Crucifixion, in order to reinforce Christian social behaviour; similar teaching is seen in Islam.[36]

4.1.4 Which is which?

As should now be clear, in each of the three cases above there is a rational scenario and an irrational, culturally driven one. A secular, reasoning, and reasonable society should aspire to avoid the cultural scenarios, which lead to needless fears and trauma, false hopes and inappropriate social actions. A reasonable *religious* society should limit its aspirations for children's upbringing to instilling core societal values; it should eschew alarmist or extremist exploitation, including the exploitation of society's concern for the young (and hence children's protected status).

But how can we know whether any given protest or policy demand or instilling of fear is rational, or culturally driven? So, for instance, how do we know whether the fear of climate change, constantly instilled into our children (Section 4.4.2), the consequent children's School Strike for Climate,[37] and the dramatic aspirations expressed by Greta Thunberg,[38] fall into the reality bracket or the cultural one? Is Thunberg's pitch to the UN[39] as reality-based as that of Malala Yousafzai, the girl who was shot in the head by the Taliban for campaigning for female education?[40,41] Is the nature of the School Strike for Climate ultimately as material and justifiable as the 1963 Children's Crusade again racial segregation in Birmingham, Alabama?[42] Large sections of society enthusiastically supported the school strikes and Greta Thunberg; they would surely answer 'yes' to the latter questions. But how do detailed comparisons actually pan out?

4.2 Reality or culture? Children in charge

An article in *Commentary* magazine, entitled 'Child soldiers in the culture wars', notes:

> The value proposition represented by politically active children is obvious. Sensitive subject matter that withers under dispassionate scrutiny thrives when that kind of analysis is taboo.[43]

This value can be cashed in by cultural or political causes.

In addition, a popular – but obviously false[44] – idea holds that children, by virtue of their innocence, have special insights or veracity. This significantly enhances their persuasiveness, and helps create an emotive smokescreen, which can amplify the irrational in our perceptions.

Below I examine whether reality or culture dominates the pitches made to authority by three prominent girls, each asking for signficant aid and major social change. To help penetrate the emotive smokescreen, some key questions to be asked are:

1. Is the child's moral platform sponsored by a culture?

2. If yes to Question 1, is the child's pitch rooted in, or driven by, the culture's main narrative?

3. Does the pitch represent a current or future wrong? Fears about harms already occurring are less likely to be cultural (although the harms could still have their roots in, or be wrongly ascribed to, fairy tales). Fears about future harms are more likely to be cultural.

4. Does the child dictate a specific solution (and timescale)? In culturally driven pitches to society's leaders, children may do so even when addressing issues of great complexity.

5. If yes to Question 4, and whatever the responses to Questions 1 and 2, does the solution seem irrational?[45] Cultural solutions generally are.

6. How big (in terms of behavioural and infrastructural change) is the ask? Cultural asks can be astronomical.

The level of respect shown to authority is also interesting; fervent cultural belief ultimately respects no authority above its own.

The answers to these questions help tell us whether the delivery of the pitch by a child is merely an extra push to an already sound reality-based proposition, or is a means to guarantee the invocation of cultural fears.[46]

It is of particular interest that all three girls are sponsored by cultures (the answer to Question 1 is 'yes' in all cases). However, this by no means indicates that *their pitches* are cultural, which is why we need the other questions.

4.2.1 Malala Yousafzai

In her pitch to the United Nations[41] advocating education for children (especially girls) and protesting the extremism, bias and poverty that closes this down, Malala Yousafzai, makes very clear that she is a religious adherent. She starts out with thanks to God and later cites inspiration from Mohammed and Jesus Christ (among others). So, she is indeed sponsored by a culture (Question 1). Yet when we come to Question 2, it is also clear that her case isn't mainly driven by religious narrative. Indeed, her own victimhood was a result of an (extremist) interpretation of religious narrative; as Malala herself puts it, a 'misusing of the name of Islam'. Her promotion of the supreme value of knowledge, and her pleas for peace, prosperity, universal free education and the protection of rights, are consistent with her religiously-framed principles, but these aims are nevertheless largely secular, and certainly not the product of culturally (religiously) instilled fears. The harms central to Malala's pitch – principally poverty and a lack of education for females – are current[47] rather than future (Question 3), and her campaign for a rejection of prejudice and for developed nations to pull their weight nevertheless avoids calls for specific solutions or dates by which major progress must be achieved. The answer to Question 4 is therefore 'no', so Question 5 is not applicable. However, with her implied goal of assisting *current* sufferers, her ask is still a big one: she is looking for a major acceleration of existing efforts, and new initiatives. Nevetheless, regarding Question 6, her demands are certainly not astronomical.[48] Finally, she shows respect to the leadership she addresses.[49]

4.2.2 Greta Thunberg

Greta Thunberg is a young campaigner whose activism began at age 15, when she started spending her school days outside the Swedish parliament, calling for stronger action on climate change. Initially attracting attention because of her youth and blunt speaking, she has criticised world leaders for a lack of action on the issue, speaking directly to international conferences, including the UN and national parliaments. Thunberg leads the children's School Strike for Climate movement, which is shown in Chapter 13 to have a presence across nations that matches national *cultural* attitudes. She articulates the emotive core narrative of climate catastrophism – the certainty of an imminent global climate disaster. Thunberg's words at the UN leave no doubt that her pitch is driven by this narrative:

> I speak on behalf of Climate Justice Now...Our civilization is being sacrificed...Our biosphere is being sacrificed...You say you love your children above all else, and yet you are stealing their future in front of their very eyes...We need to keep the fossil fuels in the

ground...we are running out of time.[50]

The answers to both Questions 1 and 2 are therefore clearly 'yes'. Notwithstanding some secondary claims of current harms,[51] Thunberg's pitch mainly concerns much greater (and 'irreversible') damage in the near future (Question 3). Regarding Question 4, she *does* dictate a solution – keeping fossil fuels in the ground – and also a timescale.[52] Whether this is rational (Question 5) is subjective; as justification she takes climate catastrophe – the 'sacrifice of civilisation and the biosphere' – for granted in her short UN speech. And though her pitches to authority in the UK[53] and France[54] suggest that the IPCC has confirmed a high certainty of global climate catastrophe in the next few decades, IPCC science doesn't in fact support such claims (see Section 5.2.2). Applying this benchmark, Thunberg's solution is geared to address an emotive invention, and not reality. Her pitch is indeed irrational, so Question 5 is a 'yes'.

Thunberg's ask (Question 6) is indeed astronomical, requiring humanity's largest behavioural and infrastructure upheavals since the industrial revolution, maybe since the invention of farming, and on essentially a crash timescale.[55] Whatever policies mainstream science might call for, it does not justify this radicalism. She shows no respect to the leadership she addresses,[56] and has been known to claim that they are (institutional) liars.[57] For those with an emotive conviction to the idea of catastrophe, this no doubt *seems* irrefutable.

4.2.3 Nongqawuse

In 1856, the Xhosa nation in South Africa, whose lifestyle and economy were largely based on keeping cattle, was under great pressure. There had been a century of serious colonial encroachment, a fatal lung disease (brought out of Europe) afflicting many of their cattle, and internal political rivalries as the nation struggled to deal with their difficult situation. In April 1856, a young girl brought a prophecy of salvation to the Xhosa leadership. Variously described as between 14 and 16 years old at the time, Nongqawuse[58] was the niece and adopted daughter of a councillor of the king. The prophecy had been communicated to her by 'the spirits of two ancestors', she said, going on to explain that to alleviate their troubles, the Xhosa

Table 1. The pitches of Malala, Nongqawuse, and Greta.

	Very likely reality	Malala Yousafza	Nongqawuse†	Greta Thunberg‡	Very likely cultural
Question 1: Child morally sponsored by a culture?	No	Yes	Yes	Yes	Yes
Question 2: Pitch driven by culture's main narrative?*	n/a	No	Yes	Yes	Yes
Question 3: Current (C) or future (F) wrong?	C	C	C	F	F
Question 4: Specific solution (and timescale)?	No (no)	No (no)	Yes (yes)	Yes (yes)	Yes (yes)
Question 5: Irrational solution?	n/a	n/a	Yes	Yes	Yes
Question 6: Astromomical ask?	No	No	Yes	Yes	Yes

*No to Question 2 negates yes for Question 1. †Although Nongqawuse's pitch concerned a current wrong, her solution was highly irrational. ‡Greta's score matches the cultural criteria exactly.

must kill all of their cattle, destroy their stores of grain, and cease cultivation of crops. In addition, new houses and enclosures must be built. Essentially nothing 'contaminated' must remain. Upon full compliance, unsullied replacement cattle would be resurrected from the dead, the granaries would be replenished, and the European settlers would be swept away. In time, albeit not across the entire nation – some chiefs resisted – the prophecy gained majority support. Several hundred thousand cattle were killed (of which the meat couldn't be eaten) and much food was destroyed. The nation soon descended into famine and chaos. The Xhosa homeland population dropped by three quarters (~78,000), the result of starvation (~40,000) and emigration to the European-run colonies in search of wage labour.[59] Xhosa independence, already weak, was lost.

The full story is highly complex, but this doesn't stop us addressing the same set of questions. Nongqawuse was indeed sponsored by a culture (Question 1): one of spiritual beliefs, including ancestor worship, blended with elements of Christianity. As her pitch came directly from the 'ancestors', who appeared to her by a river, Question 2 is a 'yes'. The fears articulated were current (Question 3), and she dictated a solution and timescale (Question 4). Her irrational solution could only ever have made things far worse, to the point of mass fatalities (Question 5). Notwithstanding the complexity and some disputed secondary aspects,[60] historians view the Xhosa cattle killings as the millennarian response of a stressed society (there is more on millennarian cultures in Chapter 15).[61] Finally, for Question 6, sacrificing the economic basis of the entire nation can only be viewed as an astronomical ask.

4.2.4 Summary of cases

These straightforward checks are not without value judgements.[62] Nevertheless, the summary in Table 1 indicates that Malala Yousafza's pitch is reality based, and Greta Thunberg's is cultural. And although the wrongs in Nongqawuse's society were current, her pitch was cultural too (and with a highly irrational solution).

Malala's pitch places her as an ambassador for the wronged, presenting serious pleas but not astronomical demands. And yes, it was spurred by (positive) emotion and had some cultural (religious) context, but this was not an emotive tale of urgent existential threat.

Greta Thunberg and Nongqawuse, on the other hand, are both essentially prophets of salvation, demanding full and strict compliance to a narrow cultural (and astronomical) 'solution', which they say is the only way to escape dire calamity. In Nongqawuse's case, the solution was a cultural invention; in Thunberg's, the emergency itself is a cultural invention (the idea of global catastrophe being unsupported by climate science; see Section 5.2.2). In both cases, society's leaders were primed by the prevalent cultural narrative.[63] When this narrative was reiterated in distilled form from the mouths of innocent girls, fears and hopes were invoked, including the spectre of guilt for those who dared to reject the pitch – the power of children to invoke adult guilt in such pitches has been noted by Thunberg herself.[64] These cultural weapons proved sufficient to override objectivity, and to unite a significant body of the public.[65]

4.2.5 Pitch characterisation

The pitches of both Thunberg and Nongqawuse are essentially millennarian,[66] a well-known cultural form that is examined in Section 15.5. From their positions as socially protected mouthpieces for their respective cultural entities, they urge 'renewal': a swift elimination of their societies' main means of sustenance and success. They also have an equivalent emotive conviction that the old ways are somehow contaminated, and must be rejected. In dictating absolutes, both girls effectively *command* rejection rather than plead their cases,[67] although they are both, in truth, only rehearsing the text of long-emergent cultural narratives.[68]

The millennarian angle comes across strongly in the Nongqawuse case, and in Thunberg's it potentially explains why many ardent adherents of climate catastrophism reject emission-free nuclear power,[69] or natural gas as a 'bridging' solution, or indeed anything that smells even vaguely of pre-renewable energy infrastructure.[70] Some within the culture of catastrophism now seem to sense that this 'isn't quite right' and have started to shift position; even Thunberg herself has recently come out in favour of nuclear power[71] (while maintaining an antipathy to capitalism[72]). The issue may eventually cause a heretical split in the culture.[73]

Thunberg is acting out the same role as Nongqawuse. Without her, someone else – not necessarily young and female – would appear instead;[74] those who most closely identify with the culture and can most effectively wield its core narrative to command others will rise to prominence from amongst the mass of adherents.

While much of the mainstream media has lauded Thunberg – or at least not explicitly criticised her – the millennarian angle has not gone wholly unnoticed;[75] there are some articles on the dangers of adults over-reacting to messianic children in general, and to Thunberg in particular,[76] on the taboo around challenging a schoolgirl and the gross over-simplification of the issues that results,[77] and the irrational response of adults (UN leaders included) to an uninformed demand from an (inappropriately) scared child.[78]

4.3 Reality or culture? Children protesting

In the last section, I looked at three cases of children making demands of society's leaders, and used a series of questions to assess whether each pitch was cultural or reality-based. These same questions can be used to assess the nature of mass movements of children. In this section I compare Greta Thunberg's School Strike for Climate movement to the 1963 Children's Crusade in Birmingham, Alabama.

The Children's Crusade was part of a wider campaign to desegregate the city and bring national attention to racial discrimination.[79] Although non-violent (in which techniques the participants were schooled), the use of children was considered controversial by many, including some adults within the desegregation campaign itself. In the end, campaigners were gambling that the protected social status of children, the shaming of authorities, and the emotive reactions of wider audiences would make the tactic worthwhile. Nevertheless, the children's involvement was risky – to the young people involved, and also to the campaign itself if their involvement was perceived cynically. President Kennedy certainly disapproved, although he added that just grievances must be resolved.

Churches allowed their premises to be used as bases and the protesters were supported by their faith, so the Crusade certainly had a cultural sponsor. However, as with Malala's pitch, its aims were secular. Importantly, children were themselves wronged, and in the here and now, although their parents and the whole community of colour was suffering in the same way. In other words, these children were not pawns; directly and indirectly they had suffered injustice and their resistance was genuine. They called for negotiation, but ultimately a specific solution too: the end of segregation (the issue scope was rather narrow). As seen in retrospect and even at the time, this was certainly not an irrational ask, and definitely not astronomical, although it required significant behavioural change from an empowered minority.

Table 2 sets out the answers to the Questions 1–6 for the Chil-

Table 2. The pitches of the Children's Crusade and School Strike.

	Very likely reality	1963 Children's Crusade	School Strike for Climate	Very likely cultural
Question 1: Child morally sponsored by a culture?	No	Yes	Yes	Yes
Question 2: Pitch driven by cultures main narrative?*	n/a	No	Yes	Yes
Question 3: Current (C) or future (F) wrong?	C	C	F‡	F
Question 4: Specific solution (and timescale)?	No (no)	Yes (yes)	Yes (yes)	Yes (yes)
Question 5: Irrational solution?	n/a	No†	Yes	Yes
Question 6: Astronomical solution?	No	No†	Yes	Yes

*No to Question 2 negates yes for Question 1. †No to Questions 5 and 6 negates Question 4. ‡Based on unfounded projection.

dren's Crusade, alongside equivalent answers for the School Strike for Climate, which are identical to those set out above for Greta Thunberg, given that she is their main spokesperson.

Would everyone give the same answers? Possibly not; it is unrealistic to assume we can be wholly free of bias. Yet most people would surely agree that the School Strike children are largely privileged, and they are not suffering current wrongs; their fear is based around a fairy tale that adults have told them, about huge and *future* harms.

They face no risk of serious opposition or disapproval (although this does not on its own invalidate their cause); indeed, there has been much approval from adults and authorities globally. In other words, these children are pushing at a cultural door that has already been left ajar. Pushing on an open door seems like a paradox for a protest. However, in cultural terms it is perfectly understandable; because their motivation is subconscious, they do not take on board that the door is open. They simply *believe* that there is an existential threat and have to take action to save themselves.

None of this implies that the children involved in the School Strike are pawns of some conscious adult agenda; as Greta Thunberg herself notes,[80] many are anxious or sad or angry or scared (and genuinely so; see Section 4.4.2). That said, they may subconsciously act as pawns of the culture of climate catastrophism,[81] proselytising on its behalf and even helping to provide it with a socially protected route into power.[82]

In contrast, the children in 1963 faced a very heavy-handed response: water-cannon, dogs, and jail. They were pushing against a closed door, which we know retrospectively (and really, even at the time) was bolted shut by a prevailing sub-culture in the locale. They certainly weren't proselytisers, acting out the pitch of their own culture and seeking to capture authority; they weren't demanding conformance to an arbitrary emotive and existential narrative. Despite deploying their social advantages as children, they represented both equity and reason, countering a long-entrenched sub-culture of racism. History has smiled on their gamble back in 1963, but such an outcome was not foreseen at the time.

4.4 Cultural susceptibility and fear in children

The approach set out above helps to tell us whether children's pitches to authority or mass protests are reality-based or cultural. In this section, I examine, for the cultural case, why children are so liable to take up these roles in the first place. I also outline how they may suffer psychological harm from cultural pressures. In particular, they may believe the threats expressed in the fairy-tale narrative of a new culture are genuine. This risk is particularly pronounced in the case of climate catastrophism, which uses the authority of science as a cloak to hide its cultural nature.

4.4.1 Children and cultural susceptibility

As we have seen, humans have exhibited cultural behaviour for an extremely long time, probably since before we were even *Homo sapiens sapiens*.[83] The most familiar cultural entities are the main religions, thousands of which may have been experienced by mankind over the millennia. The ubiquity of religious belief has led cognitive scientist Justin Barrett to argue that children have a default 'affinity', or instinct, for the concept of God or gods;[84] a religion-shaped hole if you will, just waiting to be filled by a matching social 'shape' within the child's environment.[85] Barrett's theory doesn't reject a significant role for religious indoctrination, though he makes clear his view that this isn't an exclusive or even a primary enabler of belief take-up. Instead, he suggests that indoctrination has a secondary role, supplying details of the particular belief system, and acting to reinforce its place in children's minds.

Barrett also speculates that significant effort and re-framing can 'force-fit' concepts other than divinity into this 'hole', for example the Theory of Natural Selection, or even the concept of government. However, no reasonable framing can imbue Natural Selection, for instance, with the existential fears, the hopes of redemption, the emotive cocktails, and of course a deeply felt identity that make cultures built around God – or gods or spirits – successful. (And even if, with a lot of indoctrination, Natural Selection could be forced into such a religion-shaped hole, the result would probably be a new faith, not a scientifically literate generation of children.[86])

As already noted, secular cultures are in essence religions with different details, working via the same mechanisms and featuring a similar range of emotive narratives that bypass reason to trigger identical behaviours. Hence, during child development, the culture of climate catastrophism, with its visions of apocalypse,[87] carbon sin, and tenuous hope of salvation based upon articles of faith, such as renewables,[88] will slip very easily into such a 'hole'.

Such emotive narratives are now part of school and home life in many societies.[89] Indoctrination of children (albeit not typically the aggressive sort) is ongoing, through endless repetition by teachers,[90] parents,[91] peers,[92] and the media.[93] And it works. For instance, according to Dominic Lawson in *The Times*:[94]

> Such claims are having the desired effect of terrifying children into supporting the aims of Extinction Rebellion: Thunberg is one of those children.

Such indoctrination can lead to children acting in ways normally seen as intolerable, for instance grossly denigrating a leader,[95] (this may be a cultural 'rite of passage'[81] or ritual[96]), or indeed anyone who is perceived (rightly or wrongly) to be blocking the cultural 'solution' for global catastrophe. Worse, parents, who have ultimately allowed this to happen, end up having to go along with it.

With the passage of time, climate catastrophism is changing from a culture that is learned in adulthood, and thus merely modifies existing beliefs, into a 'received' culture that is absorbed during childhood – just like religion.[97] As noted in Section 3.2.4.1, this creates much stronger and more enduring beliefs, more morally legitimised fears, more hopes too, and far less opportunity for reason to prevail. This genie won't go easily back into the bottle.

4.4.2 Interpretation of cultural fear by children

The purpose of fears conjured up in cultural narratives is to provide in-group reinforcement, via a 'moral' map. However, the threats are not real, they are not meant to trigger the same intensity of reaction as fears based in reality;[98] they aren't supposed to be taken literally.

In the main, this is exactly how things work: cultural adherents tend to strongly discount threats transmitted via cultural narratives – for example, responses to warnings of hellfire and damnation are far milder than one would expect if these threats were being taken as real. Our brains appear to have a system for subconsciously recognising a cultural narrative as such, although we don't yet understand how it works.[99]

However, sometimes the system goes wrong, producing *real* fear. For example:

(a) When a new rising culture or a cultural variant creates unfamiliar emotive narratives.

(b) In children, who lack experience of distinguishing culture from reality.

(c) When conditions such as Asperger's syndrome disrupt the proper interpretation of cultural narratives.

Point (a) results in some fearful adults, especially when – as in the case of climate catastrophism – a secular culture uses the authority of science as a cloak to hide its irrational nature. Both (a) and (b) result in many fearful (neurotypical) children, but over the longer term (years), this concern may settle down, as they learn to process the narrative correctly; a small number may even lose their belief entirely.[100] As for (c), sometimes people with these conditions are unable to properly process subtleties in social communication. For instance, they may take certain metaphors literally. In the same way, they may take cultural narratives literally, resulting in real fears.[101] When a child with Asberger's encounters a new and rising culture, all three effects are in play, and the resulting fears will be intense. The indoctrination outlined in the last subsection is therefore probably doing serious harm.

Greta Thunberg provides us with an excellent example of how cultural narratives can go wrong. She *correctly* identifies an apparently major issue in relation to this effect, which is the stark discrepancy between the extreme Catastrophe Narrative as communicated by authority sources (see Chapter 5), and the lack of corresponding urgency that one would expect if this narrative was *literally true*. Her conclusion, of 'adult irresponsibility', almost certainly arises from two main perceptions: firstly, that mainstream science supports the Catastrophe Narrative, and secondly, that everyone who's told her about imminent catastrophe *and its undoubted scientific support*, simply can't be lying.

However, there's an alternative conclusion. No doubt unknown to Thunberg, mainstream science does *not* support the idea of certain global climate catastrophe in the next few decades. However, all those people who told her about it are not lying. They are simply culturally, emotively, and honestly, *believing*. Just as she now does. More generally, the vast majority of people who propagate the Catastrophe

Narrative, including its claim of support by mainstream science, are not lying either. Thunberg's assumption that adults and governments are behaving irresponsibly in the face of certain global catastrophe, likely comes from *literal* interpretations of a cultural narrative that in reality is only meant to be a flag for cultural membership.

Moreover, there is plenty of evidence that many children are taking the claims of climate catastrophism literally. What constitutes scary information, and how exposed to it children should be, is subjective – believers and nonbelievers in strong cultural narratives will hold differing views – yet the *actual manifestation* of scared children is now commonplace. Examples have been reported in *The Evening Standard,*[102] *Euro News,*[103] *ABC News,*[104] *Treehugger,*[105] *The National Observer,*[106] *The Spectator,*[107] and, regarding young adults too, in NBC News[108], *The Washington Post,*[109] and Reddit.[110] Psychologists and guardians have accepted that the phenomenon is an important one.[111] Excerpts from three of these sources are set out below:

> Stella Brazier, 14, burst into tears when asked about her decision to attend [a climate protest]. 'It just upsets me so much because I just don't know if they [politicians] are going to do anything', she said. 'What's going to happen to humankind, what's going to happen to the whole world?'
>
> ABC News (May 2019)

> Here, for example, is 10 year old Zane: 'The reason I climate strike is because the Earth is burning before our very eyes'. According to Hannah, from Birmingham, 'there is no point in going to school if we have no future'. Lottie tells BBC Breakfast 'if we don't strike now then we are getting educated for a future that we don't know is going to exist in the way it does now'. These are quite disturbed statements. There are children who really do seem to think they, along with the rest of humanity, are about to die as a result of climate change – an irrational fear with no basis in science.
>
> *The Spectator* (February 2019)

> For Amy Jordan, 40, of Salt Lake City, a mother of three teenage children, the [climate] report caused a 'crisis'. 'The emotional reac-

> tion of my kids was severe', she told NBC News. 'There was a lot of crying. They told me, "We know what's coming, and it's going to be really rough"'...Cindy Chung, 17, of Bayonne, New Jersey, is an activist with iMatter, a network of high school students who advocate for environmental measures on a local level. She struggles to understand how people, especially adults, can continue with business as usual. 'It wasn't our choice to be born into a doomed world', she said. 'All this terrible stuff can happen by 2030, and I won't even be 30 years old. It's so frightening'.
>
> NBC News (December 2018)

Believers in imminent global climate catastrophe might claim, to paraphrase an article at *Vice.com*,[112] that it's a moral failure to succumb to despair. Or maybe even that it's a useful discomfort that will help the cause. But they cannot claim this issue doesn't exist. However, a big majority of social scientists and psychologists seem to be believers themselves, so their advice is unhelpful at best,[113] and likely exacerbating. For instance, social scientist and psychology scholar Renee Lertzman advises grieving for 'how f*** bad it all is'.

And there may be a much less noble side to the inculcation of cultural fear in children. Psycho-analyst Rosemary Randall notes:[114]

> ...Climate change makes most adults working on it feel powerless...By focusing on the weakest members of society and influencing them, the not-very-powerful adults make themselves feel better at the expense of the absolutely-not-powerful children. By making them act, we prove that we are not as powerless as we feel.

For context, Randall fully believes that climate change is a very serious and urgent global problem.

4.5 A cultural spiral

As we have seen, children have an instinctive affinity for cultural templates. Subject to indoctrination in the classroom and other social settings, many have been captured by the culture of climate catastrophism. The result has been a new army of proselytisers, their efforts amplifying the Catastrophe Narrative and disseminating it further. Genuine fears of global catastrophe are widespread, as many people, and especially children, have not internally realised the cultural nature of the narrative; the pressure upon society to act in millennarian fashion has increased accordingly.[115] Child prophets have emerged, declaring the existence of a climate emergency and demanding a crash programme of decarbonisation. Their youth protects them from criticism and gives them an aura of truth and moral purity that has mass appeal. This has helped the culture to win over (culturally primed) national and global leaderships; there is almost no official challenge to Thunberg's climate polemics.[116] Again, the result is further penetration of a fear-laden cultural narrative throughout society.

In other words, through children, climate catastrophism has created a social feedback mechanism, steadily intensifying the pressure on society to take irrational action. Yet ultimately, the culture does not possess agency, let alone sentience. It works purely via emotive selection and the consequent engagement of long evolved behaviours. And in the process, it is inappropriately scaring millions of children.

Chapter 5

THE CATASTROPHE NARRATIVE

5.1 An emergent and dominant narrative

As noted in Section 3.2.3, an emotive narrative is a critical feature of cultural entities – it is the vehicle or carrier of the culture. Successful narratives are emergent – the story that survives is the one that best attracts new adherents. Cultural narratives typically incorporate an existential threat for some people, or possibly for everyone, and are necessarily false (Section 3.2.3.2). Policing of the narrative (Section 3.2.5.2) maintains its most effective memes, protects it from attack by sceptics or competing cultures, and encourages a sense of certainty among adherents, especially about the expectation of salvation.

So, we expect the culture of climate catastrophism to spread on just such a vehicle; I will call it the 'Catastrophe Narrative'. Its existential threat is only too apparent – the certainty of imminent global climate catastrophe unless there are swift and dramatic emissions cuts.

As we shall see in this chapter, the Catastrophe Narrative is promoted by all levels of political authority, from presidents and prime ministers (and the UN elite) downwards, most religious leaders, economists, celebrities, an array of charitable organisations, many businesses, experts and influencers of every type, along with large numbers of campaigners at grass-roots level. While propagation is more intense in some organisations – for instance, universities and environmental charities and, in some nations, on the political left – it is unarguable that the Catastrophe Narrative has a huge influence upon society overall. In later chapters we measure the consequences of this influence; it is the dominant factor in shaping public attitudes to climate change.

While the Catastrophe Narrative is often called out by sceptics, and its nature as briefly outlined above is sometimes acknowledged even by those on the orthodox side of the climate-change issue, it is rarely studied, nor opposed in a manner that might realistically arrest its spread or attenuate the high levels of cultural belief that result; indeed, those who propagate it frequently win plaudits. This chapter at least makes a start on an analysis.

Like narratives from older cultures, for instance religions, the Catastrophe Narrative is in essence a fairy tale. It is *not* supported by mainstream science, and owes its success only to emotive engagement; it convinces irrationally, rather than rationally. As well as trampling upon science, it undermines all other attempts to maintain 'objectivity at social scale', such as the law, and democracy. At a minimum, it creates major bias wherever it propagates, and may even create a complete disconnect between public understanding and reality. These detrimental properties rely upon a populational structure of emotive subsidiary forms, or 'variants' (Section 3.2.3.4).

Over recent decades, many variants of the Catastrophe Narrative have emerged. These typically provoke multiple emotional reactions, an effect heightened by demands for great urgency. Such powerful 'emotive cocktails' undermine objectivity and optimise further propagation, so winning new converts to the culture of climate catastrophism. This chapter details the main narrative variants, and explains how they work. An online companion file, the CN-Archive, provides further examples, along with a fuller analysis.*

5.2 Propagation by authority level and function

Most of the CN-Archive entries are categorised by narrative variant type. However, there is also some categorisation by the authority level or social function of the propagator. The first two groups in the archive are examples of more generic quotes by first-rank authorities (Group 1: leaders or leading politicians of powerful countries, major international organisations such as the UN, religious leaders

* Weblink provided in Chapter 1. All the quotes set out here take their numbering from CN-Archive entries; some are also supported by an endnote, in which a direct weblink and sometimes further information can be found. In such cases, the CN-Archive entry numbers will be found in the endnote.

and so on), and second-rank authorities (Group 2: leaders of smaller countries, local politicians, NGO leaders and spokespersons, economists and so on). The propagation of the Catastrophe Narrative by scientists, who in theory should be more objective, is a case of special interest; there is a separate group of quotes for them too.

5.2.1 Propagation by first- and second-rank authorities

The Catastrophe Narrative has been propagated by many of the most powerful and influential figures in the West throughout the twenty-first century. CN-Archive Group 1 supplies 39 example quotes taken from 26 first-rank authorities. Three of the shorter entries are reproduced in Table 3, to provide a flavour. While based only on English language sources, as is the case for the whole archive, this group nevertheless includes the leaders, ex-leaders, and candidates for leadership of eight key Western nations, along with cabinet ministers, senior UN officials, the Pope and UK royalty. It covers the first 18 years of this century.

Throughout the examples, the narrative is framed in highly emotive terms and as being of the utmost urgency, which hugely increases its likelihood of being re-transmitted.[117] It is also clearly global in

Table 3. The Catastrophe Narrative from societal leaders.

Source	Quote
Amina J. Mohammed, UN Deputy Secretary General.	'We are the first generation to experience the impacts of climate change, and we are also the last that can prevent a catastrophe for people and the planet.' [148]
François Hollande, President of France.	'Never have the stakes of an international meeting been so high, since what is at stake is the future of the planet, the future of life.' [149]
Barrack Obama, as Democrat presidential candidate.	'This is the moment when we must come together to save this planet. Let us resolve that we will not leave our children a world where the oceans rise and famine spreads and terrible storms devastate our lands.' [150]

scope ('the planet'), and imminent catastrophe from global warming is unequivocally attributed to human greenhouse gas emissions, with hope of salvation held out through the apparently simple act of stopping them. Regarding policy, the *certainty* of impending catastrophe is often cited as the main reason to act.

Propagation of the Catastrophe Narrative is also visible among lesser-ranking politicians, leaders of less influential nations, NGOs, economists and influencers. Group 2 of the CN-Archive supplies 28 example quotes from 26 such sources. Propagation of the narrative also occurs at lower levels; claims of global climate catastrophe are ubiquitous, although the high volume and frequent difficulty of proper attribution makes detailed study of these levels very time consuming.

These quotes demonstrate the overwhelming official backing for the Catastrophe Narrative. Over many years, world authorities have primed international publics with the cultural fairy-story of certain catastrophe (absent dramatic action), featuring the ultimate existential stakes of 'all life', or indeed 'the planet' itself.

5.2.2 Propagation by scientists

In a 2016 book,[118] Peter Jacobs – a climate scientist and expert on climate communication – finds no merit in the claim 'that catastrophic anthropogenic global warming is the mainstream scientific position'. This should not surprise readers. As noted several times already, mainstream science, in the shape of the working group chapters of the IPCC's Fifth Assessment Report,[119] does not support claims of a high certainty of imminent global climate catastrophe. Indeed, when climate sceptics say (inappropriately in my view) that mainstream science supports 'CAGW' (a frequently used initialism, standing for catastrophic anthropogenic global warming), objections on climate blogs and other science-orientated forums tend to be vociferous.*

Steve Koonin,[120] from 2009–11 the Under Secretary for Science, Department of Energy, in President Obama's administration, emphasises this reality too.[121] He points out that many statements from the IPCC technical reports 'belie the canon of climate catastrophe'. And of the Sixth Assessment Report of 2021 he says:

* The use and misuse of the term 'CAGW' is considered further in Section 5.5.

> ...you don't find the words existential threat, climate catastrophe, climate disaster, at all; you find the words 'climate crisis' exactly once, and that's not a scientific finding but it is in reference to the way in which the media have amped-up their coverage.

Therefore, in addition to the narrative's central falsehood of a climate catastrophe, it also encapsulates a secondary falsehood, namely the incontrovertibility of mainstream science's support for the idea.* Nevertheless, this doesn't mean that scientists don't propagate the Catastrophe Narrative; far from it. Some do so vociferously, playing an important role in helping climate catastrophism to hide its true cultural nature. The involvement of these individuals is therefore of great interest. It is extensively covered in Section 5.4.

5.2.3 Propagation by functional areas of society

All manner of organisations have been attracted to the climate domain. There are often tangible benefits for active participation, and the costs are slight: merely extolling the Catastrophe Narrative and hence signalling cultural conformance. Nevertheless, the motivations of these bodies are not typically cynical; they simply think they're ascending to the moral high ground, or doing 'the right thing'.

As might be expected, however, the Catastrophe Narrative finds a home more readily in some organisations and parts of society than in others; places where its values resonate better – the press (catastrophe sells) and green NGOs are obvious examples. Environmental science is another – its personnel are not climate scientists and very likely are as vulnerable to the cultural message of climate catastrophe as any ordinary member of the public, perhaps even more so given the care for nature in their calling. Examples of Catastrophe Narrative propagation from environmental scientists are provided below.

Other key conduits for propagation are less obvious; the caring professions, for example. Medical organisations were likely first attracted into climate catastrophism because of the purported implications for their professions (coping with mass casualties). But in many cases their involvement now seems to have gone much further, turning into strident advocacy for action. The moral authority such

* See Endnote 3 of the CN-Archive for a long list of examples where Catastrophe Narrative explicitly includes this claim.

organisations have with publics gives their campaigning great force. Further research into this area would be fascinating, but is beyond the scope of this book. The CN-Archive (Group 8) has several representative quotes from authorities in the medical field.

5.3 Main narrative variants

Cultural narratives evolve, typically spawning numerous variants along the way. Over time, a narrative population mix emerges that best exploits our emotions. These in turn will depend upon our worldviews, which cultural narratives may nevertheless modify in their favour. There is no intelligence or agency involved; it is a simple and 'blind' process that works via differential selection of variants, balanced against overall cultural group benefits.

The most familiar example of this narrative evolution in climate catastrophism is the successful replacement of 'global warming' as the umbrella term of choice, with 'climate change'. Other variants that appeared, but proved less successful, were 'climate weirding' and 'global heating'. The reason for the success of 'climate change' is clear. Although 'global warming' has a stronger link to the original trigger for cultural emergence, 'climate change' offers far more narrative possibilities, enabling the culture to appropriate almost any changes in the physical world (and not just 'weird' ones) to its service.

Some narrative variants are very blunt, representing a kind of head-on charge at emotive engagement; these are typically more successful if they come from the kind of higher authorities who might get away with it. Lesser authorities may need to adopt a more subtle approach.

A majority of the examples in Groups 1 and 2 of the CN-Archive are in the basic and blunt category. A few of the shorter examples from these groups are set out in Table 4. The CN-Archive includes links to all sources in order to see context, but an important feature of emotive narratives is that they are frequently propagated shorn of full context, because they can be more effective in this form – Ban Ki-Moon's clock ticking metaphor is a good example.* They are there-

* Around the ticking clock idea, the text implies a world that can be 'thriving' when the young of 2015 inherit it, and suggests that 'the private sector' and 'developed economies' will survive perfectly well; the soundbite doesn't convey this.

Table 4. Blunt emotion from high-profile individuals.

Source	Quote
Ban Ki-Moon, UN Secretary-General.	'...the clock is ticking towards climate catastrophe.'[151]
Emmanuel Macron, President of France.	Macron said that without a reduction of carbon dioxide emissions and pollution, there will be no more Earth.[152]
Jan Peter Balkenende and Tony Blair, Dutch and UK prime ministers	'We have a window of only 10 to 15 years to take the steps we need to avoid crossing catastrophic tipping points.'[153]
Stephen Hawking, high-profile physicist	'We are close to the tipping point where global warming becomes irreversible. Trump's action could push the Earth over the brink, to become like Venus, with a temperature of two hundred and fifty degrees, and raining sulphuric acid.'[154]

fore much more likely to persist in this form too, turning up in the Catastrophe Narrative of others, albeit sometimes in modified form.*

Not all examples of the Catastrophe Narrative contain the words 'catastrophe' or 'catastrophic'; a little over half of the Group 1 examples do so. The others invoke similar meanings with alternative words, or even imply a future that is worse still: a 'global catastrophe' might well have many survivors, yet some expressions imply that there would be *no* survivors.[122]

More subtle Catastrophe Narrative variants are named after their emotive leverage strategy. These variants can occur separately or mixed together, and include (in no particular order):

- emotively overwhelmed conditionals
- fear plus hope
- engaging anxiety for children
- moral association

* This aspect is considered further in Section 5.3.6, and Endnote 107 is also relevant.

- agenda incorporation
- terminal metaphors
- merchants of doubt
- the voice of innocence
- attribution reinforcement
- emotive bitters
- survivalist

I touch on these briefly in the following subsections.

5.3.1 Emotively overwhelmed conditionals

By including caveats or conditional phrases alongside claims of climate catastrophe, narratives can present a surface impression of balance; they seem to be much more reasonable. For any fully rational narrative, this would weaken or limit the claims. However, if the overall pitch is sufficiently emotive, and especially if there is spuriously high confidence of global catastrophe elsewhere in the message, the caveats can easily be overwhelmed in the listener's mind. In other words, such a contradictory and emotively asymmetric presentation won't correct the false representation of mainstream science. In fact, the overall effect may well boost rather than inhibit cultural narrative propagation; the caveat serves as a protection against accusations of lying, and apparent reasonableness is an asset.

Here's an example of an emotively overwhelmed conditional (with my emphasis on the conditional):

> US President Donald Trump is the 'enemy of the people' for hampering efforts to reverse *potentially* catastrophic increases in carbon emissions, Jerry Brown said Monday, blasting White House environmental policy after signing a bill that will move the state toward 100 percent clean energy use by 2045. 'Trump is not just AWOL on climate change, he has designated himself saboteur-in-charge', Brown said in a telephone interview, citing the administration's actions against California's emissions standards, electric-car mandates and clean-power rules. He has designated himself basically enemy of the people. I'm calling him out because climate change is a real threat of death, destruction and ultimate extinction.
>
> Jerry Brown, Governor of California (September 2018)[123]

The emotive content throughout, is self-evident. 'Ultimate extinction' contextualises 'death' and 'destruction' as being relevant to essentially everyone; the phrase featuring these descriptors forms a completely spurious take on mainstream climate science.

Note that Hawking's 'could' in Table 4 does not even count as an emotively overwhelmed conditional, because no action by Trump could possibly have the outcome described. The statement is simply false, but the unquestioned scientific authority of its originator hugely increases its public appeal, and makes the message far harder to challenge.

The CN-Archive (Group 3) provides 12 further examples, from a range of authority sources and influencers.

5.3.2 Fear plus hope

As climate communicators noted some years back,[124] if fearmongering is overdone, there may be a backlash in the shape of disbelief about global warming. Nevertheless, fear memes proliferate in the Catastrophe Narrative, but often in a form that manages to avoid too much backlash. This is because 'emotive cocktails' – as noted above, more sophisticated narrative variants that invoke multiple emotions – can neutralise or offset negative reactions, while retaining their emotive persuasiveness.* One such effective cocktail is 'fear plus hope', a familiar feature of religious narratives. The 'hope' part of the narrative works to negate the backlash against the inevitable 'doom' that is explicit in the 'fear' element. And given 'hope' is an emotion too, the two parts together almost certainly increase the overall persuasiveness of the message. Here is an example of fear plus hope from economist and *New York Times* columnist Paul Krugman:

> So what's really at stake in this year's election? Well, among other things, the fate of the planet. Last year was the hottest on record, by a wide margin, which should – but won't – put an end to climate deniers' claims that global warming has stopped. The truth is that climate change just keeps getting scarier; it is, by far, the most important policy issue facing America and the world. Still, this election wouldn't have much bearing on the issue if there were no pros-

* Simpler fear memes still survive in the culture because, while they create disbelief in some, they also create belief in others; cultures are naturally polarising.

> pect of effective action against the looming catastrophe... Salvation from climate catastrophe is, in short, something we can realistically hope to see happen, with no political miracle necessary. But failure is also a very real possibility. Everything is hanging in the balance.
>
> Paul Krugman, economist (February 2016)[125]

Using the positive emotion of hope (from a mythical 'salvation') to aid the digestion of a false narrative of existential threat, is in no way justifiable, particularly when done to promote a political or policy platform. This is true no matter what the policies' merits.

The CN-Archive (Group 4) provides 12 further examples from a range of authority sources and influencers.

5.3.3 Engaging anxiety for children

Care for children is a powerful instinct that is easily aroused. Claims of serious threats to the young will thus lend power to a narrative and encourage its retransmission. No doubt the fears leading to such claims are genuine, especially when expressed by those who talk about their own children or grandchildren, which is exactly why their messaging is so emotive and so convincing to others. This emotive power can overwhelm fact and rationality. Irrational fears may become infectious, and mass guilt can be triggered, with people not wanting to be seen as failing in their responsibilities for the young.

As a result of this powerful instinct, some people treat the most nebulous possibility – even a highly unscientific rumour – that climate change will harm our children as a major and certain threat against which all-out action must be taken. This is often framed as insurance, in case there is some reality behind the possibility. However, this is actually a gross misuse of the precautionary principle, which does far more harm than good, not least to children (who, as noted in Chapter 4, will become inappropriately scared). Mixing strong anxiety for children into a false Catastrophe Narrative that already features no, improper, or emotively overwhelmed conditionals, results in a powerful persuasive effect.

Here is a short example of engaging anxiety for children:

> I won't let anyone to take us backward [sic], deny our economy the benefits of harnessing a clean energy future, or force our children

> to endure the catastrophe that would result from unchecked climate change.
>
> Hillary Clinton, US presidential candidate (November 2015)[126]

Mentioning children implies a timescale for the anticipated catastrophe; in other words, that it will happen during their lifetimes, and with the implication of 'before they grow old'. Sometimes, 'the next generation' or 'grandchildren' might be used instead. But while the former dilutes emotive impact somewhat, and the second hedges the timescale, they are all using the same psychological lever.

Genuine feelings of anxiety for children will tend to be strong, resulting in expressions of concern that can be excessive, or even lurid. This can look so over-the-top that some will inevitably conclude a cynical motivation to stoke fears, but this is unlikely for the great majority of cases. Here is an example, from a letter in response to the question, 'how do you feel about climate change?':

> Sometimes I have this dream. I'm going for a hike and discover a remote farmhouse on fire. Children are calling for help from the upper windows. So I call the fire brigade. But they don't come, because some mad person keeps telling them that it is a false alarm. The situation is getting more and more desperate, but I can't convince the firemen to get going. I cannot wake up from this nightmare.
>
> Stefan Rahmstorf, climatologist (2016)[127]

The CN-Archive provides further examples from a range of authority sources.* The Rahmstorf example above is not an isolated one. A number of other scientists, in principle professionally objective, also engage anxiety for children when they disseminate the Catastrophe Narrative; see Section 5.4.1.

5.3.4 Moral affront

If an issue can be established in people's minds as being fundamentally moral, complexity and legitimate questioning can often be steamrollered under 'moral affront'. This behaviour is another with

* See Group 1 examples i), n), u) ii], y), Group 3 example m), plus Group 5 examples aa) to ac).

deep roots in our evolutionary past, and helps with in-group reinforcement. Rather than the long process of navigating complexity, the group simply places certain activities or opinions beyond the pale, with affront directed at anyone indulging in them; it's a shortcut, in other words.

But shortcuts can be inappropriate, particularly in our complex modern world, where many different social groups have overlapping requirements, where costs and benefits may be far from obvious, and where moral issues are entangled with science, and so with scientific uncertainties. Simple 'one size fits all' solutions may have serious unintended consequences. Sometimes there just *is* genuine complexity, and it needs to be considered carefully, rather than forcefully declared off-limits, if an equitable solution is to be arrived at.

This is particularly the case for the climate domain, where an immature science is still grappling with a highly complicated system, and where the societal benefits of fossil fuels have to be weighed in the assessment. Consequently, framing swift and dramatic emissions reductions as a moral imperative will likely cause more problems than it solves. But whether or not this turns out to be the case, the argument for the fundamental morality of emissions reductions is based on twin falsehoods: a high certainty of imminent global catastrophe and the suggestion that mainstream science backs this claim.

Nevertheless, there are many and varied instances of Catastrophe Narrative that forge just such a moral association. Here is an example:

> It is crystal clear to me that we are talking about moral issues here. We are not merely talking about how to make ourselves feel more comfortable. We are talking about what we owe to our fellow human beings. Given the scale of the threat, given the fact that it weighs most heavily on those least able to protect themselves, my inclination is to say that if this question of whether carbon emissions is not a moral question then I do not know quite what is.
>
> Rowan Williams, Archbishop of Canterbury (2015)[128]

Establishing the morality of actions to avoid a certain global catastrophe doesn't necessarily have to come through direct use of the actual word 'moral'. Legal equivalents, such as 'just' or 'justice', or religious equivalents, such as 'sacred', are both common. An alternative is to associate dissenters with behaviours that society judges to be *immoral*. This can be explicit, through use of terms such as 'criminality' or 'greed', or implicitly through the deployment of a 'guilt' label, or even implying that those who question dramatic action are mentally deficient. Any of these will heighten the emotive persuasiveness of the narrative variant and increase its chances of propagation. Morality is a powerful force; resisting instinctive reactions to something presented as moral, and instead trying to think rationally and objectively, can be hard.

Examples of all of the above and others are represented in the CN-Archive; a summary of their critical elements regarding morality, extracted verbatim from quotes, is below:

- G1 m) i] 'deeply immoral'.
- G2 m) 'no greater crime against humanity'.
- G2 z) 'justice requires'.
- G3 b) 'global leaders have been guilty of willful denial'.
- G5 ac) 'some mad person keeps telling them that it is a false alarm'.
- G5 ba) 'sacred duty'.
- G5 ce) 'When we inflict our greed upon nature, nature sometimes explodes'.
- G6 n) 'work for a moral revolution urgently needed for a sustainable relationship with nature'.
- G6 v) 'This is state terrorism-sanctioned corporate terrorism, carbon terrorism and climate terrorism'.
- G7 da) 'My frustration with these greedy, lying bastards is personal. Human-caused climate disruption is not a belief'.
- G7 db) 'how they could have sacrificed the planet for the sake of cheap fossil fuel energy' (implied greed).
- G7 fa) 'current generations have an over-riding moral duty to their children and grandchildren to take immediate action'.
- G7 hc) 'How can you ignore the severe sickness of someone you are so intricately connected to and dependent upon' (implied callousness, and in this metaphor 'someone' is the planet itself).

Fuller text for the 'crime against humanity' example is given below:

> Climate change is accelerating far faster than expected, to the point where it now represents an existential threat to humanity, that is a threat posing permanent large negative consequences which will be irreversible, an outcome being locked in today by our insistence on expanding the use of fossil fuels... Already one of the world's largest carbon polluters when exports are included, Australia is complicit in destroying the conditions which make human life possible. There is no greater crime against humanity.
>
> Ian Dunlop, Australian businessman (2018)[129]

5.3.5 Agenda incorporation

This variant comes from activists outside the climate domain – promoting anything from veganism to anti-capitalism – and typically features claims that their cause forms part of the solution to global warming, or at least that their opponents are making things worse. Catastrophe Narrative repurposed in this manner re-energises the activists' movement with the momentum of climate catastrophism.

Using the Catastrophe Narrative in the service of another agenda is as unjustified as using it purely in the climate domain, given its lack of scientific support. Nevertheless, it's probably rare that agenda incorporation is done cynically. People already subject to one cultural belief, or at least a deeply and probably rather emotively held view, will have a lower threshold for belief in the Catastrophe Narrative too, if this appears to align to their cause; providing the strongest justification imaginable, namely the 'fate of the planet', makes it virtually irresistible in such circumstances. As with Catastrophe Narrative propagation generally, agenda incorporation variants are typically pushed both passionately and honestly, adding to their potency. However, they are also polarising, because some people opposing the other causes will inevitably suspect cynicism.

Here are two examples of agenda incorporation, the relevant agendas being anti-capitalism and anti-fracking:

> We are here today to voice our deep concern at the dramatic effects of climate change in the world to date. These are threatening our

> existence and the existence of Mother Earth...Mother Earth is getting dangerously close to its end...the capitalist system is responsible for that.
>
> Evo Morales, President of Bolivia (2015)[130]

> Global temperatures are currently only 1 degree warmer than they were 100 years ago, which has already proved to be hugely destructive and has resulted in a huge increase in natural disasters over the past decade, and the last year in particular. Within 20 or 30 years – well within most people's lifetime – the atmosphere's temperature is likely to raise by 2 degrees. Although this was generally considered a 'safe' temperature, the events of the past year have shown that the destructive effects of temperature increases are much more serious than most scientists expected. If we carry on at the current rate of increasing emissions, then apocalyptic temperatures are likely to be reached, with much of the Earth becoming uninhabitable and billions of people displaced...At a time when we should be rejecting the use of fossil fuels (coal, gas and oil), a UK-wide 'dash for gas' makes no sense.
>
> UK 'Frack Off' website (2018)[131]

While anti-fracking does have a direct overlap with the climate-change domain, this doesn't make citing a high certainty of imminent global climate catastrophe any more legitimate.

Agenda incorporation can also form the basis for longer-term cultural alliances, although these can be fraught with difficulties for the participants. For example, if anti-capitalists cite climate catastrophe to support their position, they might be seen as behaving cynically, to the detriment of both cultures.

The CN-Archive includes further examples, including several more featuring anti-capitalism, and others involving anti-Trump, anti-Brexit, and 'one world' governance agendas.*

5.3.6 Terminal metaphors

Terminal metaphors compare manmade climate change to real-life

* G3 c) anti-Trump, G5 dc), de), df) and dg) anti-capitalism, G5 da) anti-Brexit, G7 aa), g) 'one world' governance with complete (global) social realignment.

scenarios (or sometimes fantasy ones) having a terminal outcome (i.e. death), or at least a very high probability of such an outcome in the absence of urgent action. Such metaphors emerge because they are simple and carry a focused emotive punch. So, for example, Earth might be portrayed as someone dying from a dire disease (standing for anthropogenic climate change). The great simplicity of such metaphors opens the door wide for bias because all scientific caveats or real-world trade-offs are usually dropped – their translation into the metaphorical form is too difficult, and even where this is attempted, they tend to lose their restraining influence. Hence the resulting emotive message is that Earth, or humanity, or 'all life', simply dies, which is just Catastrophe Narrative in another (and arguably even worse) guise.

Some invocations of terminal metaphors are caveated elsewhere in the texts, so that, read as a whole, the document ends up contradictory, emphasising a high certainty of catastrophe at one point, but only a small possibility at another. As with emotively overwhelmed conditionals, the emotive part of the text – the metaphor – will win out over the more objective caveat within public perceptions. Moreover, as noted above, emotive narratives are frequently retransmitted shorn of context anyhow. Hence the core Catastrophe Narrative escapes into the wild without any accompanying restraint. Below is a (short) full example of a terminal metaphor:

> If [an atmospheric carbon dioxide concentration of] 400 ppm was a blood alcohol reading then we would be heading for an inevitable car wreck.
>
> Peter Whish-Wilson, Australian senator, The Greens (2016)[132]

A variety of terminal metaphor examples from a range of authority sources are included in the CN-Archive; a summary of their key elements (extracted verbatim from quotes), is shown below:

- G1 v) 'suicide'.
- G2 e)ii] 'a giant car heading towards a brick wall'.
- G2 u) 'drunk driver and inevitable car wreck'.
- G3 l) 'global warming is now a weapon of mass destruction'.
- G4 b) 'We are careering towards the edge of the abyss'.

- G5 ac) 'children in a burning house with no help'.
- G5 ca) 'suicidal'.
- G5 ga) 'shiny new car driving too fast on a wet, curvy road, heading straight for a crowd of pedestrians'.
- G5 gb) 'a runaway train headed over the climate cliff as we stoke the engine with more coal to increase its speed'.

Further examples from scientists are listed in Section 5.4.

5.3.7 Merchants of doubt

Although fossil fuel companies have frequently been accused – both formally and informally – of systematically peddling disinformation about climate change, the evidence has so far withered under scrutiny.[133] This is in stark contrast to the strong case made that tobacco companies knew about the risks of smoking long before they admitted it. Yet whether or not someone can substantiate this so-called 'merchants of doubt' proposition, insinuating that manmade carbon dioxide emissions will cause imminent global climate catastrophe, just as certainly as smoking causes lung cancer, is highly inappropriate. Mainstream science only claims a similar certainty about 'atmospheric warming'; it does not support the idea of global catastrophe.

The merchants of doubt variant explicitly or implicitly draws parallels between the two cases. Having raised the spectre of certain global catastrophe, it steers the resulting emotive reactions towards a scapegoat in the shape of the fossil fuel industry, which typically is castigated in the strongest terms. Understandably, this message generates anger at the perceived misjustice, and thus an even greater impetus to propagate the narrative further. None of this implies that the originators of the merchants of doubt variant are being deceptive; as noted above, generally speaking the propagators of the Catastrophe Narrative passionately believe it is true, despite all of its contradictions and issues. We are all subject to the influence of emotive cultural narratives.

The merchants of doubt variant is highly attractive to those adherents of climate catastrophism who can't comprehend why, after decades of effort and ubiquitous messaging, there is still widespread scepticism within publics. It is much easier for these people to believe an emotive meme that places the blame on corporate saboteurs with nefarious motives than it is to delve into the complex – and uncomfortable – reality of cultural belief in, and innate scepticism of, the Catastrophe Narrative. Indeed, this would require that they themselves give up the notion of global catastrophe.

The quote below is an example of the merchants of doubt variant (it also includes anxiety for children and moral association), from a letter in response to the question 'how do you feel about climate change?' There are further examples in the CN-Archive.[134]

> Public indifference and individual short-sightedness aside, I am furious that politicians like [Australian Prime Minister Tony] Abbott and his anti-environment henchman are stealing the future from my daughter, and laughing about it while they line their pockets with the figurative gold proffered by the fossil-fuel industry. Whether it is sheer stupidity, greed, deliberate dishonesty or all three, the outcome is the same – destruction of the environmental life-support system that keeps us all alive and prosperous. Climates change, but the rapidity with which we are disrupting the current climate on top of the already heavily compromised environmental health of the planet makes the situation dire. My frustration with these greedy, lying bastards is personal. Human-caused climate disruption is not a belief – it is one of the best-studied phenomena on Earth. Even a half-wit can understand this. As any father would, anyone threatening my family will be on the receiving end of my ire and vengeance. This anger is the manifestation of my deep love for my daughter, and the sadness I feel in my core about how others are treating her future. Mark my words, you plutocrats, denialists, fossil-fuel hacks and science charlatans – your time will come when you will be backed against the wall by the full wrath of billions who have suffered from your greed and stupidity, and I'll be first in line to put you there.*
>
> Corey Bradshaw, climatologist, University of Adelaide (2014)[135]

5.3.8 The voice of innocence

The voice of innocence variant is technically a subcategory of anxiety

* I have corrected a minor typo in the original text.

for children, but one that is delivered from the mouths of children themselves. It is important enough to categorise separately. While generating all the normal reactions expected from anxiety for children, the variant gains additional impact from the fact that adults typically feel guilty when morally upbraided by children.

Because of Greta Thunberg and the School Strike for Climate movement, the voice of innocence has become much more common in recent years.* However, there is some danger that its use will provoke a backlash, because although they are scared, the children's fears are ultimately not their own, instead merely reflecting the emotive cultural narratives with which they've been bombarded throughout their lives. As a result, a proportion of those on the receiving end of the variant will not feel guilt, but rather will be moved to a suspicion that the children's apparently mature concepts and phrasing are the result of their being schooled by adults.

However, despite the probability of a backlash, it is still a potent persuader. Cultures are 'blind'; they work through subconscious selection, and variants that are successful in propagating will still form part of the cultural canon, even if they also build up some longer-term harm for the culture. In other words, they accrue both harm and benefit simultaneously, but through different audiences.

In the example below, children claim that the planet is slowly dying due to the actions of adults, although apparently not slowly enough to give them a chance to grow up. They say they must drop their childhood pursuits and dedicate themselves to saving the planet.

> We are the kids, and we will fight to save the world...We are kids, and we see our schools float away in rising water. We see the ice melting, and starving polar bears in our land. We see our water wells drying out. And we see black smoke killing people. But we will fight to save the world. We see kids, and we see typhoons hitting our home. We see big people cutting down trees, and we feel how the air sometimes is hard to breathe. We see how the forest is burning. But we will fight to save this world.

* The School Strike for Climate only gained momentum after the 2018 *Climate Etc.* guest post on which this chapter is based, so the CN-Archive doesn't include any related narrative variants. However, Chapter 4 is devoted to the role of children in catastrophic climate-change culture, explicitly including Thunberg.

> We are kids, and we like to play...We like to draw, play football and read books. We like to sing and dance. But now we will fight to save the world. We are kids, and we have to pay for mistakes that we haven't made. You adults are giving us a world in chaos, and we are scared. But our future children should live in a better world than this. So we will fight to save the world! We are kids, we're still young and have a lot to learn. But this is the last chance to save our planet, and we don't have the time to grow up. Our planet is dying slowly. This has to end. That's why we're here to tell you adults: act like us kids – and fight to save the world!
>
> Statement by children attending a climate conference (2015)[136]

5.3.9 Minor variants

A few minor variants are worthy of a brief treatment.

5.3.9.1 Attribution reinforcement

Attribution reinforcement attempts to translate a certainty of imminent global catastrophe into a certainty that extreme weather events can be wholly or partially attributed to manmade climate change. Since the former claim is a cultural falsehood, the latter cannot be true either, whatever the current or future state of attribution science (one might kindly call it 'not yet mature'). See the introductions of Groups 5 and 7 in the CN-Archive for more on this complex variant, along with examples.

5.3.9.2 Emotive bitters

Over-reliance on emotion, rhetoric or fallacy, particularly if aligned to cultural positions, is not so much communicating a message as manufacturing (a false) one. With this in mind, if a largely rational speech contains even short Catastrophe Narrative cues, it is still a propagation of that narrative, and the emotive content will strongly flavour the overall speech and inculcate bias in the audience. In other words, like their namesake for drinks, a drop of emotive bitters achieves a great deal of bias. Several examples of this variant are provided in Group 5 of the CN-Archive.

5.3.9.3 Survivalist narrative

Many cultures have fringes that invest far more in the 'doom' component of their core narrative than the salvation component. In the case of climate catastrophist culture, this fringe is fond of the survivalist narrative variant, which suggests we should prepare our bunkers against the coming climate apocalypse. Such variants maintain an undercurrent of fear but are typically limited in spread; innate scepticism from within the culture restricts them. They only benefit the culture if they remain as an undercurrent; wider acceptance would likely damage the overall public profile of the culture. An example quote is provided below; see the CN-Archive for the full text.

> But any which way, barring miracles, this civilization is going down. It is time we stopped engaging in the absurd contortions and pretences of 'climate-optimism'. It's time now for climate-realism. That entails not only an epic struggle to mitigate and adapt, an epic struggle to take on the climate-criminals, but also starting to plan seriously for civilizational decline and collapse. This planning, for the sake of brevity, means thing[s] like: planning for greater self-reliance; building community; crafting values and a spirituality for a more local and Earthly future; creating seed-banks; learning to grow food; and getting yourself and your loved ones a knife-proof vest.
>
> Dr Rupert Read, former Green Party and Extinction Rebellion spokesperson[137]

5.3.9.4 *Ironic narrative*

For the ironic narrative variant, see Section 7.1.1 of the CN-Archive.

5.3.10 Coda

Two minor points are worth making. Firstly, in the narrative soup that exists in the public domain, all the variants described above may blur into each other or fully combine. Secondly, some variants (notably some of those in Group 2) feature local climate issues rather than global ones. It is therefore harder to see that they are contradicting mainstream science and are thus cultural. However, the idea of global catastrophe typically remains present, albeit implicit.

5.4 Narrative variants from scientists

Perhaps surprisingly, there are a number of scientists who propagate the Catastrophe Narrative, including environmental experts and a small minority of climate scientists (although that title seems to have a very broad definition). Arguably, their expressions are still more emotive – lurid even – than the examples in the previous section, most of which came from non-experts. Their words are therefore actually *less* objective but, given their source, still carry the authority of science.

Group 6 of the CN-Archive covers 30 examples of scientists using basic forms of the Catastrophe Narrative. Group 7 has another 26 quotes covering the main variants described in Section 5.3. Given scientists use all the same variants as non-scientists, this strongly suggests that their utterances owe far more to the culture of climate catastrophism than to anything that emerges from their studies.

5.4.1 Most popular forms for scientists

My brief review suggests that engaging anxiety for children and terminal metaphors are both very popular among scientists. I don't know why this is, but in the former case I note scientists' occasional reference to their own children in several of the extracts. Given that the Catastrophe Narrative is represented as certain and unequivocal hard science and also promotes a fearful situation for children, it could be that their rational belief in science, and possibly the fraternity of scientists, is actually amplifying their emotive personal concerns rather than helping them fend off a cultural narrative; the highest emotive concerns for parents are most often about their children.

An example of engaging anxiety for children from a scientist is set out below (capitalisation in original). See the CN-Archive for more examples from scientists.

> Global warming must not be allowed to continue as would happen by stabilizing CO_2 and temperature at present levels. Greenhouse gas buildup MUST BE REVERSED, and CO_2 reduced to levels of around 260 ppm, below pre-industrial levels. The technologies to do so are proven, cost effective, and capable of being rapidly ramped

> up, but are not being used on the scale needed due to lack of serious policies and funding to reverse global warming and stabilize the climate system at safe levels. THAT IS WHAT AOSIS* AND UNFCCC MUST ACCOMPLISH IF WE ARE TO PRESERVE OUR PLANET'S LIFE SUPPORT SYSTEMS FOR FUTURE GENERATIONS. The solutions are already in hand. Let's all get serious and stop stealing our children's future!
>
> Thomas Goreau, president of the Global Coral Reef Alliance, and member of the Jamaican delegation to the UNFCCC (2009)[138]

An example of a terminal metaphor from a scientist follows (note 'deadly', and 'might be able to save the people'). This comes from a letter written in response to the question 'how do you feel about climate change?':

> It makes me feel sick. Looking at my children and realizing that they won't have the same quality of life we had. Far from it. That they will live in a world facing severe water and food shortages, a world marked by wars caused by the consequences of climate change. It makes me feel sad. And it scares me. It scares me more than anything else. I see a group of people sitting in a boat, happily waving, taking pictures on the way, not knowing that this boat is floating right into a powerful and deadly waterfall. It is still time to pull out of the stream. We might lose some boat equipment but we might be able to save the people in the boat. But no one acts. Time is running out.
>
> Katrin Meissner, Associate Professor, Climate Change Research Centre, University of New South Wales (2014)[139]

Further examples of scientists using terminal metaphors are included in the CN-Archive; a summary of their minimal critical elements (extracted verbatim from the quotes), is shown below:

- G6 b) '...climate is a battalion of intergalactic smoking missiles'.
- G6 c) '...by driving global warming we are unleashing hell'.
- G6 d) '...very fast train heading for the wall'.
- G6 g) '...the climate dragon is being poked, and eventually

* Alliance of Small Island States.

the dragon becomes pissed off enough to trash the place'.

- G6 h) 'Imagine a giant asteroid on a direct collision course with Earth. That is the equivalent of what we face now'.
- G6 t) '...automobile driving with bad brakes toward a cliff in the fog'.
- G6 v) '...unaddressed man-made climate change is...state terrorism, sanctioned corporate terrorism, carbon terrorism, climate terrorism'.
- G6 y) '...playing Russian Roulette with the future survival of human civilization' [traditionally this is just a one in six chance of death when using a six-chamber revolver].
- G7 hb) Earth suffering a 'dire' illness to have a 'shortened life' as 'the pain and illness unfold'.
- G7 hc) Earth as a sick person, who is 'slipping away from us'.
- G7 ea) '...biblical portent of Noah type floods'.
- G8 a) comparison to World War III (very probably not terminal for everyone, but assuming it's nuclear and truly a world war, terminal for large swathes of humanity and on a timescale far shorter than anything mainstream science proposes as remotely likely for climate change).

5.4.2 Scary science competition

Scientists seem to propagate some of the scariest forms of the Catastrophe Narrative. A few examples are set out below (some are cut down from originals that can be found in the CN-Archive):

> Imagine a giant asteroid on a direct collision course with Earth. That is the equivalent of what we face now, yet we dither taking no action to divert the asteroid.
>
> James Hansen, former head of NASA's Goddard Institute (2017)[140]

> The world faces catastrophe unless global warming and this Arctic [methane] release can be stopped. Unaddressed man-made climate change is set to exacerbate an already worsening climate genocide and cause 10 billion avoidable deaths this century leaving a predicted only 0.5 billion of humanity alive.
>
> Gideon Polya, biochemist, author, activist (2015)[141]

> About a decade ago I realized we were putting the finishing touches on our own extinction party, with the party probably over by 2030. During the intervening period I've seen nothing to sway this belief, and much evidence to reinforce it.
>
> Guy McPherson, Professor, University of Arizona (2011) [142]

> If you're like me, climate change keeps you up at night on a regular basis. It's not so much that we're still on track for the worst-case global warming scenario, or that the survival of countless species – not to mention civilization as we know it – hangs in the balance, but the quiet understanding that our kids are going to feel some of the worst impacts in just a few brief decades... For natural pessimists, the inexorable destruction by climate change leads to thoughts that fall along the lines of this Jezebel headline, which asks: '*Why Would I Ever Want to Bring a Child into this Fucked Up World?*' Because really, why the hell would someone of procreating age today even consider having a baby? It feels like an utter tragedy to create new life, fall in love with it, and then watch it writhe in agony as the world singes to a crisp...
>
> Eric Holthaus, meteorologist and journalist (2015)[143]

Note that Holthaus invokes extreme versions of both terminal metaphor ('singes to a crisp') *and* engaging anxiety for children. Furthermore, as well as citing concern for youngsters alive today, he also attempts to harness our strong desire to procreate, suggesting that we should not do so until we've put in the effort to prevent imminent global catastrophe. Rahmstorf's example of children burning, from his quote in Section 5.3.3, is also an extreme form of engaging anxiety for children from a scientist. Hansen, meanwhile, uses a terminal metaphor.

Of the scientists who disseminate the Catastrophe Narrative, most are not climatologists. The few who do have domain expertise necessarily place themselves in opposition to the much more sober position put forward in the technical chapters of the IPCC reports. These publications, they say, are 'notoriously conservative'[144] or even politically compromised. For an example of their position see the 2018 publication *What Lies Beneath – The understatement of existen-*

tial climate risk, with a foreword by climate scientist Hans Joachim Schellnhuber.[145,*]

Yet their position on the fringes is rarely explained to the public and, for most listeners, the labels they carry – scientist, professor, meteorologist, oceanographer, or whatever – project scientific authority, thus adding to the perception that the narrative of certain imminent global climate catastrophe is underwritten by 'the' science.

Propagating the Catastrophe Narrative is essentially a form of advocacy, and indeed some of these scientists call for 'prevention of global catastrophe', rather than limiting themselves to the normal bounds of science.

5.5 The Catastrophe Narrative and 'CAGW'

As stated earlier, the initialism CAGW, standing for catastrophic anthropogenic global warming, is controversial – it is perceived by many in the climate domain as a 'snarl word' – an insult and a strawman argument combined.[146] But while it is sometimes used in this way, the term does have real applicability in certain cases.

A typical use would be to say 'The IPCC are just CAGW merchants', and a case can certainly be made that the expression is apposite when directed at the IPCC leadership, who commonly propagate the Catastrophe Narrative. The same could be said of the non-mainstream climate-scientists, authority figures or organisations who do so. However, it is invalid to apply the term to mainstream climate science – the actual science collated by the IPCC – which, as noted throughout this publication and as specifically cited in Section 5.2.2, *does not* support climate catastrophism.

5.6 Motivations for using the Catastrophe Narrative

Belief in global climate catastrophe and propagation of the Catastrophe Narrative, in any of its forms, in no way implies deliberate manipulation is in play. The terms used earlier in this chapter, such as 'inappropriate' or 'illegitimate', do not automatically imply dishonesty, or indeed illness or mental deficiency or any other dysfunction

* Schellnhuber himself has deployed agenda incorporation, terminal metaphor and other Catastrophe Narrative variants (see the CN-Archive).

among believers. As already explained, in general, adherents to a culture honestly, indeed passionately, believe the narrative they propagate. It is just that they've been *emotively* convinced of its truth, rather than having been persuaded through reason and logic. Indeed, this is the great power of such narratives. This is not to say there won't be some instances of greed or cynicism or disingenuousness, along with examples of noble cause corruption.* In any endeavour, particularly large ones, there will be some bad apples, but honest belief is the norm.

5.6.1 Are the motivations of scientists different?

While we can measure the attitudes of international publics to climate change, as is done in Chapters 8–10, we can't do this for scientists because they aren't identified as such in surveys, and they would be a very small demographic if they were. Nor are they surveyed separately. So, should we simply trust them? Should we pay more attention when it is scientists propagating the Catastrophe Narrative?

It is not obvious that we should. The high degree of specialisation in modern science means that expert knowledge in one narrow domain doesn't necessarily grant any insights in another, such as the atmospheric physics involved in core climate science. Increasing specialisation also makes it harder to challenge the status quo, and orthodoxies can therefore become entrenched (see Section 7.8). Scientists are also human, possessing the same cognitive capabilities, and hence biases and limitations, as everyone else.[147] In other words, they are just as likely as anyone to have emotive reactions to cultural narratives, and their use of the same highly emotive variants as laymen seems to confirm that some are simply cultural believers too.

Moreover, as briefly noted in Section 5.4.1, scientists may have a personal investment in 'the fraternity of science', which may make them more likely to become adherents of a culture that claims scientific backing, even when such claims are false, as they are in the case of climate catastrophism. In other words, many scientists – from whatever discipline – would be very uncomfortable expressing doubts about climate scares, because they feel this betrays colleagues,

* Where dishonesty does not stem from cynicism and lack of belief, but because the belief is so fervent it even overrides normal behavioural bounds.

even though doing so would in no way represent a betrayal of science. This may result in turning a blind eye, rather than actually propagating Catastrophe Narrative.

With all these issues in mind, the small minority of *climate* scientists who believe in the possibility of catastrophe are of special interest. As we have seen, this group is on the fringes, typically ignoring the more balanced interpretations from their mainstream colleagues. Is their often vocal propagation of Catastrophe Narrative due more to objective climate science, or to cultural influence? There is no way to tell for certain. But the emotive delivery, the policing of narrative, the denigration of other authorities, and the labelling of opponents as 'deniers', are all *cultural* behaviours.

5.7 Narrative ambiguity

As noted in Section 3.2.3, to the extent that narrative variants are emotively stronger, they are also more polarising. This means that there is a useful role for variants that are emotively weaker; while less appealing to ardent cultural believers, they also invoke less disbelief – less backlash – thus helping to recruit people who are less ardent, but nevertheless still culturally supportive, and in larger numbers. In short, they are less polarising. We can illustrate this point by looking at the relative penetration – the frequency of use – of some very simple terms about climate change that often feature in the Catastrophe Narrative, using Google Ngram Viewer (penetration being a proxy for support); see Figure 1. This shows that the term 'climate crisis' has a much greater penetration in the media, and its use is growing faster, than terms such as 'climate catastrophe' or 'climate chaos'.

I think this is because 'climate crisis' is a much more ambiguous term – less explicit, less visually evocative, and so less emotive – than the others. Many things in our world are crises, so the term is more mundane, and its use will be resisted less. It might imply catastrophe, but it might equally mean something more akin to the position of mainstream climate science. From the point of view of the culture, this ambiguity is hugely helpful. The term is comparatively innocuous, allowing the variant to slip beneath people's defences. But once it has inserted itself, it can still insinuate the idea of an existential threat into their minds. This works because the underlying context is still set by the rest of the Catastrophe Narrative.

The terms 'climate change' and 'global warming' are much more prevalent, dwarfing the penetration of the more emotive terms shown in Figure 1. Theoretically, these expressions don't imply any catastrophe or emergency or even crisis, unless such words are explicitly appended. And believing in 'climate change', represents a much lower threshold to people than believing in 'catastrophic climate change'. However, in just the same way as 'climate crisis' sneaks under the radar, even when used alone these less emotive terms still imply catastrophic outcomes, and it is clear from public attitudes, as measured later in this book, that people the world over largely react culturally to the topic of climate change (which is also to say, subconsciously), even where the idea of catastrophe is only implicit.

Note that terms that don't explicitly include the word 'climate', aren't neccessarily comparable to the expressions in Figure 1. For instance, expressions like 'saving the planet' from 'global catastrophe' would probably have referred to the threat of nuclear holocaust in the 1970s and 80s.

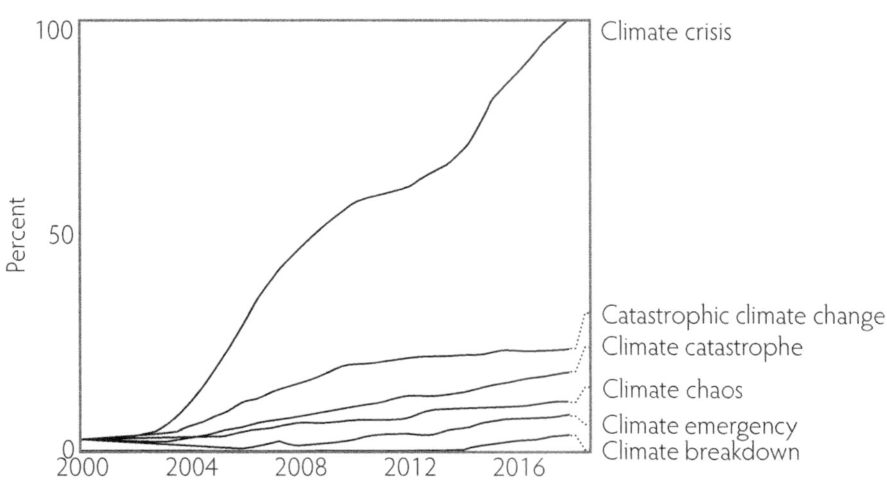

Figure 1. Penetration of simple terms about climate change.
Sampled June 2022.

5.8 Summary

This chapter has shown that climate catastrophism maintains a healthy population of narrative variants, a known feature of cultural entities. While all are rooted in the same central theme, thus maintaining coherence, the rich variety of sub-narratives allows a multiplicity of social angles to be exploited simultaneously, and more emotive hot buttons to be pressed in more people, maximising narrative propagation and thus the spread of the culture. The variant population also provides a rich base for continued evolution, for instance enabling the culture to execute major changes of direction should it be challenged by a sudden change of circumstances. The swift onset of Covid-19 in 2020 as a competing emergency represented exactly such a challenge, but one that climate catastrophism may yet leverage to its advantage.

Reading through the CN-Archive will give the reader a deeper insight into the depth and variety of Catastrophe Narrative.*

* The CN-Archive contains entries up to late 2018, when the *Climate Etc* guest post forming the basis of this chapter was published. The Catastrophe Narrative has proliferated since then.

Chapter 6

DEMONISATION AND DENIALISM

6.1 Demonisation

Adherents of strong cultures systematically demonise critics of their cultural narrative. To some extent, society has become aware that this happens in religious or extremist political conflicts, leading to a backlash. The process of demonisation is not prevented, but may be much less potent. However, if a culture is able to disguise and/or legitimise what it is doing, demonisation can still occur relatively unhindered; 'below the radar' so to speak. In recent times, the popular framing of 'denialism' has emerged as a means to do just this. Cultural adherents can use the 'denier' label to demonise people en-masse; not only without social objection, but often with social approbation. It allows them to disguise the true nature of their behaviour, in part because the pejorative framing has been legitimised by a scientific paper,* which has given it a veneer of respectability and led to its spread across academia. In this chapter, I examine the paper and expose the flawed framing of the term denialism.

6.2 Denialism

Some academics and authority sources, worried about an apparent rejection of science by large segments of the public, have suggested that this is the result of denialism. Two prominent proponents of this view are econometrician Pascal Diethelm and public health expert Martin McKee. Their paper entitled 'Denialism: what is it and how should scientists respond?'[155] has been called 'the standard scientific work on denialism'. It is certainly widely cited across the scientific literature,[156] popular science works, and both medical and public

* There is no suggestion that this was the authors' intent.

forums. Its principles form the core of the Wikipedia page on the subject.[*,157] The paper is intended as a tool with which to identify and resist denialism.

The words 'denier' and 'denialism' existed prior to the paper's publication in 2009. However, its appearance appears to have encouraged their more widespread use, and promoted a particular framing that has now become the essence of the popular understanding of these terms. In this way, it has lent academic legitimisation to their deployment as pejoratives within public discourse.[158]

The framing assumes, among other things, that denialism is a well-understood phenomenon, that it is easily detectable via straightforward tests, that it is typically the result of greed or psychological flaws or, according to a variant of the framing, dishonesty. Are these assumptions justified?

6.3 Recognising the criteria for denialism

Wikipedia asserts – using Diethelm and McKee in its support – that it is possible to recognise denialism because it has common features across topic domains. The five criteria for its recognition are:[159]

- *Conspiracy theories* — dismissing data or observations by suggesting opponents are involved in 'a conspiracy to suppress the truth'.
- *Cherrypicking* — selecting an anomalous critical paper supporting their idea, or using outdated, flawed, and discredited papers in order to make opponents look as though they base their ideas on weak research.[†]
- *False experts* — paying an expert in the field, or another field, to lend supporting evidence or credibility.[‡]
- *Moving the goalposts* — dismissing evidence presented in response to a specific claim by continually demanding some other (often unfulfillable) piece of evidence.[§]

* The page has been considerably augmented since my *Climate Etc.* post. However, the core structure and its inheritance from Diethelm and McKee 2009 and Hoofnagle's blog post, i.e. the relevance here, appears to be the same.

† This is number 3 in D&M2009, and some sources point to cherrypicking of data too.

‡ This is number 2 in D&M2009.

§ In D&M2009 this is framed more as an impossible standard of proof rather than a moving

- *Other logical fallacies* — usually one or more of false analogy, appeal to consequences, straw man, or red herring.

In other words, denialism is framed as the knowing use of fallacy, citing of false experts, and invocation of conspiracy theories; identifying it is apparently as straightforward as matching the narrative of the target individual or organisation or social group against these criteria.

However, a simplistic criteria-matching approach of this kind is problematic. As we will see, there are ambiguities in Diethelm and McKee 2009 about the underlying nature and scope of denialism. These lead to confusion about how their criteria can legitimately be applied and, if they are met, what this says about the motivation of the people being assessed. The paper's lack of guidance in these areas suggests the authors were unaware of the ambiguities and their consequences.

6.4 Applying the criteria in practice

6.4.1 The nature and scope of denialism matters

According to Diethelm and McKee, denialism (as they conceive it) is found across society at every scale: from the 'rejection of scientific evidence' by large segments of populations, down through the mass media, 'networks', 'front organisations' and corporations, to 'a few commentators' working to generate doubt in others.

This raises a fundamental issue: in a social conflict about a science-related topic, the motivations of people at the largest scale – so for substantial proportions of the public – are typically subconscious, because deeply held beliefs will dominate. However, as we shall see, Diethelm and McKee emphasise *conscious* motivations, such as greed, which might occur at the individual end of the range. The authors' suggestion that the same simple test can reliably detect denialism at both ends of this range, even though they imply underlying motivations that are fundamentally dissimilar, raises serious questions about the utility of their conception of the phenomenon. They do not address this puzzle, and while acknowledging that different

target, but the essence is the same (a moving target can never be reached).

manifestations of scale are all part of 'a larger phenomenon of denialism', and that 'denialism is a process', they don't explicitly define or describe this phenomenon or process. I return to this issue of the nature of denialism in Section 6.5, which explores causation.

Meanwhile, the ambiguity of scope raises another serious and much more practical difficulty: to whom should the criteria be applied? This is not such a simple matter as it seems.

Diethelm and McKee don't explicitly define a 'side' within a conflicted domain such as climate change, or creationism versus evolution. However, their approach seems to recognise that sides are inherent in such debates, in that they suggest that any individual or social group or organisation rejecting an overwhelming scientific consensus is in denial (so on the wrong or denialist side), while those who accept it are on the correct (or science) side. This means they are also effectively acknowledging that each side might consist of a mixture of groups, organisations and individuals. The fundamental problem this causes when attempting to apply the criteria is best understood by following thought experiments. I start with the most straightforward cases: single individuals and single organisations.

6.4.2 Denialism in individuals and single organisations

Diethelm and McKee claim that if 'some or all' of the criteria in their list are met – with the behaviours described exhibited 'in a concerted way' – this is sufficient to establish denialism. This is problematic. Firstly, they do not set out what constitutes 'concerted'. More importantly, neither do they explain what they mean by 'some'? Is a majority (three) enough; or how about two? Most of the behaviours they outline are common in public engagement on socially conflicted topics.* Therefore, even three criteria would probably catch masses of people, suggesting this threshold is not very discriminatory.

However, setting this aside, let us focus on a single individual. Even seasoned contributors to a socially conflicted topic probably wouldn't engage an expert, false or otherwise, if they were acting in a personal capacity (although they could certainly quote one who had been hired by someone else). But if, say, three or four of the remain-

* The exception is that the public wouldn't engage experts. Also, for the cherrypicking point, they wouldn't discuss 'papers', but one could substitute 'lines of argument' instead.

ing behaviours were found in someone (and assuming this detection was done without bias), then it would be reasonable to conclude that something untoward had been found.

For single organisations instead of an individual, the 'false expert' criterion comes back into play, but apart from this, the same argument holds (provided the criteria are tested against official, considered outputs of the organisation, and not against the personal views of individual employees).

However, whether that which was detected amounts to a psychological phenomenon of 'denialism', with all the very negative connotations that term entails, is moot. This depends upon causation, to which I return in Section 6.5. For example, if someone's behaviour is rooted in the subconscious – perhaps driven by a strong cultural bias – then it is not the result of dishonesty, and a pejorative framing would be inappropriate. If, however, they believe their opponents to be correct, but consciously decide to contradict them anyway, then a case could be made that dishonest denial, as promoted in the popular framing, had been detected.

6.4.3 Denialism across sides

The fundamental problem anticipated at the end of Section 6.4.1 arises because this strict application only to single individuals or single organisations does not appear to be how Diethelm and McKee intend their criteria to be used. Although they actually give no formal guidance, their examples imply that the criteria are to be applied to multiple individuals and/or organisations or groups who are perceived to be on the same side of a debate. If these disparate subjects collectively exhibit enough ('some or all') of the behaviours between them, then denialism would apparently be established.

This is an approach that is beset with difficulties. As noted above, nowhere do they define what a 'side' is; they seem to assume that this requires no further explanation. And they give no clues as to how strong they feel the links between individuals or organisations would have to be to conclude that they were indeed acting in concert, and therefore actually constituted a side, or at least a coordinated part of one. In essence, they absolve the tester from having to establish that the subjects are operating in concert at all. The side is what the tester says it is, regardless of whether the tested subjects feel that they are acting in concert, or indeed actually have any significant level of co-operation. The tester could satisfy just one of the criteria from each of five organisations not acting in concert, and conclude they'd detected denialism!

This issue cannot be ignored, especially since Diethelm and McKee have invested the term 'denialism' with intent (for example their reference to greed as a possible motivator). For instance, a side could consist of a group of organisations and individuals bound by a conscious and coordinated nefarious intent; a conspiracy, in other words. Such a case requires demonstration, with hard evidence. Alternatively, a side could be bound by a shared *culture*, based ultimately upon *subconscious* coordination (see Section 6.7). But this too would need to be demonstrated, perhaps via a cultural analysis. In summary, the criteria are meaningless unless it can be demonstrated that the chosen targets constitute a coherent side. We cannot simply say that individuals or organisations or groups must form a side because they are 'denying'; this is what the test is supposed to be telling us!

6.4.4 The criteria applied to a typical 'side'

Both sides of contested domains attract participants with motivations that have little or nothing to do with the evidence. Their input tends to divert discussion away from pertinent facts and to polarise the debate. If they focus narrowly on technical issues (pseudoscientifically[160]), their impact on the debate may be modest. But if they promote arguments with wider policy and societal implications, this will draw more people into the debate and inflame passions, which brings existing biases to the fore, in serious cases leading to noble cause corruption.[161]

Diethelm and McKee claim to be able cut through this complexity; they say that the espousal of conspiracy theories and the use of logical fallacies are both reliable criteria with which to detect denialism, and can therefore be used to detect which side must be wrong on a socially contested issue. However, even a brief consideration shows that this cannot be true.

Consider a contested issue that features a largely evidential position, which is opposed mainly by religious believers. The religious side has a strong cultural alliance with a political party, X, which is thus drawn into the debate on the same side (religion is a culture, and the 'tribal' expression of politics, likewise). This sparks a reaction from X's political opponent Z, which weighs in on the evidential side. However, by default, Z does not deploy evidential arguments, but instead resorts to its accustomed cultural (tribal) weapons, for example claiming that 'folks who support the X party or (via association) oppose the evidential position, have inferior brains'. And indeed, fire is returned in the same manner. This scenario is depicted in Figure 2.

So, both sides of this debate feature groups whose motivation is purely cultural and nothing to do with the evidence, and both will indiscriminately use conspiracy theory, logical fallacies and demonisation; even the 'right' side is inextricably entangled with a morass of cultural behaviours, both offensive and defensive. In fact, such groups may form the majority of the support on both sides of the debate. In this case, useful information and genuine truth may disappear entirely – both sides are wrong, the public doesn't understand the technical issues, and these aren't even discussed anyway because everyone is engaged in a cultural slugging match.

Because both sides end up using fallacy and conspiracy, Diethelm and McKee's use of these as tests for denialism will fail, or at best be ambiguous. The same can be said of two of their other criteria, namely cherrypicking and false experts. How, then, do we tell who is right? Unfortunately, Diethelm and McKee give only one item of further guidance on this question, namely the existence of an 'overwhelming scientific consensus'. In other words, they argue that the 'right' side must be the consensus side. However, they do not acknowledge the difference between a scientific and a social consensus, or that the latter can pose as the former.[162] A socially enforced consensus (arising from a culture, or groupthink) can completely hide the state of the science, even from other scientists, if they are outside the relevant area of expertise. It can also increase the chances that scientists who *do* have appropriate domain expertise will straddle the rift between sides, or maybe even end up mostly on the 'wrong' side. Authoritative, apparently settled scientific consensuses have been overturned many times;[163] scientists and policymakers are not separate from

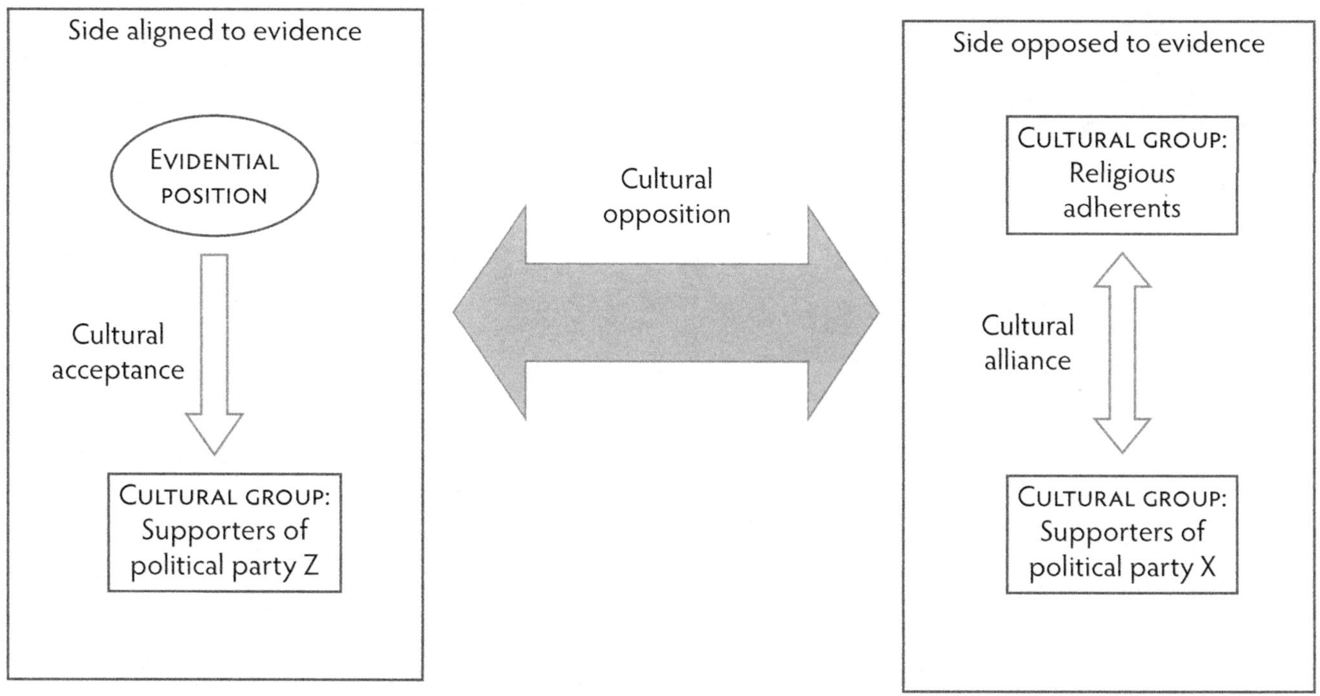

Figure 2. 'Sides' and cultural groups in a socially conflicted topic.

society, and they are subject to the same biases as everyone else. In the climate domain, for instance, there is plausible evidence that at least some scientists have biases that stem from the Catastrophe Narrative.[164]

Diethelm and McKee provide no method to distinguish a social consensus from a scientific one, or to determine the maturity of the latter (see also Section 6.4.6). For example, in the public domain and among public authorities there is a dominant cultural consensus on the certainty of global climate catastrophe, which claims backing by science, despite contradicting the mainstream scientific consensus (and sceptical scientific opinion too). In the public dispute over climate change, Diethelm and McKee's criteria would be ambiguous at best, and their backup guidance – the existence of an 'overwhelming consensus' – would fail.

6.4.5 Cherrypicking, experts, and the avoidance of bias

As noted above, two more of Diethelm and McKee's tests – cherrypicking and use of false experts – are likewise problematic. Where the main motivator is cultural bias, one would expect most cherrypicking to be unintentional and subconscious, so probably subtle in nature. But in a complex domain mired in claims and counterclaims, it can be difficult to identify even blatant cases without fairly extensive domain knowledge. Similarly, the picking of 'discredited papers' is subjective – it depends upon believing those who did the discrediting and their reasons for doing so, which implies a prior judgment that itself can only be based upon significant domain knowledge, and so may also be biased. Indeed, the very allegation of cherrypicking could itself be a cherrypick, if for instance this only presents an unfavourable part of the original case. In other words, the knowledge required to identify cherrypicking is itself domain dependent, thus tending to thwart objectivity.

It is the same story regarding experts. To know if an expert is 'false' requires domain knowledge. What they are paid and by whom is not, on its own, a definitive criterion (or even a major one; ideological bias often motivates people much more than money, although the two can also be aligned). Moreover, in a contested domain, navigating often labyrinthine funding paths can be almost as complex as evaluating the topic evidence; members of the public certainly don't have time to do this and, in all likelihood, nor will most domain authorities. Worse, the interpretation of funding network influences is itself subject to bias and polarisation.[165] Opposing networks are to be expected in a contested domain, and no simple rule of thumb – 'scientists paid by industry are less reliable', say – helps us cut through the complexity. Government scientists and university employees have just as much skin in the game as people funded by industry.

Finally, where a strong culture is at work in a conflicted domain (without conflict there probably wouldn't be 'denialism' anyhow), there is evidence that the more domain-knowledgeable individuals are, the more polarised they are too.[166] Whether this effect continues up to the level of true 'experts' would be hard to determine, but there is also much anecdotal evidence of highly polarised experts. This is all very problematic when hoping to resolve disputes by calling up the opinions of the more knowledgeable; it may make things worse!

So, absent some novel methodology (Diethelm and McKee do not suggest any) we have a fatal recursion. Correctly identifying cherrypicking and false experts implies domain knowledge that is reasonably deep, and necessarily unbiased. This in turn means already knowing – despite the confounding factor of a highly polarised environment – which side is in fact 'speaking to truth' and which is 'denying'; but this is essentially what we were meant to be finding out in the first place! Or in other words, the domain knowledge that's needed to investigate these characteristics brings with it domain bias, which may lead to erroneous judgement.

6.4.6 The standard of proof – a more useful criterion?

So, four of the five test criteria are not reliable; the kinds of behaviours they target can be seen on both sides of a conflicted issue. But what about the other criterion: 'Moving the goalposts'? Diethelm and McKee didn't actually call it that; they framed the point slightly differently, saying that denialists ask for an impossible standard of proof. However, the two definitions are equivalent; constantly moving the goal outwards must eventually reach a standard of proof that's impossible to attain. Diethelm and McKee imply that there should be some stable and realistic threshold of proof for any contested question, to enable the 'right' side to be objectively recognised. As examples, they cite four main domains: HIV/AIDS (denial that the former causes the latter), creationism, smoking/cancer, and climate change. I look at each of these domains below.

The first of these seems to have a definitive threshold; if one can independently replicate the development of AIDs from HIV, then proof has been achieved. Unfortunately, replication is only too easy. The most tragic cases were the children who developed AIDS after receiving HIV in blood transfusions.[167] Then there were the three laboratory workers (with no other risk factors) who, in separate incidents, developed AIDS after accidental exposure to cloned HIV. Or the six patients who picked up HIV from an infected Florida dentist – four developed AIDS and three died. All these cases were confirmed by genetic sequencing. Up to December 1999, of 56 health care workers in the United States with occupationally acquired HIV infection (and in the absence of other risk factors), 25 developed AIDS. Extensive research since then has revealed the life-cycle of HIV, and the development of AIDS in untreated patients can now be predicted.

For climate change, however, there isn't even a single domain of argument. The public debate is almost entirely divorced from the scientific issues, and for the most polarised case (the US), there is no more a standard of proof for certain global catastrophe in a few decades' time than there is for an international hoax. These are tribal positions, both of which are false.

Scientific debate, meanwhile, centres on the value for climate sensitivity – the warming expected from a doubling of carbon dioxide concentrations – and the damage that might occur for any particular value. But the IPCC hasn't been able to narrow a relatively large range for climate sensitivity in 30 years, and the damage functions used estimate the resulting costs necessarily integrate many disciplines (including economic models – hardly a beacon of successful predictions), greatly increasing the range of potential outcomes. To make matters worse, scientists outside the mainstream push the climate sensitivity range out further still, and in both directions.

But while Diethelm and McKee cite two scientific questions that occupy much of the conversation on climate blogs – global temperature records and the role of climate models[168] – they rather miss the key point: if there is a wide range of climate sensitivities (and associated damages), how do you even set a standard of proof? Or, looked at from the other direction, if a standard of proof had been determined, then the question of climate sensitivity would have been solved already.

In practice, the IPCC justifies the range by reference to 'multiple lines of evidence'. This may be perfectly acceptable, but it doesn't break down to a clear criterion of 'proof' for any particular outcome, and doesn't preclude legitimate debate about any of the plethora of factors that make up those lines of evidence. Nevertheless, because of cultural effects, the one damage scenario that is beyond any plausible range still dominates the public domain: a certainty of imminent and global catastrophe.

The secondhand smoke debate is home turf for Diethelm and McKee; they're acknowledged experts. But others in the field have robustly criticised their work – not only the paper we're discussing here, but also their studies on secondhand smoke – complaining of selection bias and the use of rhetorical devices and defamation.[169] As a novice in this domain, how do I know which experts are false, or if in fact neither is false but that the science is immature?

So, the secondhand smoke debate features a division of expert opinion, and as usual there is defensive behaviour on *both* sides. It is therefore unsurprising that any threshold of proof for the domain* is itself contested. Determining one isn't a matter of straightforward replication, but relies upon complex social and medical statistics that have to be slowly and painstakingly accumulated; this means there will be ample opportunity for results to be affected by researcher bias. Clearly, even Diethelm and McKee's home turf is problematic when attempting to detect denialism using their proposed tests.

Like the HIV case, proof of evolution over creationism seems like a very safe bet. Familiar issues, such as the increasing resistance of diseases to antibiotics, allow us to actually perceive evolution in action. Yet what would this contested domain have looked like, say, 20 years after Darwin's publication of *The Origin of Species*? What

* Properly a sub-domain.

supporting evidence was available then?[170] I submit that while the relevant criteria for proof may be obvious now, they were not so back then, even to the educated elite.[171] Therefore, if the correct evidential goalposts and hence the 'right' side can only be confirmed long in retrospect, a standard of proof for current debates will often be unreliable; we cannot be certain of where on the timeline of science emergence we actually stand.[172]

This all suggests that objective determination of a stable and achievable standard of proof is not a simple matter. The only case manifest via instant replication,[167] and so requiring no consensus, is HIV/AIDS. For many domains, a stable standard of proof simply reflects the maturity of the relevant science. If immature (and the long time necessary to collect and analyse social trends or medical or climate data can impede maturation), standards of proof will be contested, just like everything else. And they could legitimately move too, so will not be easily and objectively pinned down.

6.4.7 The failure of the five criteria

So even the most hopeful of Diethelm and McKee's criteria fails to provide us with a reliable means of identifying the side that is, overall, denialist. And because the behaviours targeted may appear on both sides of long-conflicted domains, do they truly define denialist activity anyhow? Can both sides be denialist? And assuming one side is indeed denialist overall, this still implies that some of its supporters are legitimately motivated. Together, such questions probe beneath the immediate problems with Diethelm and McKee's simplistic test criteria to reach more fundamental issues, of which the principal one is: what is the main cause of denialism?

6.5 Causation

6.5.1 Bad actors?

Diethelm and McKee suggest that the proximal cause of denialism is 'a few commentators', who they frame as bad actors: people who 'sow doubt' in defiance of an apparently overwhelming scientific consensus. In practice, such a consensus might not be scientific, even if it presents itself as such; it could in reality be cultural or a product of groupthink. This is the case for the consensus on certain climate catastrophe, which has led to widespread but inappropriate finger-pointing at supposed bad actors (through the merchants of doubt meme; see Chapter 5).

However, we can still ask ourselves whether Diethelm and McKee's view is that people who are persuaded into disbelief by bad actors become deniers themselves? Or should one consider such people the victims of deniers? In other words, are the bad actors merely agents behind a wider denialism, or are they the only 'true' deniers? This ambiguity isn't resolved, although the authors do say that the 'few commentators' are one part of 'a larger phenomenon of denialism'.

Another issue with the bad actors suggestion is that, in some domains at least, it is simply inapplicable. Creationism (opposing evolution) is certainly not driven by a 'few commentators', and historically at least, completely the reverse. One hundred and fifty years ago, essentially everyone believed in creationism, because everyone subscribed to one religious faith or another; a majority of the world's population still does. The legacy of those earlier times is that opposition to the science of evolution is inherent in large sectors of the public in some nations, including the US.*

On the other hand, at least in the West, it is generally accepted that the term 'denier' legitimately applies to anyone who subscribes to the view that the Holocaust didn't occur, not just a smaller number of bad actors who actively promote this view. This suggests that the same rule should be true of other domains, especially as parallels are often drawn between Holocaust denial and denialism scenarios in general, which Diethelm and McKee point out.

This ambiguity about scope adds still more confusion to the nature of denialism as perceived by Diethelm and McKee. However, despite contradictory elements, the stronger implication from the paper appears to be that denialism is not just a highly selective characteristic of 'a few commentators', but a wider phenomenon encompassing all those who take issue with what is an (alleged) overwhelming sci-

* Notwithstanding the recent innovation of 'God guided evolution' as a compromise position.

entific consensus.* I follow this interpretation from now on.†

While members of the public may indeed cherrypick and use fallacious arguments in support of their worldviews, as a practical issue it would be impossible to properly evaluate many of them according to the five criteria, even if these were reliable. However, this doesn't stop people from attempting to do so, applying Diethelm and McKee's general ideas (including the ambiguities) and the spirit of their test, for example as seen at Wikipedia. Those targeted may be anyone who opposes consensus views,‡ or indeed anyone who questions the particular views and biases of the 'tester', ranging from single individuals up to large social groups. In such circumstances all nuances about applicability and scope of the test are lost; indeed the 'test' may never be more than a brief thought-experiment. Hence spurious accusations of denialism become legion; the label is simply too useful a way to demonise opponents. Diethelm and McKee cannot be held responsible, but they have at least contributed inadvertently to giving the pejorative framing a veneer of academic respectability.

6.5.2 Diethelm and McKee's proposed primary causation

Diethelm and McKee have only a short paragraph dealing with the distal causes of denialism. This is disappointing; when boldly stating that a complex social phenomenon can easily and reliably be isolated, implying at least that it is well-defined, a reasonable grasp (or at least a theory) of causation is surely a prerequisite.

The relevant paragraph simply states that denialist motivations are: 'eccentricity' and 'idiosyncrasy' (with both apparently encouraged by 'maverick celebrity status'), 'greed' (with corporate oil and tobacco as examples), and 'ideology' or 'faith'.[173] A major problem from a social-psychology point of view is that these are very different motivators, each with contrasting power, scope, and resultant behaviours.[174] Lumping them together, without considering their fundamental differences, suggests that the authors have barely considered

* Diethelm and McKee don't address the idea that groupthink or a cultural narrative can imitate a scientific consensus.

† Especially when I consider if there may be a more useful conception of denialism, outside of Diethelm and McKee's framing. See Section 6.7.

‡ And of course the consensus does not necessarily have to be a scientific one; it could be cultural or groupthink.

the critical issue of cause.

At this point, it's worth noting that Diethelm and McKee's paper is in essence a distillation of a series of blog posts by Mark Hoofnagle, a medical doctor and prominent writer on denialism.[*,175] They have added specific examples from different domains, along with a few theoretical nuances,[176] but their 'five characteristics of denial', as shown in Section 6.3, are the same as Hoofnagle's. However, Hoofnagle has more to say about the cause of 'denialism', which he mainly and clearly attributes to dishonesty, although without analytical justification.[177] Diethelm and McKee conspicuously drop this claim, along with Hoofnagle's strong hints ('cranky', 'delusional') that mental illness is also a causal factor.[178] They are wise to do so.[179] Dishonesty is not a prime social driver, and for instance it couldn't seriously power the behaviour of the 45% of Americans that Diethelm and McKee say reject the evidence for evolution, or produce the significant public minorities who exhibit similarly strong resistance in other domains. Nor is there any justification for concluding such large numbers of people are 'cranky' or 'delusional'.

All this highlights that Diethelm and McKee's explanations of causation have run aground, as a result of their ambiguity of scope. Some items on their list – for example eccentricity – stem from personal psychology (as indeed does dishonesty, as proposed by Hoofnagle), and therefore can't explain mass phenomena. However others, such as faith, are mass phenomena, and stem from social psychology at the largest scale; they are not generally compatible with 'a few commentators' as causal agents. This seem contradictory. We would not expect a universal 'denialism' to have two entirely different psychological causes at different scales.

As noted above, I assume that if denial exists, it must be a social phenomenon, capable of shaping attitudes across a significant section of society. It could potentially be universal, with explanatory power at all scales – from the very small (groupthink) to the largest (a major culture) – thus capable of accounting for the great majority of what we see within the different domains.† A mass motivator

* Hoofnagle is properly cited by Diethelm and McKee, so I am not suggesting any wrongdoing.

† A social phenomenon would also explain the situation where bad actors are the visible

rules out dishonesty, eccentricity, idiosyncrasy and celebrity status (although they can be secondary or tertiary factors). Greed is not usually primary either, although at an institutional level and perhaps aided by the dishonesty of influential individuals, it can be a significant factor in some 'denialism' phenomena. Even then, like corporate dishonesty,* it may just be a proxy for a toxic culture; in other words it is still a group phenomenon.

This leaves two remaining possible motivations from Diethelm and McKee's causation list, namely 'ideology' and 'faith', which we already know are powerful cultural drivers. If these are indeed involved, then a whole new world of implications is opened up. But surprisingly, beyond the two words, Diethelm and McKee say almost nothing; I address this omission in Section 6.7.

Diethelm and McKee's list of assumed causes appears to have been 'grabbed off the street', with Hoofnagle's claim of dishonesty and crankiness removed, but with no theoretical framework inserted in its place. Their framing is therefore unsound; it appears not to be based on a properly characterised phenomenon of 'denialism' with a specific cause.

6.6 Diethelm and McKee has little utility

The flaws outlined above result from a lack of underlying principles. Framed in the manner of Diethelm and McKee, Hoofnagle and Wikipedia, there is no solid phenomenon of 'denialism' to actually test for. It isn't possible to objectively evaluate their five criteria, and even it was, we still wouldn't be able to reliably tell which side was right.

'Denialism', as envisaged in the paper, largely comes down to the use of a handful of fallacies, which in fact is a subset of a venerable and much longer list,[180] some dating back to classical times. Diethelm and McKee neither add to this list nor to our understanding of how such devices affect people. Moreover, people use fallacies all the time, and if they do so systemically or excessively, it is readily detected even by the uninitiated.

Diethelm and McKee provide no means to objectively discern

part of a wider social group that remains hidden because it is held to be beyond the pale.
* See Endnote 179.

why people are deploying such devices either. In the complex sphere of human society, there are many motivations, only *some* of which would imply that they were inappropriately, and indeed knowingly, opposing genuine and proven scientific facts. Typically, the motivation is subconscious, especially when passion and deep bias dominate.

6.7 Identifying denialism

6.7.1 Does denialism actually exist?

If we are going to establish whether denialism really exists, we need to return to the issue of causation. We have already eliminated three of the factors that Diethelm and McKee say might motivate it. The remaining two – ideology[173] and faith – indicate the presence of political and spiritual cultural entities respectively. Could denialism, if such a thing exists, be motivated by strong cultural influences, and if so could we formulate a more objective test for its presence?

Fortunately, there is a great deal of research on cultural behaviours, from which we know, for instance, that when an individual's cherished values are threatened by new notions, they will typically react defensively, dismissing or even aggressively challenging them.[181] At the same time, they will tend to unquestioningly accept notions that protect their values. In other words, they exhibit strong bias; they lose objectivity. In principle, the values involved don't necessarily have to be cultural ones, but are much more likely to be cherished and robustly defended if they are.

Such defensive behaviours may be induced by a threat from any strong consensus, whether scientific or cultural.* If there is indeed a universal phenomenon of denialism, then it makes sense to look for it within this defensive behaviour. Interestingly, science journalist Michael Specter, whose writings on denialism have been another important contribution to the popular framing of the term,[182] almost gets to the same position when he says that 'fear of technology' is the main cause, because such fear essentially stems from cultural defence. However, he then veers away from a universal cultural explanation

* And notwithstanding the fact that these arise by very different means.

and heads back to similar territory to Diethelm and McKee, saying:

> Corporations, wrapping themselves in the mantle of progress but all too often propelled by greed, have done more than religion or even Luddism to inflame denialists and raise doubts about the objectivity of science.

Interested readers can find further details in the endnotes.[183]

6.7.2 Denialism as cultural defence

If denialism is motivated by cultural defence, we should expect a great deal of similarity between deniers and other people who are just defending their cultural values, as they have done throughout history. This being the case, what would a denialism motivated by cultural defence look like? (Only for the purposes of the list below, to distinguish this from the popular framing, I'll call it 'denierism' as exhibited by 'denierists'.)

- *Denierists would mostly not be dishonest or mentally ill*: One reason cultures are so powerful is that they are not driven primarily by dishonesty, which is a relatively weak force; overall, cultural belief is both passionate and honest. Hence most denierists would not be mendacious; they would be defending the truth as they see it. Similarly, they would not be mentally ill.
- *Denierists would exhibit a range of behaviours:* Cultural defence is not black-and-white. In response to alien values, it may produce resistance – and there may be an extreme fringe – but also compromise to various extents. So denierists should exhibit a range of behaviours too, although if they are only a subset of cultural defenders, they may exhibit a correspondingly smaller range of behaviours.
- *Denierists would defend the values of allies*: Just as with the defence of nations, cultural defence calls upon alliances. Hence powerful and complicating alliance effects will be in play, such as those described in Section 6.4.4. This means that denierists may defend not only their own core cultural values, but those of an allied culture too.
- *Everyone would be a denierist*: No one is free of cultural influence, hence in theory we'll all exhibit denierism about some-

thing.

- *You can be a denierist on one conflicted issue, but objective on another*: Cultural defence is domain orientated. People can be hugely biased in one domain, but perfectly objective in another, so one cannot assume denierism will cross domain boundaries.*
- *Denierists will sometimes be right*: Our instincts can't tell if an invader is cultural or evidential but, since cultures are all built around fairy tales, rejecting them would undoubtedly be right.† An obvious example is a reaction against extreme ideological propaganda.
- *Detection of denierism does not imply detection of evidence rejection*: Cultural defence can be triggered by both social and scientific consensuses, so detecting denierism doesn't mean that evidence has been rejected. It may just be that a culture has been detected. This is especially tricky if a scientific consensus has become intimately associated with a cultural position, in the way described in the example in Section 6.4.4.
- *Etcetera*: Cultural effects are many and varied; denierism will likewise exhibit complexity.

It is clear then that 'denierism' – cultural defence raised against a perceived threat from a consensus of some kind – in general looks nothing like denialism, in the popular perception of the term as framed by Diethelm and McKee and others. In light of this, a more suitable question to ask than at Section 6.7.1 ('Does denialism actually exist?') is: 'Does cultural defence better explain the widespread resistance that we see to various consensuses?'

6.7.3 Cultural defence versus popular denialism

Resistance to consensuses is seen at all scales, but this does not necessarily mean that a single phenomenon is responsible for all cases. However, a phenomenon that can explain behaviours across most of the scale would make for a more parsimonious model. Cultural defence therefore looks a strong candidate, because it can be mounted

* Although when there are two allied cultures, the denierism may cross boundaries.

† In the rest of the book I use the term 'apt', rather than 'right' to refer to situations in which the cultural defence mechanism of 'innate scepticism' correctly detects a culture. See Chapter 7.

by anything, from members of a small group (either a splinter of a larger culture or just a small groupthink entity), up to large sections of society (adherents of a major cultural entity such as a religion). It may also explain some cases that are often assumed to be individually motivated. For instance, resistance to some consensuses is often said to be the result of corporate greed, which in turn is put down to the individual greed of influential managers, but in reality this could be a kind of toxic groupthink.*

While Diethelm and McKee touch briefly on cultural motivations, namely 'faith' and 'ideology', their emphasis remains very much on individual motivations. However, these do not scale well: dishonesty, greed, idiosyncrasies or psychological flaws would not inspire large sectors of the public. So, resistance to a scientific consensus, say, would be seen among smaller numbers of people, and they would exhibit weaker levels of motivation. They would therefore be much easier to combat than the large numbers who stubbornly oppose such a consensus in reality.

So on this score, cultural defence looks a far better candidate. This is not to say that individualistic motivations are absent, only that they are peripheral. Socially conflicted topics will inevitably attract some greedy or dishonest people seeking advantage, but they are not the cause of the conflict; they are a symptom, and will turn up on both sides, not only on the one that the passage of time proves to be wrong. However, in the popular framing, such people are typically seen as causal.

While failing to identify a single cause that could explain resistance at different scales, the framing put forward by Diethelm, McKee and Hoofnagle has other weaknesses too. A tendency to emphasise good and bad motivation fosters a black-and-white paradigm,[184] a problem further entrenched by their claim that simplistic criteria can distinguish one side from the other. The real world is more complicated than that. For instance, cultural defence may be triggered by a new consensus that challenges cherished values. This 'apt' innate scepticism, as I term it, is exactly what is exhibited by most of the

* For example, the case of corporations resisting the scientific consensus on tobacco may stem from a group motivation to protect their corporate 'tribe'.

public who are sceptical of claims of certain global climate catastrophe.* Such people are not 'bad'; they have correctly identified a false consensus (because it is cultural). And even in cases where cultural defence is triggered mistakenly, which is to say that people's innate scepticism is 'inapt', and that they are therefore wrong, it still does not follow that their motivation is bad. As an example, people who are sceptical of the science of evolution are motivated by their honest religious belief. So there are a host of 'shades of grey' between the black and the white of good and bad motivations. Once again, cultural defence better explains these real-world characteristics, which the popular framing doesn't effectively address.

In the majority of domains, most of those who resist a consensus are honestly motivated, but this may not always be so. For instance, Western Holocaust denial† at least superficially resembles the popular framing. But cultural defence can lead to a wide range of behaviours, and it is therefore quite possible that some of them, at the extreme fringes, might correspond to such obnoxious behaviours. However, most people exhibiting cultural defence will not be like this at all; instead they are free of nefarious motivation and their behaviours are far more moderate. And it is the failure of Diethelm and McKee's test to distinguish the honest and the nefariously motivated that does most damage. Their naïve five criteria will pick up honest cultural defenders, often on both sides of a socially conflicted domain, and flag them as deniers. As a result, all will be stigmatised, tarred with the same brush used for real extremists.

6.7.4 What should we do about it?

It's possible that behaviours reflecting the popular framing of denialism could never be isolated from cultural defence, never meaningfully distilled into a 'denialism' that's worth the pejorative name. Nevertheless, we need more reliable and constructive methods of objectively clarifying contested domains than the inevitable dishing out of labels – greed or crankiness or lying – to people who do not deserve them.[185] This in turn could lead to more productive ways of

* Innate scepticism and its trigger conditions are explored in the next chapter.

† In some Arab countries this has a very different character, being much more disconnected from European history, and indeed motivated by 'vanilla' cultural defence, in this case from the perceived threat that the state of Israel represents to Arabic/Islamic culture.

combatting resistance to genuine evidence and reason. Some progress on this has come from Joshua Rosenau, the head of the US National Center for Science Education. He doesn't attempt specific tests for denialism, but agrees with the causation proposed here, pointing out that resistance to scientific evidence is 'about deep fears and core personal identity', where identity is linked to a 'social group'. Rosenau proposes a promotion of the relevant science from in-group members or leaders, to show that 'rejection of [the] science is not a prerequisite for membership'.

Whatever the answer, we should always determine first whether the fears expressed by those resisting scientific evidence are apt or inapt – in other words, whether a culture is in play or not. That doesn't tell us what (science) is right, but it may tell us who is wrong (recall that all cultures are wrong). It may also help in scenarios where (correct) science and a culture may occupy the same side, with resistance prompted by a (false) cultural narrative that is promoting the science, rather than the science itself. And, as we will see in Chapter 7, we should always bear in mind that an innate scepticism of groupthink or cultural dominance, which society lauds, shares the same roots as innate 'denialism', which society stigmatises.

6.8 This framing of 'denialism' thwarts the authors' intent

Diethelm, McKee and Hoofnagle have, laudably, long fought against anti-science factions. Diethelm and McKee wanted to provide health professionals with the tools to do so. Hoofnagle wanted a means to combat invalid emotional arguments. However, their priors and the weaknesses of their work[186] have created enormous problems for the enterprise of science and, more importantly, for society as a whole. Hoofnagle's emphasis on individual psychology, and Diethelm and McKee's vague, unsupported list of causes, have diverted attention from social-psychological explanations in general, and cultural causation in particular. Their focus on the personal, and their spurious claim of being able to distinguish 'correct' and 'denialist' sides, have inadvertently given academic legitimacy to anyone who wants to call out opponents as being ruled by greed, or as psychologically impaired or systemic liars, or almost anything else. Accusations of 'denialism' have been turned into a powerful weapon for policing cultural consensuses – notably climate catastrophism – and protecting the fairy-tale narratives at their cores. A cloak of psychological and academic 'wisdom' hides the fact that such accusations are nothing more than out-group demonisation – a feature of cultural entities since the dawn of mankind.

Lacking proper definition, Diethelm and McKee's framing of denialism has spread as a negative meme that is useful to any emotive position.[187] As a consequence, across many contested domains, misunderstanding, stigmatisation, fear and other emotive reactions have been amplified at the expense of reason and science. The casual and common use of the word 'denier' to equate legitimate questioners with those who deny the Holocaust, is the inevitable result. Back in 2012, Hoofnagle partially acknowledged this problem,[188] and even the Wikipedia page recognises it too. In a paper published only a year after the one examined here (2010), Diethelm and McKee complained that accusations of 'denialism' were being used by a 'wrong' side. This too was inevitable. But I doubt they, or anyone else, will get the pejorative 'denialist' genie back into a bottle anytime soon.

Chapter 7

INNATE SCEPTICISM

Public resistance to an alien culture is often ascribed to belief in a competing one. However, I suggest here that it is the result of a different mechanism, which I call 'innate scepticism'. Although a competing belief will amplify innate scepticism, it is not a necessary condition for resistance. Properly understood, innate scepticism is key to explaining what's really happening in culturally conflicted domains. In this chapter, I probe its origins and nature, while in Chapter 8, I measure its presence in the climate domain.

7.1 Asking a fundamental question about scepticism

In an article entitled 'Science and the public: debate, denial, and skepticism',[189] social psychologist Stephan Lewandowsky and colleagues ask a profound question:

> What characterizes the public response to scientific discoveries that are 'inconvenient', or threatening to one's lifestyle, livelihood, or deeply-held beliefs? Is it debate, denial, or skepticism?

Unfortunately, the important issues this question raises are sidestepped by the authors. Let us consider two examples.

The first one arises from the authors' framing of the question in terms of the public reaction to 'scientific discoveries'. Categorising a negative public reaction as 'denialism' might be appropriate when dealing with 'discoveries'; in other words, where the science is mature and findings can be considered factual. But public attitudes are likely to be conditioned by scientific hypotheses as well as discoveries; in other words, by science that is immature, and far from settled. In some cases, the maturity of the science may itself be in dispute. The framing of the authors effectively sidesteps such distinctions, hence missing much of the complexity behind public responses.

The second one is that, as we saw in the last chapter, it can be far from easy to distinguish 'denialism' from scepticism. The authors claim it is possible to do so with 'relative ease', citing three main sources in their support. However, the first is the Diethelm and McKee 2009 paper that we examined in Chapter 6, which provides neither an objective test for denialism nor any social theory of causation for it. The second paper, McKee and Diethelm 2010,[190] while somewhat wider in scope and acknowledging some role for cultural factors, still supplies no underlying theory and also sticks to the same flawed test criteria. The third, a book by Naomi Oreskes and Erik Conway,[191] focuses chiefly on the climate-change domain and on some particular scientists and personalities too, offering little social theory to explain denialism and settling instead on what amounts to an allegation of nefarious conspiracy.[192] This is similar to motivations proposed by Hoofnagle (dishonesty), and Hoofnagle, Diethelm and McKee (corporate interests and greed).

The article has been robustly criticised by people working in the climate domain[193] (as have Lewandowsky's methods in other works on 'climate psychology'[194]). From the perspective of this book, the concern is that the authors don't seem to have realised that a need for tests to distinguish 'denialism' from scepticism suggests that the two phenomena are fundamentally related. Had they done so, and pursued this line of thinking into the climate domain (which is their main focus), they would surely have come across the cultural nature of the narrative of certain climate catastrophe, and legitimate* resistance to it. This in turn might have prompted Lewandowsky to reconsider his rather unconventional view that higher uncertainty in climate science can only mean more likelihood of disastrous outcomes,[195] a position that appears itself to be the result of significant 'leakage'[196] from the cultural narrative of climate catastrophe.[197]

Despite these problems, the authors make some excellent recommendations in regard to easing social conflict around scientific issues,[198] and raise deep and pertinent questions about the nature of public scepticism. As we shall see, when lifestyle, livelihood, and

* Stemming from cultural defence, as noted in Chapter 6.

especially deeply-held beliefs are threatened by science, cultural resistance – scepticism and 'denial' – is indeed aroused in the public.[199]

To assess the full picture, it is just as important to ask the opposite question too: how does the public react to science that is 'convenient' or encouraging to lifestyles, livelihoods, or deeply-held beliefs? In other words, not all support of evidential positions on conflicted science topics is due to rational understanding. But this question too is sidestepped by Lewandowsky and his team, as we will see.

7.2 The origins of scepticism

7.2.1 The two faces of scepticism

Barry Marshall, the doctor who discovered the role of *Helicobacter pylori* in causing ulcers and stomach cancer, said in an interview with *Discover* magazine (emphasis mine):

> I presented that work at the annual meeting of the Royal Australasian College of Physicians in Perth. That was my first experience of people being totally *skeptical*. To gastroenterologists, the concept of a germ causing ulcers was like saying that the Earth is flat.[200]

To the detriment of patients, it took some years for his theory to be accepted,[201] even after the famous demonstration in which he infected and then cured himself; nevertheless, plausible evidence of the involvement of bacteria had been accumulating over a far longer period. Indeed, a hypothesis very similar to Marshall's had been developed over a century earlier by Böttcher and Letulle, but failed to prevent scepticism of bacterial causation from becoming dominant.

So why does scepticism appear to have two faces; often healthy and indeed crucial to scientific inquiry and progress, yet at other times an unhealthy impediment? To get a better understanding, we should go back to the beginning and look at its origins.

7.2.2 How old is scepticism?

In accordance with the general theme of this book, in most of this chapter I will restrict myself to looking at the responses of the non-expert public to disputed issues. So for now, we'll leave behind gastroenterologists and climate scientists and all the other experts. We'll return to them briefly later.

When members of the public respond in a sceptical manner to a narrative, be this religious or political or scientific in origin, their response is not typically shaped by knowledge of Socrates or Pyrrho or Descartes or Hume. Such philosophical sceptics have been around at least since the Sophists in the fifth century BC,[202] yet they cannot be the engine driving public scepticism, which can be detected much further back in time. For example, it is argued that there is scepticism in the works of Homer,[203] who may have lived centuries before the Sophists, or alternatively may not have been real at all – it is quite possible his stories were collections inherited from oral tradition stretching still further back. *The Harper's Songs*, originally from the Middle Kingdom of Ancient Egypt (c. 2000 BC), certainly contain scepticism, mainly in the form of doubts about life after death and in contradiction to the dominant religion of the time.[204] Even more explicitly sceptical on that subject is *The Immortality of Writers,* a text dating to the end of the Nineteenth Dynasty, in around 1190 BC.[205]

These examples don't prove that ancient *publics* were highly sceptical, but none of the texts are exclusive to intellectual elites.[206] Moreover, the personal art and graffiti of ordinary people in the ancient world confirm the presence of a robust sense of scepticism. I particularly like the private drawing of an Egyptian tomb-painter, who humorously[207] depicts his god-pharaoh as a mouse driving a chariot pulled by a dog.[208]

Detecting scepticism this far back depends upon the evidence from written and other records.[209] However, we can infer something about prehistoric scepticism in a more roundabout way:

- from observation of systemic scepticism in the oral societies that survived into the 19th and 20th centuries;[210]
- from the fact that scepticism seems to stalk every religion (indeed every social consensus[211]), and in turn, that religious practices of one form or another have been around for a very long time.[212]

It is reasonable to hypothesise that these patterns have always held, and hence to conclude that scepticism is very old indeed. It is probably old enough to have been part of our evolutionary processes.

Of course, there was no formal science to be sceptical about so long ago. But public reactions aren't different depending on whether challenging narratives happen to be religious or scientific or philosophical. The public is, almost by definition, largely uninformed; they aren't experts in any of these domains. Their reaction must therefore arise from factors that are largely independent of detailed knowledge.

7.2.3 Where does scepticism come from?

The Oxford Online Dictionary defines scepticism as: 'A sceptical attitude; doubt as to the truth of something'.[213] However, this straightforward text hides an apparently fiendish complication regarding public scepticism. Lacking domain knowledge, how do the uninitiated know whether their doubts are well or ill founded? How could they even begin to tell? The answer appears to lie in evolutionary theory, an observation that accords extremely well with the suggestion of scepticism having ancient roots, as noted above.

Deception, and the ability to detect it, is not unique to humans. It has evolved in many species, including apes, octopuses, fireflies and even plants.[214] There has always been a continuous deception-versus-detection 'arms race', both within and across species, as philosopher David Livingstone Smith notes (regarding the former):[215]

> Deceptive maneuvers have been studied in many species of flora and fauna, including our closest non-human relatives, the chimpanzees, who are capable of sophisticated tactical deception of one another...Just as the proliferation of reciprocal altruism encouraged the evolution of deception, so intra-specific cheating facilitated the evolution of cognitive mechanisms for discriminative altruism and cheater detection, leading to an escalating co-evolutionary 'arms race' in which ever more sophisticated methods of deception were matched by ever more sophisticated methods of detecting and safeguarding against deception.

This arms race has resulted in the development of complex strategies and interactions, especially in humans.[216] Indeed, as noted by evolutionary biologists Luke McNally and Andrew Jackson, the same pressures selecting for group co-operation, a key characteristic of humans, appear to have selected for deception too:[217]

> ...ultimately, our ability to convincingly lie to each other may have evolved as a direct result of our cooperative nature.

There are two levels in the human arms race. The simplest is where individuals deceive, and other individuals attempt to detect this deceit. The more complex situation is where cultural groups deceive, via their fairy-tale narratives, and individuals attempt to detect this group deceit. The interaction at each level is very different, but both feature a common key concept: a point at which people can sense there's a problem, yet are unable to prove that a deception is taking place, so choose to 'withhold judgment'.[218] This suspicion of deception, 'doubt as to the truth of something', is instinctive, a scepticism that arises from skills long honed by the evolutionary arms race.

Individual deception and detection are relatively well understood. Science writer Michael Shermer lists signals that people can use to detect deceits:[219] nervousness, excess control,[220] apparent rehearsal, inconsistency of a story over time, and an increase in such signals when the subject is under cognitive load. These clues might be up- or down-weighted depending on existing levels of trust in the messenger but, most importantly, none of them have anything to do with the information being conveyed to the audience: no domain knowledge is required for these instinctive skills to work!

But what if the concept in question isn't presented to us by an individual, but by an entire cultural group? What then does the arms race look like? What clues to deception do we search for?

7.3 The nature of innate scepticism

7.3.1 Scepticism of group deception

As in the individual deception/detection case, the conditions for an arms race are satisfied for groups as well: there are incentives for group deceptions to work, and there are incentives for individuals to detect them. The relevant groups are cultural entities, which, as we saw in Section 3.1, are ultimately a product of cultural group selection and gene-culture co-evolution.[21] We also saw that altruism towards other members is a critical feature of such groups and, in order for it to function, members need to know who is part of their group and who is not. This information does not come from biology, but from cultural signals – behavioural cues that are associated with a 'consensus', a necessarily false narrative (see Section 3.2.3.2), acceptance of which indicates 'correct behaviour' in the group. The fairy tales propagated by religions, those dominant cultural narratives throughout history, are examples of such powerful collective deceptions. The incentive for deploying them is that they are hugely advantageous to groups.

However, there are also incentives for people to detect group deceptions. For example, detecting a rival cultural consensus helps prevent adherents drifting away to other groups. Another less obvious incentive is to enable cultural adherents to identify when their own native culture is becoming too burdensome; if adherents become abject slaves to an unrestricted and eventually grotesque cultural narrative, their productivity as a group will fall. In both cases, individuals need clues to tell them that a narrative is cultural, which is to say, a group deception (although the trigger-level for scepticism will be much higher in the second case).

An example of such a deception that potentially covers both of these scenarios is an extremist ideology, with its cultural narrative disseminated via state propaganda (as noted in Section 3.3.3, extreme ideologies are also cultural). Stephan Lewandowsky and his co-authors have confirmed that a 'stable personality trait' of scepticism boosts our resistance to this type of misinformation – deceptions propagated by those with a strong cultural or worldview bias.[221] This ability may be even more useful when cultural signals are more subtle than the propaganda of extremist regimes, but are nevertheless still harmful.

Most adherents of a culture are honestly motivated; their belief is subconscious, so they're unaware of the deception in the cultural narrative they are propagating. As a result, Michael Shermer's list of clues for detecting individual deception, outlined in Section 7.2.3, is no help when trying to determine whether a narrative is a fairy tale pitched by cultural adherents or a rational proposition. Nevertheless, there are still clues to be gleaned. I suggest that a cultural entity may be detected if the narrative and the way it is pitched is:

- too coherent and coordinated (cultural consensuses are instinctively policed, creating a stark uniformity).
- too certain (belittling or bypassing challenges, including uncertainty).
- too forceful (suppressing other views).
- too emotive (both positive passions, and negative ones such as fears and worries).
- too arrogant (demeaning and/or demonising dissenters).
- too universal (claims of universal applicability are tenuous, yet often boldly stated).
- too existential (threats are exaggerated – 'bogeyman' stories).
- too conveniently helpful to a (different) culture.*

Similar to the individual case, trust in the messenger will also be a factor in the assessment.

I propose that an evolved ability to detect such clues is the origin of public scepticism of cultural deceptions. This is in accord with the historical examples of scepticism among inexpert publics set out in Section 7.2.2. If they are sufficiently sceptical of a narrative to think it may be cultural, and therefore a deception, large swathes of publics will 'withhold judgment'. Moreover, and most importantly, just as with detection of individual deceit, they don't need significant domain knowledge in order to do so.

Detection of individual deception is not relevant to this book and not pursued further. However, 'innate scepticism' – the instinctive detection of cultural group deceptions – is critical. And as we will see for the climate domain in Chapter 8, we can measure its presence.

7.3.2 The importance of pre-existing beliefs

Although innate scepticism works independently of detailed domain knowledge, pre-existing social values and aspirations can work to suppress or enhance it. For example, if the messaging from a rising cultural consensus aligns well with individuals' pre-existing values, the clues set out in the last section will prompt very different behav-

* I am referring here to the idea of allied belief, which is considered in Section 9.3.

iours. Instead of provoking scepticism, they will have the effects shown in Table 5, in each case, closing down someone's ability to detect the deception; the clues are simply far less obvious to them.

However, if their pre-existing values are unaligned with the rising cultural consensus, or actively opposed to it, then it is very likely that innate scepticism will be triggered, and strongly so in the latter case.

Some of the pre-existing values that modulate the expression of innate scepticism in this way will themselves derive from collective deceptions. In other words, the expression of innate scepticism also depends upon whether adherents of an existing culture see a newer one as a rival or an ally (or both simultaneously – see Section 9.1.5). This is what is behind the final list item in Table 5: highly aligned values will likely mean that the new culture is perceived as an ally.

So innate scepticism can be undermined, as the powerful influence of cultures on human societies makes clear.[222] However, significant scepticism about cultural beliefs always seems to persist; the one seems never to have been found without the other.[210] This much

Table 5. Effect of messaging that aligns with pre-existing values.

Message is:	Effect
Highly coherent or certain	Confirms the comforting value-framework upon which social identity is based, or at least does not challenge it.
Forceful, emotive	Evoked emotions will resonate, hence thwarting the objectivity in sensing that these may be inappropriate, and also that the messaging is too forceful.
Arrogant	Demeaning or demonising of those disagreeing with the cultural consensus may seem justified, or at least a blind eye will be turned.
Universal	Claims of universal applicability will seem quite natural.
Existential	Once emotions are engaged, the existential stakes will seem justified, especially if information sources are largely trusted (which may simply mean they are sympathetically biased).
Helpful to a different cultural group	Sympathy or support for the rising culture.

is clear from the historic references above, but it is still more obvious today. Although science has amplified scepticism towards traditional religious narratives, no modern secular culture – communism or fascism, for example – has escaped battle with public scepticism either.[223] Even when a low profile is necessary to avoid official displeasure, scepticism still exists and finds a way to express itself.*

7.3.3 Innate scepticism as cultural defence

As noted in Section 3.1, cultural entities are a legacy of our evolution, and a product of cultural group selection. Moreover, they are not the only legacy. After eons living as groups, the very way we think is orientated towards our society; for instance, our brains automatically model the aspirations of others, mirror those around us, and suppress self-interest for group benefit. Our thinking is essentially social thinking.[224] As a consequence, our personal identity is tightly bound up with the values of our social groups, and in turn, throughout the ages, the most important of these values have typically been explicitly cultural; in other words, they come from cultural entities, such as the religions.

As we have just seen, if a rising culture threatens our existing values, innate scepticism is triggered. Our identities are precious to us – they are us – so a fundamental way of viewing innate scepticism is that it is a mechanism for protecting who we are. This view is consistent with the 'identity protective cognition' effect, identified in the US public by social psychologist Dan Kahan. His very strong data suggests that the opposing sides on highly conflicted science-related topics are each biased in a manner that best protects their different cultural identities.[225] In other words, the conflict doesn't represent competing knowledge positions, but competing identities.

We have passed through this ground before, in Section 6.7, when proposing that 'cultural defence' may not only be the source of 'denialism', but also of entirely appropriate resistance to cultural consensuses. As we will see in Section 7.6, innate scepticism is the mechanism for both.

Some readers might be tempted at this point to wonder why, in

* Authority displeasure renders most official sources useless for confirming historic scepticism, which is why unofficial sources of information, such as graffiti, are so useful.

the twenty-first century, we should still care about cultural belief, or rejection. For the majority of people throughout most of history and before, cultures were a huge deal. We couldn't just take them or leave them. Lack of support for one's (typically religious) dominant local culture probably meant severe social disadvantage, if not complete ostracisation. In places where two or more cultures competed, your very life might depend on which one you'd committed to, and which ones you had rejected. So instinctive behaviours reflecting this importance won't disappear overnight. In some countries, the secular cultures of fascism and communism are far from finished; the culture of climate catastrophism is global and highly intrusive; cultures based on Critical Race Theory and extreme trans-rights are just two of the newer kids on the block. And the great majority of the world still believes in one religion or another. Over a period of little more than a century, millions have twice answered the call to sacrifice themselves for what amounted to the cultural values of their home nations. And the work of Kahan and others shows that, even in everyday social and political disputes, cultural identity still matters. When expressing what is important, it seems people must either defend against insistent cultures of various kinds or promote them; on the whole, we still can't just take them or leave them. If anything, politics in the West is becoming more and more cultural; in other words, more about identities than policies.

So innate scepticism can be thought of as a cultural defence against something that threatens our identity. Equivalently, cultural identity restrains or disables any innate scepticism aimed inwards, towards our own beliefs.[226] That said, if cultures go too far or become decadent – abusing their hosts – this may well trigger an innately sceptical reaction from their own adherents (as noted in Section 3.2.2.1).

So, to return to the principle outlined at the start of the chapter, our 'lifestyles, livelihoods, or deeply-held beliefs' are readily identifiable in our responses to scientific theories or discoveries – whether innately sceptical or culturally supportive. (This is statistically identifiable at group level; we must remember that every individual is unique and not predictable.)

However, it's worth noting that different cultural narratives will challenge different deeply-held values. This means if we observe that an individual is innately sceptical of one cultural narrative – say, catastrophic climate change – we cannot conclude that they will be innately sceptical towards another – say a religious one. In fact, the same individual can offer up fierce opposition to a cultural narrative in one domain, and none whatsoever to another narrative in a different domain.*

7.3.4 Innate scepticism and truth

Innate scepticism is not about discovering the truth. Doing so isn't necessary in order to protect deeply held personal values, nor is it typically possible. For instance, for centuries, many people were sceptical of religious consensuses on human origins. However, the truth of the matter wasn't known, so until the science became clear, they withheld judgement. This stance is still the typical result of innate scepticism, in whatever domain it is expressed. Sometimes, however, the result is a modified belief, a 'bounded scepticism' so to speak, an example of which is when people express belief in God, but a disbelief in his current agents upon Earth. The belief in 'God-guided evolution', widespread in the US, and which partially resolves the conflict between science and religion regarding human origins, is another.

When cultural entities clash, the central narrative of one can become a focal point for sceptical resistance to the other. Similarly, sustained internal scepticism of an established culture (perhaps due to its decadence) can create the conditions in which a breakaway eventually arises.[227] Such schismatic movements may proclaim themselves to represent the 'truth', but in a literal sense all competing cultures are just as untrue; they're all collective deceptions.

7.3.5 Cultural belief and disbelief are separate functions

We can think of innate scepticism – an emotive rejection of culture – as 'cultural disbelief'. However, it is not just the opposite of 'cultural belief'. The two functions are separate, and act semi-independently.

The default state is for people's innate scepticism to be switched on when assessing incoming narratives. However, as noted in Sec-

* However, see also Section 7.7, item 8), the note in brackets.

tion 7.3.3, strong belief in a culture, say Zogism, will heavily attenuate its effect for narratives stemming from this culture (unless Zogism becomes decadent). In contrast, innate scepticism will be enhanced when Zogists, especially the more ardent, assess the narratives from competing cultures. Let's say that Gozism is the main opponent of Zogism. Where belief in Zogism is high, innate scepticism of it is low. And where innate scepticism of Gozism is high, belief in it is low. This simple scenario suggests that cultural belief and disbelief are mirror images of each other. However, this view is really a consequence of the simple binary nature of the example: two opposed cultures, so no entangling alliances, and no shades of grey between the low and high of either belief or disbelief.

In practice, even for a single culture, both cultural belief and disbelief exhibit a range of strengths across a typical population. In a country acknowledged as Zogist, say, only a minority are likely to be ardent adherents of Zogism, and only a minority risk the disdain of their peers by eschewing it. The rest of the population are somewhere in between. However, importantly, belief and disbelief will not always mirror each other. As we shall see later in this book, innate scepticism is also expressed as a result of the intrusion of hard realities that clash with cultural narratives. The stronger these 'reality constraints', the more innate scepticism comes to the fore. This means that for the same population, cultural belief and cultural disbelief can both be strong. However, it is different circumstances that bring out one or the other. Populational adherence to a culture is the sum of everyone's belief and disbelief across all circumstances. We can think of the contradiction between levels of belief and disbelief (e.g. both high), as 'cultural hypocrisy'.*

Further cultural hypocrisy may occur if there is also indirect belief in Zogism, from its alliance with, say, the culture of Zapism. So, when there are no reality constraints, belief comes to the fore, but when reality kicks in, disbelief takes over. There is an extreme contradiction, an almost complete cultural hypocrisy. A real-world example, supported by measurements across many national populations, and

* It is hypocrisy in the sense that adherents attempt to impose their contradictions on others.

for different strengths of cultural belief in, and innate scepticism of, climate catastrophism, is explored and explained in Chapters 8 and 9.

7.4 The entanglement of science

In the modern world, there are additional complications to the story of innate scepticism. This is because it has become entangled with the enterprise of science, at least in regard to scientific issues that have a (real or perceived) major social impact. Such entanglements occur because, at least:

1. Correct science may be associated with, or promoted or transmitted by, a specific culture.

2. Correct science may challenge values and contradict knowledge established by a culture.

3. Scientific theories often have genuine and significant uncertainty, opening a window to cultural judgments and bias.

4. Via an array of bias mechanisms, culture can divert or hijack science in a particular domain.[228]

5. Science (as an enterprise) has picked up cultural characteristics of its own.[229]

6. Strong innate scepticism about a promoted theory may motivate a pursuit of truth via science.[230]

As noted in Section 7.2.2, for uninitiated publics, innate scepticism is the only way to judge competing claims about a socially contentious scientific theory or discovery: the uncertainties, the allegations about funding and biased information sources, and all the rest of it.

This is problematical because innate scepticism is a way to detect potential group deceptions, not a way to judge scientific truth. Consider what may happen if a strong culture pitches some genuine science-based claims (Point 1). If out-groupers detect any signs of (group) deception from cultural narratives associated with the pitch, they will react to the whole thing as a threat and reject it, including the science, *even if that science is correct.*[231]

This assumes that signs of group deception are correctly detected, but that doesn't have to be the case. Consider what happens when the adherents of a culture have their assumptions or values seriously challenged by new science (Point 2). Because their values are unaligned with the scientific claims, their innate scepticism will be armed, so to speak. And from their perspective, the absolute certainty that comes with a replicable scientific result will tend to look just like the certainty that strong cultural consensuses proclaim (first two bullet points in Section 7.3.1). This may result in a false positive, an incorrect detection of group deceit and, if so, will also lead to rejection of the (correct) science.

A similar issue arises when scientists are overzealous. When this happens, the non-aligned may detect the emotion (fourth bullet point in Section 7.3.1), triggering their innate scepticism; if the zeal is excessive, rightly so; such zeal might be biasing the science. More broadly, the authority that scientists project, perhaps too arrogantly sometimes, can trigger innate scepticism in some members of the public because it looks very like the demeaning and demonisation (fifth bullet point in Section 7.3.1) that cultures employ.

In contrast, those with aligned values may exhibit a kind of allied belief when confronted with scientific claims;* although they don't understand the theory, they simply believe it 'because it is science'. For some topics at least, they treat science as a cultural friend.

The uncertainties surrounding science that is nascent (Point 3 in the list above) give more scope for innate scepticism and cultural beliefs to operate. This is true, not only for members of the public, but potentially also for the scientists involved in the field (they are not automatons, nor wholly independent of public opinion). This can unfortunately lead to cultural entities diverting scientific enterprises (Point 4), or in the worst case even hijacking them, so as to pose as a scientific consensus. Such a consensus will correctly trigger the deceit detection mechanism of innate scepticism in some people, but not those with closely allied value systems. In addition, situations where the authority of science more generally has taken on some of

* See Section 9.3.

the trappings of a culture (Point 5) may also be correctly detected.

Therefore, innate scepticism that is prompted by a science-based pitch will sometimes be apt, and sometimes inapt. In the former case, it correctly detects a deceit, or at least emotivity or arrogance that ought not to be part of science (and that deceit can sometimes be on the same side as genuine science). In the latter case, it is falsely triggered (no deceit was detected). In practice, we may not be able to tell which situation is which, but Section 7.5 looks at how we might 'disentangle' the effects of culture on conflicted science, giving useful insights into the realities, giving useful insights into the realities of social conflicts around science.

Finally, whether innate scepticism of publics is apt or inapt, it may motivate some to pursue the truth through science – in other words, innate scepticism may inspire scientific scepticism (Point 6). However, only a few will respond in this way, and in the inapt case they must eventually come to a dead end.

7.5 Disentangling science and cultural behaviours

Science frequently becomes entangled with cultural behaviours, in all of the ways noted above, especially when it is perceived to have important social implications; this sometimes results in polarised opinions and serious social conflict, which in the worst case could undermine science itself. Can knowledge of cultural mechanisms, including innate scepticism (an 'anti-cultural' cultural behaviour), help us to figure out which claims are wrong and which are right?

Typically, the more orthodox side of a socially conflicted issue will claim it is backed by science and is facing 'denialism'. Alternatively, both sides might make such claims. However, the tests cited by Lewandowsky and co-authors, as listed in Section 7.1, which are intended to resolve such claims by distinguishing 'genuine scepticism' from so-called 'denialism', provide no help whatsoever. As explained in Chapter 6, they detect only cultural behaviours that we'd expect to see on both sides of a socially conflicted science issue. Observing them therefore doesn't tell us which side is right. Worse still, the shallow nature of the tests and the lack of a theoretical grounding means that objectivity is inevitably replaced by bias, whether that of authors of the original tests,[*,232] or that of anyone trying to apply them in practice.

The tests also fail to mark the critical difference between *sides* and *groups*. As we saw in Section 6.4, each side of a socially conflicted issue is typically made up of people with a variety of motivations: their cultural beliefs, innate scepticism, a general trust in science, or, for some, a rational consideration of the evidence. However, the heterogeneous nature of a side might not always be obvious. For example, Chapter 11 explores the alliance in the US between Dem/Lib culture and climate catastrophism; because of the high-profile conflict between Dem/Lib and Rep/Con 'tribes', the fact that climate catastrophism is a culture in its own right, on the same side as the Dem/Lib tribe, is simply not perceived. However, once we understand that the sides of a socially conflicted issue are rarely monolithic, our knowledge of cultural mechanisms starts to become useful. This is because we may be able to use social data – if there is enough of it – to identify those groups who are engaged in cultural behaviours, and how they are aligned within the conflict. This data also enables us to use a more formal means of investigation than circumstantial evidence about logical fallacies or false experts or whatever.

As we have seen, strong cultural narratives are wrong by definition;[†] they are collective deceptions unrelated to reality. Social analysis can reveal the groups of people, who, by virtue of their belief in such narratives, are wrongly motivated, even if they end up on the 'right' side. Innate scepticism directed at such groups, no matter what side they're on, is entirely apt, which is to say *not* wrongly motivated. In summary – and putting it more colloquially – people engaging in cultural behaviours can safely be ignored.

Those exhibiting inapt innate scepticism are wrongly motivated too; this is an anti-cultural instinct that is mistakenly triggered.[‡] Meanwhile, an evidential position is no guarantor of truth, but nor is it a collective deception or an instinctive reaction against a collective deception (apt innate scepticism); it might be right. So, while social

* See also Section 6.8

† See Section 3.2.3.2.

‡ Based on Sections 7.3.3 and 6.7.2 regarding cultural defence, we could view this as a kind of 'denialism' but without the implications of systemic lying or pathological issues.

analysis cannot tell us what is correct science, if it can break down the sides sufficiently to see these group positions, this helps us on the road to uncovering where the truth might lie, or at least rule out those who must be wrong.

It's possible for the overwhelming majority of the public to be wrong about a socially conflicted issue – for example when both sides are engaged in a cultural slugging match, with neither actually considering the evidence. And even if one side is shown in the future to be 'right', most of the people expressing such views are still likely to have been culturally motivated rather than rationally convinced; in other words they were right for the wrong reasons.

For example, extending the picture of sides on the socially conflicted topic of climate change in the US, there are actually four cultural groups involved: religion, the Rep/Con tribe, the Dem/Lib tribe, and climate catastrophism. The latter two groups are allied, and so occupy the same side. On the opposing side, Rep/Con cultural attitudes strengthen innate scepticism against climate catastrophism. Religion influences both sides, but in different ways. Chapter 11 covers this cultural landscape in detail.

Outside the US, the breakdown of sides in the climate domain is simpler: all nations can be covered by a model involving only two cultures and their interrelationship. The measurements in Chapters 8–10 show that attitudes to climate change across national publics are unequivocally cultural, and stem from a dominant narrative on certain climate catastrophe, which people accept through cultural belief or reject via (apt) innate scepticism. The proportion of the public that accepts or rejects the narrative varies systemically per nation according to the level of their religious belief and the scenario framing (what attitude is being probed, and how). In other words, apparent 'sides' depend mostly on national religiosity. However, these sides are not in binary opposition because the relationship between religion and climate catastrophism is dualistic; different scenarios produce different 'sides' (this is fully explained in Chapter 9).

In both of the above examples, so everywhere in the world, the consensus on certain global catastrophe is a group deception; a fairy tale. Hence innate scepticism against it is apt, and is not 'denialism', in the popular framing. This is true whether the innate scepticism is culturally inspired or not, and remains true even where ambiguous or even completely false theories are offered up as alternatives to the cultural consensus, and even if this is accompanied by bad behaviour, say the use of fallacy or even demonising of opponents. All this tells us very little about 'what is right' in the climate change case, but one can say that the majorities of all publics are wrong no matter what side they're on; in the main their attitudes are not rooted in science, mainstream or any other.

Another example is the debate over evolution in the US. One side encompasses:

- an evidentiary position based upon evolutionary theory
- the Dem/Lib political 'tribe'
- those who aren't strong Dem/Lib adherents but are still innately sceptical of cultural attacks from the opposing side (and who don't know much about evolutionary theory themselves).

The opposing side consists of religious creationists, in alliance with the Rep/Con 'tribe'.*

All of the support on the incorrect side is cultural, but it's likely that most of the support on the correct side is too – rational involvement in the debate is not possible for most people because rationality needs knowledge.† The cultural nature of much of the support for the scientific evidence can be seen in Dem/Lib demonisation of their Rep/Con opponents. However, the existence of such cultural support does not affect the validity of evolutionary theory.

7.6 Innate scepticism and 'denialism'

We are now in a position to confirm the insinuation in Lewandowsky's question at the top of the chapter: members of the public will indeed resist scientific theories or discoveries that they find 'inconvenient'

* Very religious Dem/Libs and irreligious Rep/Cons will blur the boundaries, but there is still a net cultural face-off.

† In the US, the penetration of knowledge about evolution is hampered by cultural resistance to its teaching, which hinders children of Democrats just as much as it does those of Republicans. Lacking factual understanding, and against a background of intense political tribalism, much of the Dem/Lib support for evolution will be cultural.

or threatening to their lifestyles and livelihoods, or that challenge their deeply-held beliefs. Their innate scepticism will be aroused, and, if strongly so, they will engage in the debate, but not in a rational way. Meanwhile, if the science is 'convenient' for their lifestyles and livelihoods, or aligns with their deeply-held beliefs, their innate scepticism will be disabled. This results in general support for the science, but again not typically as a result of objective understanding.

Therefore, people are *not* typically 'anti-science'; an oft-thrown accusation. They are against science that challenges their values, and in favour of science that aligns with them.[233] Moreover, there's a sense in which those who end up supporting the science through cultural mechanisms are 'just as wrong' as those who resist it for the same reason. Generally speaking, attitudes can be statistically revealed across populations according to cultural memberships: political allegiances, religious beliefs, or belief in climate catastrophism. However, we must bear in mind that such analyses tell us nothing about the motivations of individuals; a single person might buck any trend.

So where does so-called 'denialism' fit in? Innate scepticism across broad swathes of publics bears almost no resemblance to the vaguely defined pathological conditions and systemic lying for nefarious purpose that are said to motivate denialism in the popular framing (as explored in Chapter 6). This framing might apply to tiny fringe groups, but the term is meaningless when describing bulk public reactions to socially conflicted science issues.

We need to confront a much more challenging reality. Our long-evolved instinct of innate scepticism is not a different reaction depending upon whether future history eventually proves it to be right or wrong. Innate scepticism of socially promoted 'science' that turns out to be both apt and correct, challenging groupthink and cultural dogma that wears the cloak of science, is (eventually) lauded. Meanwhile, innate scepticism that turns out to be inapt – mistaken – impeding public understanding and progress, is stigmatised. But the very same psychological reaction drives both cases; these are the two faces of innate scepticism.

7.7 Recap and discussion

The theory of innate scepticism in a non-expert public is as follows:

- Innate scepticism works independently of detailed knowledge of a contested domain.
- Innate scepticism attempts to detect (cultural) group deception, not pursue truth.
- Innate scepticism leads to doubt, withheld judgment, and sometimes modified beliefs.
- Innate scepticism of the narrative is suppressed by cultural belief in it, although some always remains in the population.
- *Apt* and *inapt* innate scepticism are fundamentally the same reactive behaviour.
- Innate scepticism not pitched against a cultural consensus[234] (a group deception) is inapt.
- Apt innate scepticism is a reaction to a group deception, and says nothing about what alternatives are true.
- An individual's innate scepticism cannot be assumed to cross domain boundaries (although, if domains have a strong cultural alliance, innate scepticism will more likely be aligned).
- The capability for innate scepticism arises from our evolutionary legacy; it is instinctive.
- Innate scepticism is entangled with the enterprise of science in various ways.

These characteristics have fundamental implications for all socially contentious science issues. Just one example is the likely failure of 'climate-change education' to achieve the aims of its instigators. Given the essentially cultural nature of the taught narratives, the program will cause as much innate scepticism as belief – in other words, polarisation – and *neither* side of the divide will end up with rational views.[235]

7.8 The innate scepticism of experts

Shifting our focus away from the public, what about the innate scepticism of experts? The *Helicobacter* story at the start of this chapter defies a full cultural analysis; there is rarely enough data about scientists' current attitudes, still less so for those working many decades in the past. But we still observe classic cultural behaviours among scientists, and can see, with the benefit of hindsight, that these may suppress correct science. So, our knowledge of cultural behaviour could in principle serve as a warning system, even when the social data is lacking. In that spirit, some of the questions raised by this historic example are revealing when we consider conflicted science in the modern era.

In the 1980s, the strong consensus view was that gastric disorders had a physiological basis, and were not due to infection. This dogma had dominated since the 1950s, despite there being significant historic evidence to the contrary. As a result, gastroenterologists – astonishingly – *lacked expertise in bacterial infections*. So, were they left to use the same tools as the public? In other words, did they judge Marshall's theory (and earlier ones) using their innate scepticism (inapt, as it turns out)?

These are fundamental questions. If experts become 'too challenged' – if the search for potential solutions has moved beyond their specialism or comfort zone – do they revert to the same instinctive approaches as the general public? If so, any sufficiently radical ideas that challenge scientific orthodoxy will face not only evidentiary opposition, but instinctive opposition too: innate scepticism, demonisation, comforting groupthink (culture writ small), and so on. This would represent a significant barrier to progress.

Moreover, in cross-disciplinary fields in particular, no scientist will have expertise across the board. So the process of building the big scientific picture becomes much more vulnerable to warring beliefs and innate scepticism. Culture and identity could overwhelm the truth, even when scientific evidence starts to move against the prevailing opinion. The uncomfortable question of whether the huge public culture of climate catastrophism is affecting climatology in this manner is unavoidable.

7.9 Conflation of innate and scientific scepticism

Science is essentially a rational form of sceptical inquiry. But the enterprise of science is always intertwined with powerful cultural mechanisms, one of which is innate scepticism. To understand what is science, we therefore have to identify what is cultural, including what is innate scepticism (and is therefore not rational scepticism).[236] Otherwise we'll never be certain if science is operating as objectively as it should, or is tainted by cultural influences, or maybe is even submerged by them. It is therefore important to understand the difference between rational and innate scepticism.

Debate about the nature of scepticism will be at best confused if its innate form isn't even acknowledged.* For instance, in an article in the *Huffington Post*, the head of The Skeptics Society, Michael Shermer, says that 'science and skepticism are synonymous',[237] a claim that makes the conflation of innate and scientific scepticism almost inevitable. Indeed, that Shermer recommends investigating the sources of claims to help establish truth or untruth is itself a serious slip into the territory of innate scepticism. Sources are about identity, not arguments or evidence. In other words, they're irrelevant for objective scientific scepticism.†

The article dates from 2013, so Shermer would have been unaware of Kahan's later work on cultural defence, or indeed of my own. But I think his view is representative of how scepticism is typically perceived, both then and now. Further conflation of the two kinds is apparent in the same article, and is even phrased using terms reminiscent of (religious) culture:[238]

> It is to find the essential balance between orthodoxy and heresy, between a total commitment to the status quo and the blind pursuit of new ideas, between being open-minded enough to accept radical new ideas and so open-minded that your brains fall out. Skepticism is about finding that balance.

In the sense of social conformance to particular views, science should have no need of orthodoxy and no commitment to a status quo (a social device). Hence there's no need for heresy either, nor

* Or at least similar identity-related accounts of behaviour, like Dan Kahan's.

† Although sources are relevant for a social-psychological analysis of the conflict, which is also a scientific inquiry, albeit in theory one that is independent of topic knowledge/arguments. However, they are necessary only for the purpose of establishing ownership (or otherwise) by a cultural consensus, a procedure not addressed by either Shermer's article or Lewandowsky's tests.

for any balance between these culturally defined poles. All that is required is a valid critique, which should stand or fall on its merits; on the evidence, in other words. Should that evidence not yet be obtainable, then it is acceptable – advisable, even – to withhold judgement. This is a position common to all scepticisms, so our brains are unlikely to 'fall out', and no alternative theory need be presented to hold this position.

Nor would a blind (to domain conventions) pursuit of new ideas normally represent a danger to society.[239] A much greater danger is that institutional science loses its objectivity, becomes vulnerable to cultural modes, and ends up pushing policy measures that are based on fallible innate scepticism, or even collective deception.

The Ancient Greeks attempted to formalise instinctive scepticism in order to make it independent of cultural values. However, their 'philosophical scepticism', which has evolved into today's scientific scepticism, is still challenged by cultural values, which inevitably come into play whenever science has a strong social impact.

7.10 Conclusion for innate scepticism

Social psychology asks some pertinent questions about public scepticism. But it seems that its practitioners, especially those covering the climate domain, have avoided investigating its root causes and have steadfastly sidestepped answers that might challenge orthodoxy. Perhaps they fear what they might find, and are nervous that they might be pulled away from a comfort zone of an 'approved' scientific consensus in which 'denialists' are placed firmly on the naughty step.

To fully understand culturally conflicted domains, whether climate change or any other, it is vital to recognise that bulk scepticism in publics is mostly instinctive rather than rational, or indeed pathological. Moreover, this innate scepticism – 'cultural disbelief' – is just as important as cultural belief and, as noted in Section 7.3.5, these two cultural mechanisms or functions are semi-independent, so can contradict each other to varying degrees.

Chapter 8

MEASURING CLIMATE CATASTROPHISM

8.1 Measuring a culture

Circumstantially at least, Chapters 2–5 strongly suggest that a new culture of climate catastrophism is operating in society. Because cultures tend towards social universality,* as they expand in both membership and social reach, new ones will eventually have to interact with older ones. This interaction will be particularly strong and stable if two highly successful cultures have existed alongside each other for many years. Therefore, the simplest method of demonstrating that climate catastrophism definitely is a culture is to reveal its interactions with older entities that are unarguably cultural. If a major interaction is occurring, then climate catastrophism must indeed be a culture too, provided rational or physical causes of any apparent relationship can be ruled out.

To measure this interaction, I utilise the results from a number of opinion polls of attitudes to climate change, conducted by the EU, the UN, and a variety of mainstream polling organisations, among national publics from around the world. These mainly independent sources reveal cultural relationships that are highly robust, and consistent with a single, straightforward theoretical framework. They constitute a hard-data measurement of the cultural entity of climate catastrophism.†

Note, however, that the USA is considered separately, in Chapter 11; although the same underlying cultural mechanisms are at

* See Section 3.3.2.

† All the survey source data are publicly available, and all relationships are easily replicable using only a spreadsheet.

work there, the unique level of public polarisation in US society makes the relationships more complex.

As social psychologist Dan Kahan has stated, when Americans are asked about climate change or other culturally polarised issues,[240] they respond with 'who they are' (i.e. their cultural worldview, or cultural 'identity'), not with 'what they know'. And as far as climate change is concerned, the publics of any nation don't actually *know* much at all. In other words, people give cultural responses, unless questions have been specifically geared to avoid identity issues. This is not the case in the surveys we are using, or in fact in almost any survey of attitudes to climate change, so the responses in the datasets we are looking at are indeed cultural, to a greater or lesser degree. That is not what the survey designers envisaged, but for us, it's a useful feature rather than a bug; if the questions were culturally neutral, we might see little of interest. This wouldn't mean that cultural attitudes weren't present, merely that they had not been brought out by the questions posed. To put it another way, any (climate) cultural bias in the questions that resonates with publics, and so brings forth their true attitudes, is actually helpful. If, in the climate domain, a culture is indeed dominating society, we will see mostly emotive rather than rational reactions.

An accompanying spreadsheet contains all the charts presented in this chapter, and indeed the rest of the book, along with links to all the source data. This file is referred to throughout as the 'Excel-Ref', and a link to it is provided in Chapter 1.

8.2 The lens of religiosity

With what older culture might climate catastrophism interact? We need one with global reach and which is, for its most basic values at least, consistent everywhere. It turns out that there's a perfect candidate, namely religion; there is a wealth of data on levels of religiosity – its cultural strength – for virtually all nations.

To what extent might public attitudes to climate change be influenced by religion? If they are largely grounded in objective reason, we might still expect a modest religious influence, but we certainly wouldn't expect it to be overwhelming. On the other hand, if attitudes are predominantly and inherently cultural, we should expect them to be affected by an interaction with the wholly cultural phenomenon of religion, to the extent that attitudes to climate catastrophism will be heavily or overwhelmingly dependent on national religiosity.

In other words, the religiosity of nations provides a 'lens' through which we will 'see' the newer culture of climate catastrophism, if indeed it is a cultural entity in its own right. To deploy this lens, a standard scale for national religiosity is needed. I use a very basic one, namely self-assessed religiosity averaged with the inverse of self-assessed *ir*religiosity, helping to ensure consistency across faiths. Most studies in the field use behavioural proxies – hours spent praying or the number of visits to church, say – which are also often incorporated into more complex scales. These may provide a much better measure *within* any faith about what particular levels of religiosity actually mean for adherents, but we don't need that information, and I suspect such methods give much less consistent results across faiths, because the targeted behaviours are themselves faith dependent. See Appendix A for further explanation on the religiosity scale, including links to the data sources, and details of a correction to some minor self-assessment bias in the underlying data.

8.3 A sneak preview – paradoxical results?

At the simplest level then, we can take the survey results for attitudes to climate change across an array of countries, and plot them against the corresponding national religiosities. The first survey question we will consider comes from a 2019 YouGov opinion poll,[241] which asked: 'How much of an impact, if any, do you believe climate change will have on your life?' Figure 3 shows on the *y*-axis the percentage of people in each country who gave the climate-change most-endorsing answer, 'A great deal',* with these figures plotted against each country's national religiosity on the *x*-axis. Climate-change 'most-endorsing' responses are those that are the most supportive of, and/or the most concerned about, dangerous manmade climate change.

Clearly, there is an extremely robust positive correlation between the two variables, with publics in largely secular countries, on the

* The other responses being: 'A fair amount', 'Not much', 'No impact', and 'Don't know'.

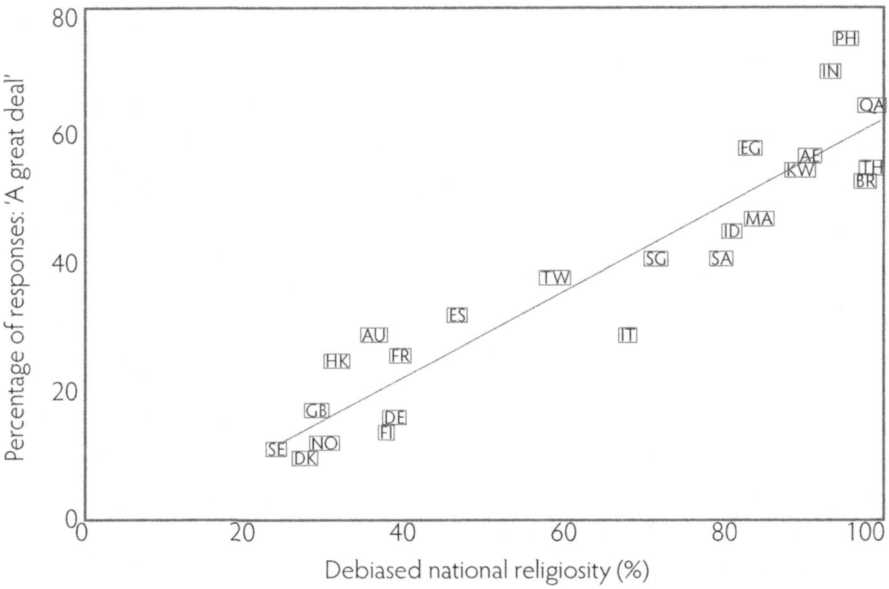

Figure 3. National attitudes to climate impacts.

How much of an impact, if any, do you believe climate change will have on your life? Proportion of respondents giving climate-change most endorsing response: 'A great deal'. $R^2 = 0.87$. A list of country codes can be found in Appendix L.

left of the figure, resisting the climate-orthodox concept that climate change will have a major effect upon their lives. People in countries that have a strong religious culture, meanwhile, are far more supportive of this idea. An explanation of the results comes later, when we've seen more series; but it is worth noting that, compared to typical social-psychological relationships, the dependence of attitudes to climate catastrophism on religiosity is very strong.

Now consider a second survey question, in which publics were asked to pick their most important six out of seventeen global issues. This is from a huge 2015 UN survey.[242] Figure 4 shows the percentage of people in each country who selected 'action on climate change' in the top six issues. There is again a robust relationship between

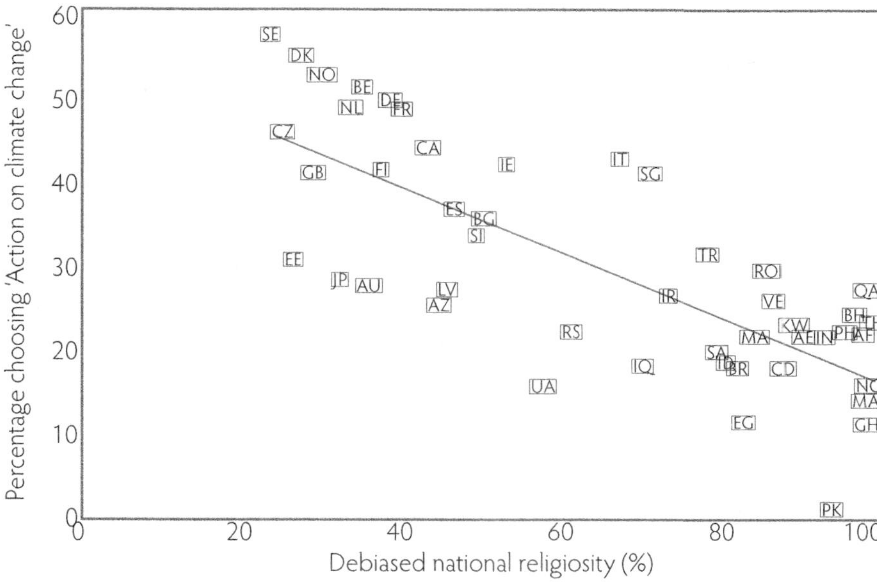

Figure 4. National priorities for climate change.

Respondents were asked to list their six most important global issues from a list of seventeen. Proportion choosing climate change as one of the six. 47 nations, $R^2 = 0.58$. A list of country codes can be seen at Appendix L.

this climate-change endorsing attitude and national religiosities – a very emphatic result – but this time it is an anti-correlation. In other words, respondents in largely secular countries gave answers that were aligned with climate orthodoxy ('climate change is the overwhelming issue of our time'), while those in strongly religious countries were far more resistant.

This result is, at first glance, a paradox. Many national publics appear to be simultaneously supportive of, and yet resistant to, climate orthodoxy. For instance, those religious countries having far and away the most concern about climate-change impacts (at the right of Figure 3), also express the least priority for climate-change action against other issues (at the right of Figure 4). And the difference in positions is not marginal; it is huge. How can this be? The answer takes a little while to unfold, across the rest of this chapter and the following one but, as we shall see, the apparent paradox is ultimately a function of the cultural nature of attitudes to climate change. To be clear, this is more than just saying that such attitudes are influenced by religiosity. The key point of this chapter, and indeed this book, is that attitudes to climate change are cultural *in their own right*.

Observant readers will have noticed that there is also some systemic secondary variance about the trend in Figure 4. The feature is discussed in Section 10.1.5.

8.4 Mapping the survey question framings

8.4.1 'How the questions are asked': framing

It's a truism that the responses to surveys about attitudes to climate change depend upon 'how the questions are asked'; in other words, how their framing influences the answers. A common take-away from this observation is that the results from such surveys are, in some sense, unreliable or erratic and must therefore have little value. This inference is made especially by those – whatever their own view on climate change – who are uncomfortable with the implications of some of the responses. However, it could hardly be more wrong.*

That different framings produce apparently capricious and contradictory responses – perhaps simultaneously supportive of and resistant to climate-change orthodoxy, as above – does not speak to 'our unfathomable brains' or some complex mix of motivations (there have been such proposals), or indeed to flaws in the surveys. It is far more likely to reflect a simple lack of comprehension about the underlying factors that determine public responses. Consider, for instance, that so long as the questions are framed in precisely the same manner, then in any given country at dates that are not too far apart, valid statistical samples of the public will respond to different surveys with the same answers (give or take the sampling errors). Such an underlying consistency of responses strongly suggests that the above factors come from a straightforward motivational model;

* They are less predictable for weakly-framed or mixed-mode questions – see later – but even these are predictable within a defined 'envelope'.

this work proves the existence of such a model, and describes its features.

In practice, it turns out that of all the different ways in which questions can be framed, only a few turn out to be important. And indeed, just two ways are relevant to the analysis here: reality-constrained and unconstrained questions. Reality-constrained questions force survey participants to consider climate change (or global warming) alongside other real-world issues, typically by asking them to rank it in a list of issues of public concern.* Unconstrained questions don't force such a trade-off; they present issues in isolation. That just two basic framings can produce a wide range of responses matching the above scenario is because, as we will see, each may come in different 'strengths', and also because some questions are a mixture of the two.

8.4.2 Unintuitive results correspond to primary framings

The first survey question, 'How much of an impact, if any, do you believe climate change will have on your life?', which generated the results in Figure 3, is unconstrained; survey participants will naturally tend to consider it in isolation of other factors. The second question, in which respondents were asked about the priority for 'action on climate change' relative to a list of other issues of major concern, generating the results in Figure 4, is reality-constrained; it brings hard reality into participants' minds by implying the need for trade-offs.

It may seem suprising but, as we shall see, all questions that probe publics about their attitudes to climate change using these two primary framings, whatever the question details, will always produce the two generic response patterns seen in Figure 3 and 4: the climate-change most-endorsing responses will correlate with national religiosities for unconstrained questions, but will anti-correlate for reality-constrained ones. This is true across the 64 nations measured, suggesting that the apparent paradox is almost certainly the result of very stable patterns of motivation in the respondents.

8.4.3 Framing strengths

As alluded to above, the 'strength' of the questions is important too;

* Another common style of reality constraint is to relate questions to personal or tax dollars that participants might be willing to pay for particular policy options.

that is, different question wordings can modify the responses to some degree (although the basic pattern of correlation or anti-correlation will remain). For reality-constrained questions, 'strength' reflects how much respondents are forced to compromise with reality. So where survey participants are asked to nominate the *x* 'most important' from a list of *y* issues (of which just one is related to climate change), strength is higher where *x* is smaller as a proportion of *y*. So, for instance, nominating the most important single issue out of 9 is a stronger reality constraint – a stronger clash with the other issues – than picking the top 3 of 12 say, which is stronger in turn than the top 6 of 17 (the latter matching the Figure 4 example). There are other factors relevant to the strength of this type of reality constraint too. For instance, for the same size of *x*, a list of *y* national issues creates a stronger constraint than a list of *y* global issues, because the former are typically closer to people's day-to-day and personal (so emotive) concerns.

The strength of unconstrained questions, on the other hand, reflects how closely 'aligned' they are with the emotive and existential values of climate catastrophism – in other words, with the vehicle of this culture, the Catastrophe Narrative (see Chapter 5) – and also how much they target the personal. Another way of thinking about strength for this type of question is as the extent to which questions are biased towards climate catastrophism, emotive content being a feature of bias, and not of rationality.

The details of how strength applies to questions of each primary framing type – reality-constrained and unconstrained – are rather different. However, the essence is the same in both cases, reflecting the extent to which they prompt a cultural response, which is to say an emotive rather than a rational one (though for strong reality constraints, which make the culture yield to reality, one could argue that rationality has leaked through).

8.4.4 A basic framing map

To elucidate the effect of framing strength on responses, I categorise questions into one of the buckets shown in Table 6. These match up well with the ways in which such survey questions are typically phrased, in particular the questions in the surveys used in this book.

Table 6. Framing map for survey questions.

Reality-constrained questions	
Fully constrained	FC
Strongly constrained	SC
Medium constrained	MC
Weakly constrained	WC
Unconstrained questions	
Strongly aligned	SA
Medium-strong aligned	MSA
Medium-weak aligned	MWA
Weakly aligned	WA

8.5 The approach for cultural analysis

If attitudes to climate change are largely cultural, analysing survey data as though it was a collection of rational, objective responses to the prime facie subject matter of each question will tell us virtually nothing. As we've already seen, it is wrong to assume that climate-change most-endorsing responses from national publics will be logically consistent; they change radically with the question framing, and this apparent paradox points to a cultural origin.

So, in order to understand the survey responses, we need to recognise their nature. The two primary framings above (reality-constrained and unconstrained) and their strength scales, are our means to decode climate survey responses from the apparently capricious, into the systemic and predictable.

However, we need more than systemic question framings attuned to the cultural. In practice, we only observed the paradox above because we were using a wholly cultural *x*-axis (national religiosity). With any non-cultural axis, and perhaps some cultural variables that aren't a decent proxy for religiosity, we'd have seen no meaningful pattern. So, we need our lens of religiosity too.

For any given question framing and framing strength, we can then plot a data series representing the survey results across nations on the *y*-axis, against national religiosity on the *x*-axis. From the resulting patterns we can develop a model of what is truly driving the responses. If the motivations of publics are not primarily cultural, using these methods won't give us straightforward and repeatable patterns that correspond to our framings.

8.6 Primary results

8.6.1 The 'Basic' series: survey questions

We are now in a position to expand on our preview, and look at a modest range of further questions taken from international surveys on attitudes to climate change (see Table 7).

Assigning reality-constrained questions to one of the buckets for strength is straightforward: the ordering is pseudo-mathematical and so fairly intuitive. However, for the unconstrained questions, some textual analysis is required. This is always going to introduce an element of subjectivity and inexactitude. However, in reality, the bucket boundaries are arbitrary because strength is a continuum, so it's not particularly important where they fall so long as the approach used is consistent.

The question for the SA series is very strongly aligned both to the Catastrophe Narrative and to the personal (which engages emotions more), because it alludes to direct individual impact from climate change. The concept of certain and global catastrophe implies that such personal impacts must occur.

The question for the MSA series is likewise strongly aligned to Catastrophe Narrative, in that it speaks to personal harm from climate change. However, its edge is taken off by two factors:

- It targets concern about harm, not harm itself, a level of indirection that will likely reduce emotive engagement.[*,243]
- The explicit mention of a timescale, albeit a long one – 'your lifetime' – may raise reasoned thoughts about when climate

* Although using 'concern' also prevents an effect whereby only 'doomsters' respond, severely reducing the response rate. See endnote 247.

Table 7. Survey questions for the Basic series.

Series	Question	Strength	Response
Unconstrained series			
SA	How much of an impact, if any, do you believe climate change will have on your life?	Strong	A great deal
MSA	How concerned are you, if at all, that global climate change will harm you personally at some point in your lifetime?	Medium-Strong	Very concerned
MWA	Because of human activities, the Earth is close to tipping points in nature where climate or nature may change suddenly, or may be more difficult to stabilise in the future	Medium-Weak	Strongly agree
WA	How much power, if any, do you think international bodies (e.g. the United Nations) have to combat climate change?	Weak	A great deal
WA1	How serious a problem, if at all, do you think climate change is?	Weak	Extremely serious
Reality-constrained series			
FC	(Result estimated from WC)	Full	Climate change
SC	Climate-change as important in 1 of 9 (global threats)	Strong	Climate change
MC	(Intuited series)	Medium	n/a
WC	Action on climate-change as important in 6 of 17 (global issues)	Weak	Action on climate change

Appendix C ties each series to a link for its survey source, and provides a reference to an associated full-data chart in the Excel-Ref.

harms could strike, which will again tend to reduce emotive engagement.

The question for the MWA series also raises the issue of impacts. However, much of its alignment strength is eroded because the public knows very little about the technical issue of 'tipping points',* nor does the question explicitly link these to certain catastrophe or personal impact, only to the much more nebulous concept of 'instability'. Also, the mention of 'nature' likely conflates manmade calamities with natural ones (such issues can occur when responses are largely emotive, not rational).

The questions for the weakly-aligned WA and WA1 series are very impersonally phrased, and don't explicitly invoke the spectre of impacts or harm. Many things in this world are 'serious problems', but that doesn't typically mean they'll invoke more than weak emotive responses.

The questions in Table 7 and their responses generate what I refer to as the 'Basic' series. These are depicted in Figure 5. We are only measuring the climate-change most-endorsing responses at this stage (listed in the last column of the table). Less-endorsing responses, and ones that are climate-change resistive to different degrees, are considered later. As with our two sneak-preview series above, for each question, the percentage of respondents across many nations who gave the most-endorsing answers, on the *y*-axis, is plotted against national religiosity on the *x*-axis.

8.6.2 The 'Basic' series: summary charts

The two charts in Figure 5 (one for unconstrained responses, one for reality-constrained responses), are summaries, showing trends only; the sheer number of data points means they would obscure each other if depicted. The full data for the SA and WC series can be seen in our sneak previews above, and for WA and SC in Appendix B. The original graphs and data sources for all series are provided in the Excel-Ref.

The nation encodings along the top of Figure 5 represent the

* There is a genuine scientific concept of tipping points, but the term has been hijacked and its meaning corrupted, to the extent that is has become a Catastrophe Narrative variant.

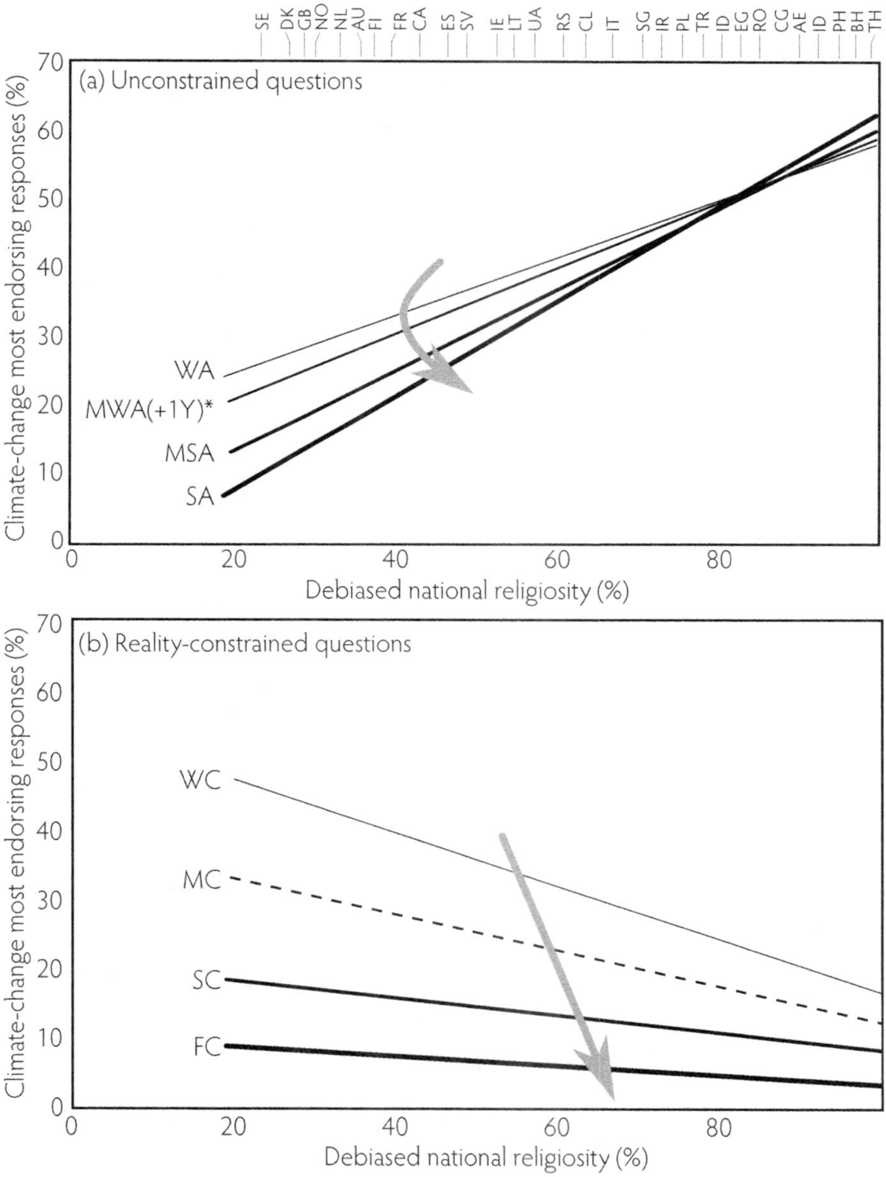

Figure 5. Main survey results.

Percentage of people per country giving most climate-endorsing response for each question, plotted against national religiosity. (a) Unconstrained questions and (b) Reality-constrained questions. Trends only. Grey arrows indicate effect of increasing strength. * MWA has been uplifted by 1% on the y-axis, to enable the RH ends of WA/WA1 and MWA to be distinguished. A list of country codes can be seen in Appendix L.

religiosities of a representative sample of nations, as reference points for readers who are unfamiliar with how this varies around the world; there isn't room to include all of the 64 nations covered by the main series in this book. Those shown are not necessarily represented in all of the Basic series, but note that the same framework encapsulates the relationships of climate change attitudes with national religiosity in all nations, in whatever series they are represented.

All of the climate-change most-endorsing responses to unconstrained survey questions strongly correlate with national religiosity. For the reality-constrained survey questions, however, the responses strongly anti-correlate with national religiosity. The grey arrows depict the nuanced changes in responses for increasing strength within each of these primary framing types. The responses to unconstrained survey questions have a higher gradient with increasing strength, and share a pivot point. Responses to reality-constrained survey questions lose some of their gradient with increasing strength, and sink down the *y*-axis.

8.6.3 Caveats, correlation parameters and source data

There are some caveats associated with these Basic series. Firstly, the MC trend (dashed) is intuited, so is not from an actual measurement; I haven't yet found a series that roughly bisects the space between SC and WC. Secondly, the Fully-Constrained (FC) trend is an estimate, calculated from WC, that is also spot-confined by five individual national measurements at the lower religiosity end.* This is because I can find no survey that provides a Fully-Constrained series across many nations; that is, one which uses the same framing to ask many national publics to nominate climate change as the most important from a long list of national issues. As WC is for nomination of the 6 most important out of 17 issues, the FC estimate is created simply by dividing the WC responses by 6. Thirdly, the trends for the two weakly-aligned series (WA and WA1), don't actually land in quite the same place. However, they're pretty close, and in this summary chart

* These spot data-points hint that FC is too high, which makes sense because the WC series is about global not national issues. But it stands for now, as there isn't enough data to make a reasonable correction. As Fully-constrained measurements have such low *y*-axis values, noise is an issue too. The spot data-points can be enabled on summary chart 'Trends' at sheet 'Main Trends' of the Excel-Ref (enable the 6th and 7th entries in the list of series).

I simplify by representing them with the same line.*

Finally, this book only measures public attitudes to climate change in order to probe the potential presence of a cultural entity. In sticking to publics, I don't mean to imply that the views of climatologists can't be influenced by a powerful culture such as climate catastrophism. For instance, the defensive manner in which the topic of uncertainty is treated within the climate-science community *might* be a symptom of cultural pressure. The revelations of Climategate strongly suggest this is the case, as do the pronouncements of at least some scientists on the subject of catastrophe (see Section 5.4). However, the attitudes of national publics are a straightforward target for hard measurement and characterisation; there are many suitable surveys. This simply isn't the case for the enterprise of science; there's virtually no useful data available on the attitudes of climate scientists as a separate demographic. So this book sticks to the attitudes and underlying beliefs of *publics*, and doesn't attempt to quantify potential cultural impacts on climate science.

The correlation parameters for the Basic series are shown in Table 8, at the end of the chapter. All series are statistically valid. Appendix C ties each series to a link for its survey source, and provides a reference to an associated full-data chart in the Excel-Ref.

8.6.4 These results are revelational!

The importance of the patterns revealed in Figure 5 can hardly be overstated. They are fundamental and revelational. Here, I pre-empt the explanations later in the book somewhat, in order to give an initial feeling why this is so. Figure 5 shows or implies:

National religiosity is a potent single predictor: The charts in Figure 5 depict national religiosity as a 'social predictor' for international (non-US) attitudes to climate change. There is a sizeable literature that examines such predictors, the search for the best ones having been ongoing for many years, but Figure 5 and the other series in this book renders it obsolete. National religiosity is a far more powerful predictor than anything published to date. Moreover, it is a single variable –

* WA1 starts in the same place as WA at low religiosity, but has a slightly steeper gradient, so ends up higher at high religiosities. The full series can be seen at sheet 'WA1+O2 and WC37' of the Excel-Ref, chart WA1+O2 (green data).

most studies group the predictors in an attempt to increase predictive power.

There is a single, straightforward explanatory model: The nature of the primary framings and their strength ranges come from an expectation of cultural responses. As explained in the next two chapters, the responses summarised in Figure 5, and those from the other primary series reported in this book, all fit into a simple, single framework that reflects an underlying cultural motivation. The literature doesn't appear to come anywhere close to such a straightforward (and demonstrable) explanation for the very wide variation of international public attitudes to climate change. It is mired in complexity and detail, appearing to miss the wood for the trees.

Patterns are common to all cultures: Assuming the straightforward explanatory model noted above is valid, what we're actually doing in producing Figure 5 is 'interrogating a cultural entity'. In doing this, we'd *expect* the basic patterns obtained; that is to say, the two framings should bring out contrasting groupings in the responses, and varying the strengths should introduce a pivot point for unconstrained questions but not for reality-constrained ones. In Chapter 9, it is shown that this is the case for another culture entirely, namely religion. Seeing such patterns in the climate domain confirms our cultural explanation.

Public attitudes to climate change are cultural in their own right: Although fully explained later, we can already see intuitively that the trends in Figure 5 can't result from the climate or climate exposure of nations, or from climate science or climate policy, or indeed from anything that is rational. No such factors could possibly correlate and anti-correlate simultaneously with the most-endorsing attitudes to dangerous climate change. But cultures have hypocrisy at their very hearts, so cultural causation can explain this apparent paradox.

Social science generally and the social 'predictor literature' in particular, has failed to identify inherent cultural causation for international public attitudes to climate change. The detail of how this has happened is examined in Chapter 10. For now, it suffices to say that this is an extraordinary oversight.

Table 8. Correlation parameters for the Basic series.

Series	Strength	N	R	R^2	p
Unconstrained series					
SA	Strong	24	0.94	0.87	2.2×10^{-11}
MSA	Medium-strong	26	0.67	0.45	1.7×10^{-4}
MWA	Medium-weak	17	0.81	0.66	7.4×10^{-5}
WA	Weak	24	0.89	0.80	4.6×10^{-9}
WA1	Weak	37	0.70	0.49	1.4×10^{-6}
Reality-constrained series					
FC	Full	5*	n/a	n/a	n/a
SC	Strong	16	−0.57	0.33	2.1×10^{-2}
MC	Medium	n/a†	n/a	n/a	n/a
WC	Weak	47	−0.76	0.58	4.1×10^{-10}

N, number of data points; R, correlation coeffient, R^2, coefficient of determination; P, probability value. *Confining an estimate based on WC. †Intuited series.

Chapter 9

THE CULTURAL MEASUREMENTS EXPLAINED

This chapter explains the results presented in Chapter 8. The underlying model, based on cultural causation, is straightforward and rests upon the fundamentals covered in Chapter 3, the ubiquitous presence of Catastrophe Narrative in the public domain, as covered in Chapter 5, and especially the concept of innate scepticism, as explained in Chapter 7. As a brief reminder, innate scepticism has nothing to do with rational scepticism. It is an instinctive reaction against an invasive alien culture (or a local cultural overreach). Cultural values determine whether or not it is triggered in individuals. Innate scepticism can be thought of as cultural disbelief in a cultural narrative – an emotive rejection – just as cultural belief is an emotive commitment to a cultural narrative.

If the model described here is true, similar patterns to those depicted in Figure 5 will occur when interrogating any strong cultural entity. Additional validation is therefore provided by demonstrating that this is indeed the case for a culture other than climate catastrophism.

9.1 An ambivalent relationship

9.1.1 Apparent paradox

As noted in the last chapter, there is an apparent paradox, in that those nations expressing by far the most concern about climate change (at the right-hand ends of the trends in Figure 5a),* also express the least priority for climate change relative to other issues of concern (right-hand ends of the trends in Figure 5b). The opposite is true at the

* See page p. 147.

left-hand ends of these charts, although the unconstrained and reality-constrained trends overlap more. This pattern defies any explanation based upon physical or rational factors. However, it can be explained in terms of cultural mechanics. At the heart of the matter is the ambivalent relationship of the relatively new culture of climate catastrophism with older culture, in the shape of the main religious faiths.

9.1.2 Why the unconstrained trends occur

The leaderships of all of the main faiths have now been signed up to Catastrophe Narrative for some years, and propagate it in speeches and statements of their official positions.[244] This means that their flocks are comfortable with the narrative, at least when no reality constraint intrudes. The innate scepticism of climate catastrophism in religious believers is therefore disabled; they believe in the Catastrophe Narrative too. This leads to the unconstrained response trends in Figure 5a. As we move from left to right, there is less and less innate scepticism as the proportion of religious people within national publics increases. In colloquial terms, we could say that the unconstrained questions are eliciting (climate catastrophist) virtue-signalling responses from the religious.

9.1.3 Why the reality-constrained trends occur

However, whenever any reality constraints appear, the flocks aren't buying. This situation has not gone unnoticed in the literature but, having overlooked cultural causation, mainstream researchers have found explanations elusive. At best, they allude to 'deeper attitudes' of religious adherents that are 'ambivalent, complex, and muted'.[245]

However, with a cultural model, the explanation becomes straightforward: the reality constraints are sufficient to switch the innate scepticism of religious believers – their cultural disbelief in the Catastrophe Narrative – back on. The presence of actual demands and priority aspirations associated with climate change clashes with their religiously-orientated values. This is cultural defence in action; a role of innate scepticism, as noted in Chapter 7.* Reality prompts religious adherents to sense the presence of the new culture and the

* See p. 120.

threat it represents to their deeply held beliefs; their innate scepticism kicks in strongly and their virtue-signalling responses collapse. This explains the dramatically lower responses of the religious nations to reality-constrained questions (at the right-hand side of Figure 5b), as compared to unconstrained questions (at the right-hand side of Figure 5a).

For the irreligious, innate scepticism is 'on' by default, but in reality-constrained scenarios, it is expressed in fewer of them than might be expected (for the WC series, just over half the population – at the left of Figure 5b). While the irreligious do not have strong cultural values to defend, why this expression is quite so weak for reality-constrained questions bears further discussion, which I defer to the next section. However, given this is the case, the picture for reality-constrained trends is completed. Moving from left to right, there is more and more innate scepticism, because there are more and more religious people within national publics for whom this is strongly triggered.

9.1.4 Pascalian assent

The responses of the irreligious are also different for unconstrained and reality-constrained questions, although the contrast is not as stark as it is for the religious. With reality-constrained questions, something appears to reduce their innate scepticism, which results in somewhat higher levels of climate-change most-endorsing responses than would otherwise be expected.* This cannot be due to a direct cultural belief in climate catastrophism; if this were the case, the unconstrained trends would also be higher at the left of Figure 5.

Whatever the cause, it is almost certainly still a cultural mechanism of some kind: publics effectively have no understanding about climate change,† and have been inundated by Catastrophe Narrative for decades; so it is extremely unlikely to be anything rational. There are two candidates that match this profile. The first is a general public trust in science, which would make itself felt when people are assessing relative priorities in reality-constrained questions; after all,

* Although these responses still shrink proportionally to the strength of the constraint, as anticipated.

† See Section 9.2.1.

the Catastrophe Narrative (falsely) claims it is fully backed by mainstream science. Blind trust in science may be a good bet on average, but it is still a bet.

The second candidate is something akin to Pascal's wager.* In reality-constrained circumstances, some people make a subconscious bet that, being ubiquitous and authoritative, the Catastrophe Narrative might be true. One can think of these wagers as a kind of 'mental insurance policy' among disbelievers, who are taking out cover against the possibility that the story of climate doom might not be wrong after all (albeit only when the cover is cheap; i.e. the constraints are light). In Section 9.6, I chart a similar effect regarding belief in life-after-death among adherents of a religious culture.

The two candidates may be working hand in glove; a general trust in science is bound to make a big difference to a subconscious wager placed on a narrative which claims 'the' science as its foremost authority. Some of the people who are not emotively committed to climate catastrophism – they are not direct or allied believers – nevertheless trust this aspect of the narrative; essentially, they assent to the story of 'the' science. I therefore term these people 'Pascalian assenters'; the gamble of their assent can still be regarded as a cultural mechanism, albeit a more subtle one, prompted by the powerful presence of climate catastrophism.

To pin this mechanism down further, we would like to know much more about how it occurs. For instance, is it essentially independent of the alliance between religion and climate catastrophism, stemming from a direct relationship of the irreligious with climate-change propositions? Or is it somehow dependent on the alliance? Given those exhibiting Pascalian assent are irreligious, the former would seem more likely. Unfortunately, we cannot distinguish these possibilities from the survey data presented in Chapter 8. However, it turns out that the unique situation in the US, examined in Chapter 11, sheds more light on this issue; I pursue it there.†

* The 17th century philosopher, Blaise Pascal, suggested that people should live as though God exists. If they do so, they lose little. If they do otherwise, they risk eternal damnation if they are wrong.

† See p. 173. There is further support for Pascalian assent in Online-appendix D.

9.1.5 The paradox reflects blind cooperation and competition

Even a weak reality constraint can cause the fundamental switch in attitudes among religious adherents. Compare the weakly constrained (WC) series with any of the unconstrained ones; there is a remarkable difference. The apparently robust religious-based endorsement of climate-orthodox values that emerges from unconstrained climate surveys – above 60% of the public in the most religious nations – is actually very shallow, nothing more than ephemeral virtue signalling. Support collapses to a fraction of this number when reality intrudes, for instance via the introduction of contending priorities.

This is important far beyond surveys of public opinion because, as their name implies, reality constraints do not just appear in surveys and studies, but in the real world too, with fundamental implications for climate policies that may burden publics. Support for such policies must necessarily be traded off against other demands and, as we will see in Chapter 12, therefore closely mirrors responses to reality-constrained questions.

A more generic way to think of the relationship between religion and the culture of climate catastrophism – the cause of this seemingly paradoxical attitude 'flip' – is that they are simultaneously co-operating and competing. Such ambivalent relationships are not unusual for cultural entities; they are neither sentient nor agential, working only via emotive selection, so they can blindly follow conflicting paths, and will indeed follow them if it is net beneficial to do so.

Once the Pascalian assent of the irreligious is incorporated, our model fully explains the flip in the trends of the most-endorsing responses, from robust correlation with national religiosities for unconstrained questions, to robust anti-correlation for reality-constrained ones. The nature of the responses to different strengths for each of those question framings is important too, and is examined in Section 9.4. However, we must first take a brief look at the bigger social picture.

9.2 The big social picture

In Section 9.1, I described the fundamental mechanisms behind the Chapter 8 results. However, there is a bigger social picture: the mechanisms outlined are leading us to the same conclusion as Dan Kahan (see Section 7.3.3), a scholar who is at the forefront of the 'cultural cognition' approach to analysis of socially conflicted issues. Kahan has said of public attitudes to climate change in the US: 'What people "believe" about global warming doesn't reflect what they know; it expresses who they are'. In other words, it expresses their cultural identity.[246]

However, Kahan's theory is derived from US social data only, and assumes that the relevant cultural identity is overwhelmingly about political 'tribe': Rep/Cons or Dem/Libs. His findings are now widely accepted, and in part this seems due to a rush to blame toxic US political conflict for public 'irrationality' about climate change. This naturally leads to the further assumption that non-US attitudes to climate change must be much more rational, because tribal political polarisation doesn't afflict these countries to anywhere near the same degree. However, the findings in this book suggest that everyone is overlooking the full range of contributions to cultural identity.

As we will see in Chapter 11, cultural identities in the US come from interaction of four cultures: the two political tribes, plus religion and climate catastrophism. In other countries, without the highly polarised politics, to a first approximation there remains just religion and the culture of climate catastrophism, which, as we have seen, are together sufficient to entirely explain the patterns in Figure 5.

So Kahan's conclusion that attitudes to climate change are culturally driven turns out to be universally true – it is true outside the US too, as definitively shown by the measurements in Chapter 8. However, the social sciences, perhaps blinded by the high-profile of the Rep/Con versus Dem/Lib conflict, and the US focus of most research in the area, have overlooked the presence of a 'climate culture' in its own right – hidden under the surface inside the US, but prominent everywhere else. As I show in Chapter 10, this has led to a failure to search for inherently cultural attitudes to climate change outside the US, and so, naturally, a failure to find them.

9.2.1 Climate change understanding in global publics

There is relatively good agreement across international publics that carbon dioxide is a greenhouse gas and that climate change is mostly anthropogenic. Some regard this as evidence that people are technically well-informed in the area. However, publics still know little or nothing of the impacts and costs that might be imposed on them by climate change, or by any policies to mitigate it. A rational assessment is therefore beyond them, and they have to assess climate change – the potential impacts and the priority for addressing it relative to other issues – culturally.* As a result, patterns of responses are usually either linear (Chapter 8) or inside an envelope with linear boundaries (Chapter 10) when plotted against national religiosity.

This observation prompts the question of whether strong public agreement about the anthropogenic nature of climate change is cultural too. Surveys of international publics show that in the most irreligious nations about 60% of people apparently agree that climate change is manmade,† whereas in religious places the figure is over 90%. It is extremely doubtful that religious populations are more knowledgeable on this issue than irreligious ones; I therefore propose that the ostensibly higher agreement instead reveals a cultural response. This being the case, we may ask how much of the religious response is cultural, and how much represents objective knowledge. We don't know. Given that cultural mechanisms are clearly in play, we could also ask how much of the irreligious response is cultural – for instance, Pascalian assent‡ – and how much represents objective knowledge. Again we don't know, but those giving objective responses could be significantly less than 60% of the total and perhaps much less; there may be other drivers.§

Deducing the degree of objective understanding when responses are strongly cultural (which is to say emotive) is no easy task. As noted above, researchers in the climate domain tend to take 'correct' answers at face value, and assume that the public is well informed. However, more careful consideration leads to the conclusion that they have little real understanding; their responses are emotive rather than reasoning. As an example, consider a survey that probed publics on

* A fact confirmed by many charts in the Excel-Ref.

† See, for example, the survey results from Levi 2021 (in the 'Extra' sheet of the Excel-Ref) and OECD2 (in the 'PostCovid' sheet of the Excel-Ref).

‡ See Section 9.1.4.

§ Especially because, as the text accompanying the Levi 2021 chart points out, the overall response is probably the net result of several underlying cultural components.

the question of the effect of unabated global warming on volcanoes. In the most irreligious nations, just over 50% of respondents said that more frequent eruptions were unlikely, while in the most religious ones, the figure was less than 20%.* Rationally, of course, the figure should be 100% for all nations, but in fact majorities in most nations gave the incorrect answer. Significantly, religious populations appear to be (much) less knowledgeable this time. This outcome appears to be due to the emotive conflation of disaster scenarios (technically, it's a Catastrophe Narrative resistive trend).

Section 13.3 and Figure 29 also provide an important perspective; it is clear that 'information about climate change' is in itself viewed culturally, no matter what detail it may convey.

9.3 Allied belief and core belief

Two concepts that are useful when trying to understand Figure 5 and the relationship between religion and climate catastrophism are 'core belief' and 'allied belief'; these give a more intuitive feel for what's actually going on in the subconscious minds of those responding to the questions.

The thickest trendline in Figure 5b (the Fully Constrained, FC series) represents people who still give climate-change most-endorsing responses even in the presence of extremely strong reality constraints. These are the 'ardent believers' (a familiar feature of religions), who collectively represent society's level of 'core belief' in the culture of climate catastrophism.

However, not all climate-change most-endorsing responses stem from a direct commitment to climate catastrophism. As we have seen, when questions are unconstrained (Figure 5a), such responses are also given by religious adherents whose innate scepticism of the Catastrophe Narrative is disabled by their faith. In other words, their endorsement results from a surface alliance between their faith and the culture of climate catastrophe. Hence, I term this effect 'allied belief'. The thickest trendline in Figure 5a (the SA series) represents the most emphatic belief of this type across nations, and I generally use this series as the yardstick for allied belief unless specific

* See chart OECD3 in the Excel-Ref.

strengths are discussed. In practice, the allied belief responses must surely include the core belief responses as well. But there are few of the latter, so they barely affect the figures in the former.[247]

These two conceptual belief types dovetail naturally with explanations for the effect of different framing strengths in survey questions on the corresponding responses. This is discussed in the following section.

9.4 The effects of framing strength

9.4.1 On unconstrained questions

In Figure 5a, the trendlines of responses to unconstrained questions tilt with strength, becoming steeper for more strongly-framed questions. Recall from Section 8.4.3 that a stronger framing represents a closer alignment to the highly emotive Catastrophe Narrative. This should prompt more allied belief in religious people, whose innate scepticism is disabled, as noted in Section 9.1.2. This is what we see as we move through the series from WA to SA; more of the religious express climate-change endorsing responses (although significant numbers of people, even in the most religious countries, are still unmoved[248]). Travelling towards the right on the *x*-axis, the religious represent a higher and higher proportion of publics, so this effect is heavily weighted to the right-hand side of the chart.

However, among the irreligious, the increasing alignment of questions to the Catastrophe Narrative has the opposite effect, prompting more innate scepticism (which for them is enabled 'by default'). As a result, fewer of them will give climate-change endorsing responses. Travelling towards the left on the *x*-axis, there are more and more such people, so the effect is heavily weighted to the left-hand side of the chart.

Together these two effects rotate the trendlines around a crossover point, which happens to be located at around 82% religiosity.*

* In practice, the crossover is blurred between about 75% and 90%. Because there is an asymmetrical response between the irreligious and the religious populations to the same change in framing strength, the crossover occurs nearer to the right-hand side. The *y*-axis response range from WA to SA, at the left of Figure 5, is about 3.6 times larger than on the right. This ratio for religion alone (Figure 7) is about 1.6, so I wouldn't expect anything less. At least in part, the ratio may be higher in Figure 5 because the unconstrained belief (in

Another way of thinking about this rotation is that it reflects the fact that cultures are emotively polarising, and the more so as the emotive content of their narratives increases.

Our range of unconstrained question strengths reveals that publics will react differently to Catastrophe Narrative variants that have more or less emotive content. Although preferential selection of the most emotive narrative variants is in part of how cultures emerge, as noted already,* a population of variants having different strengths is actually optimal for the culture, and sucks in more adherents; this evolutionary balance prevents the most emotive variants simply wiping out the weaker ones.

9.4.2 On reality-constrained questions

For reality-constrained questions the situation is very different, although religious and irreligious people again react differently.

Among the religious, and as noted in Section 9.1.3, the intrusion of reality triggers innate scepticism of, and collapsing allied belief in, climate catastrophism. And with religious values to defend, that scepticism is sharp and enhanced. But the strength of the constraint matters too; the greater the intrusion of real-world facts and priorities, the greater the scepticism, and the less endorsement of climate change there will be.

For the irreligious, innate scepticism is enabled 'by default', but, as with the religious, strength still matters: the more of a reality constraint there is, the less climate-change endorsement will be given. However, as a result of Pascalian assent (Section 9.1.4) or possibly because scepticism that is not religiously inspired is weaker, there is more muted innate scepticism among the irreligious. Their defence of existing priorities from the aspirations of climate catastrophism is therefore less stubborn.

As a result, a steadily increasing reality constraint produces a commensurate heightening of innate scepticism across both groups. As a result, the trendlines are pushed down towards the x-axis (Figure 5b), and also flatten out as they sink because the reality constraints main-

catastrophism) here is due to an alliance, rather than resulting from direct cultural belief, which may mean that a given increase in emotive framing produces less extra engagement (of religious adherents). This warrants further investigation.

* See p. 24.

tain the same impact ratio across both groups.

Ultimately, for the series with the strongest reality constraint (FC), only small minorities of publics across nations exhibit core belief, and still give a climate-change endorsement. However, the true line of core belief is probably in reality even lower than the FC series, because there's a limit to how strong a reality constraint can be written into a survey question that probes policy preferences.

An alternative approach, which may provide a stronger reality constraint, is to ask people how much money they'd contribute to policies that combat climate change. There are fewer surveys of this kind, but those which have been conducted suggest that even very small sums of money – almost trivial amounts – are perceived as strong constraints. Additionally, 'personal contributions' are seen as being stronger than 'tax contributions'. Any level of contribution sufficient to make the policies useful in practice is therefore going to have very little support, even in principle, and still less in reality. And certainly much less than that represented by the FC trendline in Figure 5b.

The trendlines from reality-constrained responses can't in principle cross over each other, so long as each represents a discrete hard-reality scenario. A way of thinking about these responses is that they reflect a 'negotiation' or accomodation of the cultural entity (in this case climate catastrophism) with reality. Cultural entities would never have become so embedded and ubiquitous in humans (and also co-evolved with our biology) if they couldn't, to a sufficient extent, accommodate reality; societies dominated by cultures that are too distanced from reality, too fantastical at the expense of rationality, will be more likely to fail. So cultures will typically be accommodating of reality constraints, but only in proportion to their strength.*

The term 'negotiation' does not imply that cultural entities are sentient, or even agential – as noted throughout this book, they are sustained subconsciously, through the preferential selection and propagation of an emotive narrative. The term is a useful metaphor through which we can present the cultural entity's 'point of view'

* This may not be the case for a brand-new or reinvigorated culture, which is going through a 'rampant' phase.

(analogous to the phrase 'the selfish gene'). We can also say that allied belief is the result of a cultural negotiation, but with another cultural entity rather than reality.

9.4.3 Virtue signalling and the real world

Unencumbered by reality constraints, the responses to unconstrained questions are essentially pure virtue-signalling (an intuitive term I have already slipped into the earlier parts of this chapter), either indicating direct membership of a cultural club, or sympathy with – 'associate membership', so to speak – an allied one.

In contrast, reality-constrained questions elicit responses that should correspond to real-world behaviour. It's as though respondents are actually choosing among policy options from a limited budget of their own money, or perhaps voting for particular parties that endorse climate change to a greater or lesser extent. That these responses accommodate reality to some extent does not make them non-cultural.* They still have a very robust relationship with national religiosity (a purely cultural phenomenon), and they are not what we'd expect from informed rationality, or from physical drivers (e.g. the climate or climate exposure of countries). They are still cultural –virtue signalling in other words – but negotiated to varying extents with reality.

This is important, because it means we should expect the real-world implications of climate catastrophism to have very similar patterns. In Chapter 11, I show that the actual deployment of renewable energy across nations is indeed culturally determined.

9.5 The cultural relationship changed with time

The patterns of all the survey results in Chapter 8, and those of the series presented in later chapters, all fit a single cultural model. However, there is significant inertia associated with mass cultural attitudes, so these patterns will have taken some time to establish themselves. In other words, the cultural relationship described above

* Adherents of a culture always accommodate constraints proportionally, essentially always giving way to reality. Their responses to reality-contrained survey questions therefore exhibit a gradient on a cultural axis (across nations, say). The responses of those who are not adherents may still vary with strength, but not so systemically, and they will not have such a gradient.

will have evolved quite slowly. Hence if attitudes are measured far enough back in time, an expectation of the cultural model is that the patterns would be different.

Exactly what we expect to see in 'historic' data, and how old is sufficiently historic to see it, is a complex matter. It depends on the precise unfolding of a race between several changing factors across nations. Firstly, climate change will have entered the public perception as a science topic; the associated narratives were certainly much less emotive then, and it took time for the cultural aspects of the domain to emerge and eventually to dominate. Secondly, in parallel, the alliance between the emerging culture of climate catastrophism and religious leaderships would also have taken time to be cemented; this wasn't simply decided top down by the ultimate leader of each faith, it is itself the result of an emergent process among religious elites. Thirdly, as awareness of the issue of manmade climate change, which no-one had originally heard of, spread across the world, the issue would have been framed in a different way depending on whether it was transmitted through a secular or a religious pathway.

Although the precise trajectories for these factors are all uncertain, we can reasonably expect that if we can go back far enough, attitudes should be systemically different. All the modern measurements in this book are from 2015 or after. That faith leaderships are effectively 'signed up' to climate catastrophism by this point is easy to establish.* However, their level of commitment at any point within the preceding 25 years is much harder to ascertain.

In those earlier years, commitment in religious nations would often have been weak. In such cases, it is impossible to distinguish nascent cultural (allied) belief from acceptance of what may then have been interpreted as culturally neutral scientific pronouncements. Resistive views are, however, much more likely to be cultural in nature. In some religious nations, it's possible that climate catastrophism will be culturally rejected because the relevant faith leadership has yet to establish a significant alliance (so giving the opposite of modern measurements!). In other religious nations, awareness of climate change may actually arise through an early manifestation of

* See Appendix D.

such an alliance. For irreligious nations, meanwhile, there is likely to be far less innate scepticism; initial perceptions of climate change as a science topic should overwhelm an instinctive rejection of nascent cultural catastrophism. This will also result in most-endorsing responses to unconstrained questions that are effectively opposite to modern measurements, in that irreligious nations will be positioned high upon the *y*-axis. In summary, relative to modern measurements for unconstrained responses, our expectation is for much higher *y*-axis scores from irreligious nations, and more varied scores for religious nations.

I use the 2005–2009 World Values Survey (WVS), a long-running periodic poll of global public opinion, to test these expectations.[249] This data is probably 'historical enough'; although Al Gore's film *An Inconvenient Truth* injected catastrophist memes into the mainstream media from about mid-2006, bulk public attitudes take some years to change. A problem with attempting to measure attitudes before this period is a lack of surveys covering a wide enough range of nations; looking at developed Western nations only would severely narrow the lens of religiosity we are using to view cultural developments. At the end of this window, towards the end of 2009, there was a much bigger inflection-point with the coincidence of the first heavily publicised climate 'Conference of the Parties' (COP15 in Copenhagen) and the Climategate scandal, both of which will have heightened public perceptions further, as well as shifting them to more (polarised) cultural positions too. So, our historic data should preferably end at 2009.

Among the many questions posed by the WVS survey, national publics were asked to rate the seriousness of various environmental issues for the world as a whole. Figure 6 shows the percentage of these who rated 'global warming or the greenhouse effect' as 'very serious' (which is the climate-change most-endorsing response for this survey), plotted against the same religiosity scale that is used throughout this work.

The correlation of this unconstrained series with national religiosity is negative (R = −0.46), and thus the opposite of the modern measurements. However, this may be partly because most-endorsing attitudes will be 'artificially' lower in those (typically very religious)

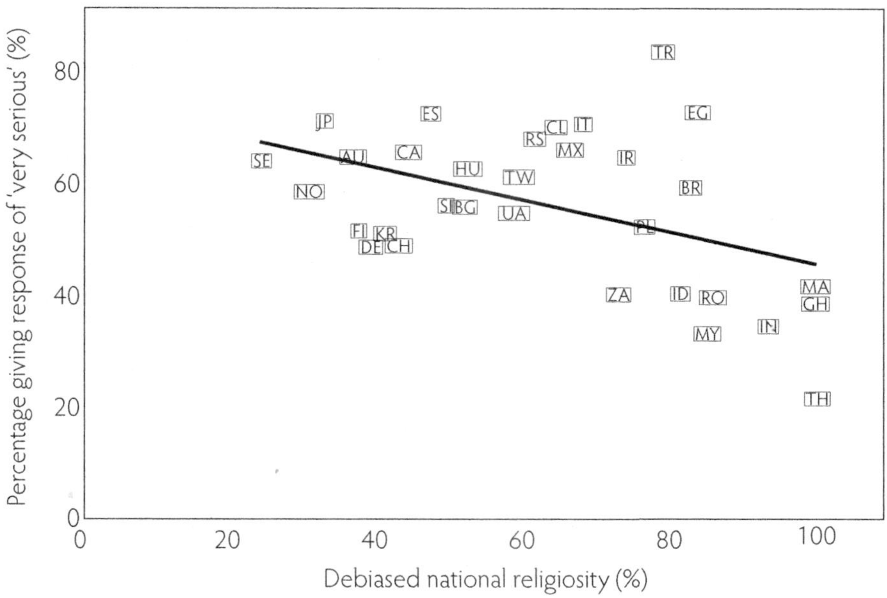

Figure 6. Before faith leaders signed up.

The survey covered 32 nations in 2005–9. See main text for details of question. $R^2 = 0.21$ All the data and sources can be found in the Excel-Ref, at the Extra sheet

countries where the spread of knowledge about climate change was still modest at the time. However, with or without such countries (Morocco, sub-Saharan Africa, India and Indonesia), the *y*-axis scores of the least religious countries are still very high indeed compared to today.* And the remaining very religious countries (Thailand, Romania, Brazil, Egypt, Turkey) have very varied scores, reflecting mixed cultural relationships that are still being formed, so with correspondingly different and changeable attitudes. Both of these factors match our expectation, hence providing further validation for the model of cultural causation.

* Except some lifted series, which are examined later; the historic series in Figure 6 should not be lifted.

Further detail on international public attitudes to climate change in this historical era, including the different nature of 'very serious' as a response back then, along with another smaller but confirmatory historical dataset, can be found in the endnotes.[250]

9.6 The same signature for a different culture

Because cultural entities arise from fundamental properties bequeathed to us by evolution, as described in Chapter 3, they all operate in the same manner no matter how different their surface features. This means that if any cultural entities are interrogated at the level of their most basic values, they should all respond in a similar way to climate catastrophism, as measured in Chapter 8. So, if we conduct surveys of attitudes among adherents of another culture – whether a religion or a political ideology – using questions equivalent to those in Chapter 8, we should always see patterns very similar to those in Figure 5.

Because all cultural entities are emotively polarising, we should see a set of response trends to unconstrained questions that rotate around a common pivot point. And because all cultural entities must negotiate with reality, we should see a set of response trends to increasing strengths of reality-constrained questions that progressively sink towards the *x*-axis and level out, finishing at a line of core belief. I suggest that these two patterns represent the 'signature' of a cultural entity.*

In order for this to work, the surveys used need to feature questions that prompt cultural reactions, and are framed as 'pure' reality-constrained or unconstrained types (i.e. not contaminated by content from the opposite type). In addition, each framing type must have a range of strengths, as set out in Section 8.4.3. Finally, the survey would need to cover representative samples of publics across many nations.

* While two such cultural signatures are demonstrated, it's hard to prove the negative case. Firstly, finding a topic that is unequivocally non-cultural, but is the subject of frequent surveys, is challenging. Secondly, unconstrained survey questions on such a topic are unlikely to feature (emotively) strong framings. Finally, we must demonstrate that a lack of signature isn't due to mixed-mode framings (Section 10.1.4), or the influence of multiple cultures, which could mix or hide the signature patterns; in other words, we must somehow show that the data is indeed unequivocally non-cultural.

The theory can be demonstrated by interrogating the old and familiar culture of religion, for which a number of suitable surveys on religious attitudes are readily available from mainstream pollsters. The question texts are shown in Table 9, alongside the framing buckets I have applied in an equivalent manner to Chapter 8.

The faith-endorsing responses to these questions from international (non-US) publics are charted in Figure 7. The nation encodings along the top of the figure reflect the positions of an example subset from the list of 44 nations that are used to construct the series in this 'religion-only' cultural view. Each nation is not necessarily represented in every series. There isn't room in the book format to put all nations across the top of the charts, and the large number of data-points involved would obscure each other if depicted. The marked example nations therefore provide useful reference points.*

Table 10 shows the main faith(s) for the exampled nations, in the same order as in the figure. As this table helps to make clear, the trends in Figure 7 are sustained by contributions from all of the main faiths, with each represented at several places on the trendlines. In other words, for basic enough values, all the main faiths act as one. While in principle each is a separate cultural entity, they have shared the same social-psychological domain (creation myths, deities, life-after-death and so on) for an extremely long time, and more is common about them than is different, despite their constant competition. As is clear from all the charts in Chapter 8, this is also true of the generic reaction of religion to the culture of climate catastrophism; that is, all the main faiths react to this new culture in the same manner. It is as though, in this regard at least, 'religion' consists of a single cultural entity.

The correlation parameters for the series in Figure 7 are shown in Table 11; all series are statistically valid and have robust or very robust correlation coefficients.

In Figure 7, we are plotting religiously-orientated values against national religiosities.† Because there's only one culture involved here, all the trends must slope the same way. However, the patterns of the

* Table 11 shows the number of data-points (i.e. nations) for each series. Full-data charts and sources for all series are provided in the Excel-Ref1.

† Using the same religiosity scale as used throughout this book.

Table 9. Survey questions probing religious-orientated attitudes.

ID	Survey question	Strength	Response
Unconstrained series			
SA	Belief in God is necessary in order to be moral and have good values.	Strong	Yes
MA	Religious practices are an important factor in the moral life of my country's citizens.	Medium	Agree
WA	How often, if at all, do you think about the meaning and purpose of life?	Weak	Often + sometimes
Reality-constrained series			
FC	We depend too much on science and not enough on faith.	Full	Agree
SC	Which of the following is most important to you? *(Your continent, Your nationality, Your local county/state/province/city, Your religion, Your ethnic group, None of these, Don't know)*	Strong	Your religion
MC	Please tell me which of the following is closer to your own point of view: *(Creationist... Evolutionist... Don't know)*	Medium	Creationist
WC	Which of the following sentences best describes your beliefs about what happens when you die?*(I believe in an afterlife but not specifically in a heaven or hell, I believe you go to heaven or hell)*	Weak	Total life after death options
WC1	In which of the following things do you believe, if you believe in any? *(Life after death)*	Weak	Yes

Links to the source surveys can be found in Excel-Ref1.

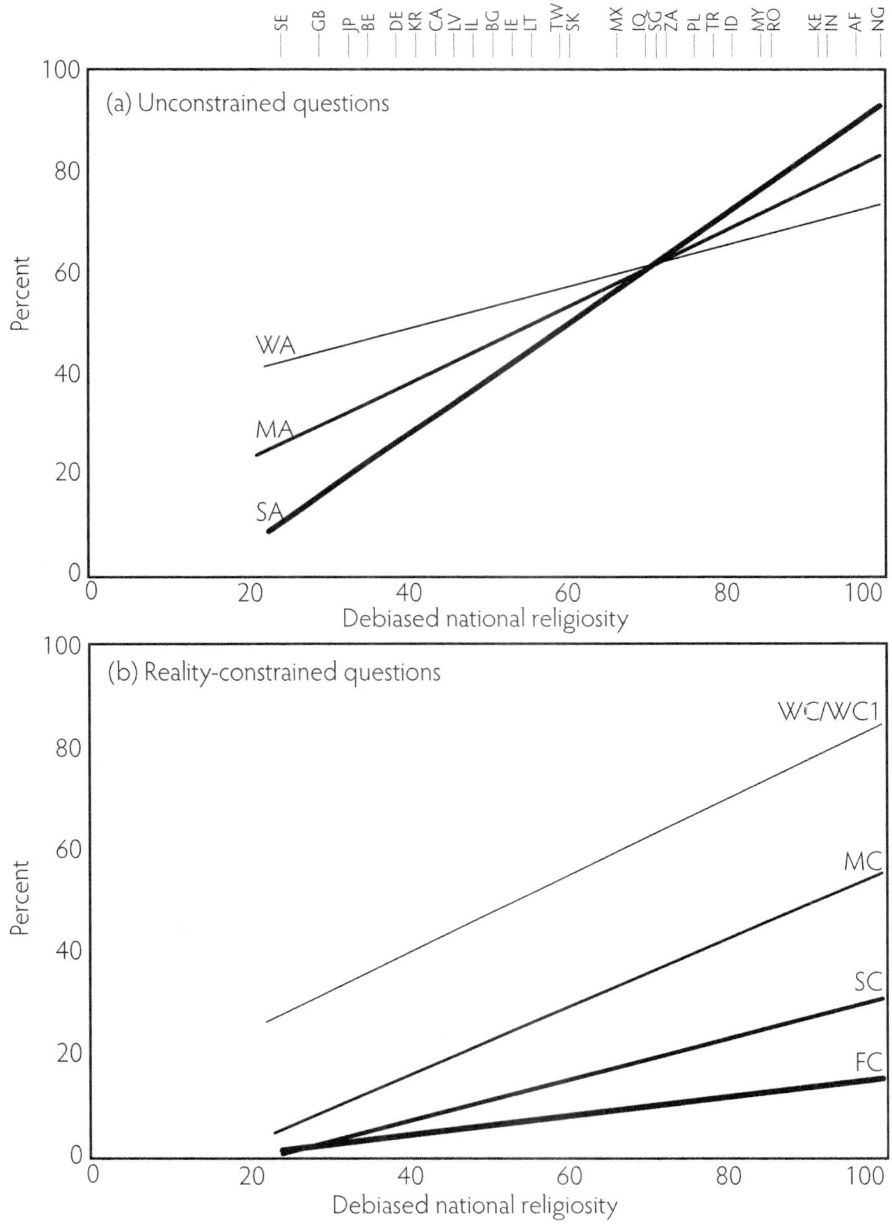

Figure 7. Interrogating another culture.

Religious-faith endorsing responses to questions about religious attitudes, plotted against religiosity. Trends only. Questions per Table 9, correlation parameters per Table 11. A list of country codes can be seen in Appendix L.

Table 10. Nations and their faiths.

	Primary faith	Secondary faith		Primary faith	Secondary faith
Sweden	Christian Protestant		Mexico	Christian Catholic	
UK	Christian Protestant		Iraq	Islam Shia	Islam Sunni
Japan	Shinto	Buddhist	Singapore	Buddhist	Islam Sunni
Belgium	Christian Catholic	Christian Protestant	South Africa	Christian Protestant	African Zionist
Germany	Christian Protestant	Christian Catholic	Poland	Christian Catholic	
South Korea	Christian Protestant	Buddhist	Turkey	Islam Sunni	
Canada	Christian Catholic	Christian Protestant	Indonesia	Islam Sunni	
Latvia	Christian Protestant	Christian Orthodox	Malaysia	Islam Sunni	Buddhist
Israel	Judaism		Romania	Christian Orthodox	
Bulgaria	Christian Orthodox		Kenya	Christian Protestant	Christian Catholic
Ireland	Christian Catholic		India	Hindu	Islam Sunni
Lithuania	Christian Catholic		Afghanistan	Islam Sunni	Islam Shia
Taiwan	Chinese folk	Buddhist	Nigeria	Christian Protestant	Christian Catholic
Slovakia	Christian Catholic				

Faiths: ✝, Christian Protestant; ✝, Christian Catholic; ☦, Christian Orthodox; ⊕, African Zionist; ☪, Islam Sunni; ☪, Islam Shia; ✡, Judaism; ☯, Buddhist; ॐ, Hindu; ⛩, Shinto; 福, Chinese folk.

reality-constrained and unconstrained trendlines exactly match our expectation. The pivot point for the unconstrained trends happens to be at about 71% national religiosity.

For the reality-constrained trends, committing to faith above (all) science is a very strong constraint indeed; only those having core belief will do so. Putting faith above national or ethnic or local identities is less strong, but is still a major commitment; almost everyone who is religious will also have at least some other loyalty of this kind. Committing to faith above the specific narrow scientific concept of evolution is weaker still. And weakest of all is a commitment to life after death, about which there is much scientific scepticism, but which can probably never be disproved.

Table 11. Interrogating another culture – correlation parameters.

ID	N	R	R^2	p
Unconstrained series				
SA	30	0.88	0.773	1.64×10^{-10}
MA	18	0.91	0.824	1.93×10^{-7}
WA	23	0.55	0.305	6.31×10^{-3}
Reality-constrained series				
FC	20	0.64	0.413	1.20×10^{-4}
SC	36	0.76	0.573	2.24×10^{-3}
MC	20	0.80	0.641	2.22×10^{-5}
WC	20	0.79	0.631	2.87×10^{-5}
WC1	30	0.86	0.732	1.70×10^{-9}

N, number of data points; R, correlation coeffient, R^2, coefficient of determination; p, probability value.

Support in the WC/WC1 series, at the left of Figure 7, is actually higher than the proportion of national populations claiming to be religious. This response is likely boosted by Pascalian assent from disbelievers (Section 9.1.4).* The same effect might also boost support in weakly unconstrained framings. The minimally emotive WA framing garners almost twice as much support as there are religious people within nations at the far left of the chart; that religion has been around for millennia would help with this kind of acceptability, even for an unconstrained scenario.

Probing religion itself is a huge topic, so any further explanation is out of scope. However, this brief examination demonstrates robustly (as per Table 11) that, across international publics, responses to both reality-constrained and unconstrained questions about religious attitudes match the expected patterns.[251] This is excellent validation of the cultural theory proposed in this book.

* Where in this case the insurance bet is on life after death.

9.7 Summary

The measurements in Chapter 8 demonstrate that public attitudes to climate change have very strong relationships with national religiosities across the globe. As we shall see in the next chapter, this is not only true for climate-change most-endorsing responses, or for the particular attitudes measured in the 'Basic' series, but for many other attitudes too, including those that are resistive to the concept of dangerous climate change.* Given that religiosity is a purely cultural phenomenon, even without further explanation we can deduce that such attitudes are themselves almost certainly cultural in nature. Indeed, nothing physical (such as the climate or climate exposure of nations), or even anything rational (such as science or technical policy), could cause the simultaneous correlations and anti-correlations of the climate-change most-endorsing attitudes with national religiosity, as shown in Figure 5.

To add to this intuitive take on the data, Section 9.6 clearly demonstrates that, provided similarly framed questions are used, the same pattern of responses is obtained when interrogating religion, which is undoubtedly cultural. This confirms that climate catastrophism is indeed a culture too.

Section 9.1 presents a straightforward model of cultural causation that explains the pattern of responses. It rests upon the concepts of cultural belief and innate scepticism, and their expression by national publics participating in the ambivalent relationship between climate catastrophism and religion. Exactly as we'd expect (and as demonstrated in Section 9.5), this relationship was very different before 2009. This is a very useful confirmation of the principles of the model. Finally, as noted in Section 9.2, these principles also chime very strongly with Dan Kahan's theories, the state of the art in cultural cognition as applied to public attitudes on conflicted topics, including climate change. However, this book moves beyond Kahan's exclusive focus on the USA, to show that public attitudes to climate change are inherently cultural everywhere.

* See Section 10.2 for a summary of all the primary series on climate-change attitudes, activism and policy presented in this book.

9.7.1 Insights from the model

What we see in Chapter 8 are the fundamental characteristics of a cultural entity, in fact, of *two* cultures, because the net effects are also a product of our lens of national religiosity. The analysis here shows clearly that the older culture of religion is interacting with a new one: climate catastrophism. The survey questions in Chapter 8 are interrogating a culture that itself *owns* the conceptual domain of the questions, in the same way that the concepts of God or life after death are owned by religious culture.

As noted in Chapter 3, cultures are ultimately in-group/out-group definition and reinforcement systems. Large swathes of publics are signalling 'in' to the 'club' of climate catastrophism, simply because their religious leaders are signalling 'in'. But when hard realities appear, their support collapses. The club of religion matters to them much more than virtue signalling about climate, whatever alliance their leaders have entered into. And, as reality bites more and more, support from all quarters – both religious and irreligious – falls away, eventually leaving only a rump of core belief within all publics: ardent believers who are still culturally committed.*

The whole scenario is the result of cultural belief, an emotive commitment to a culture, and also of innate scepticism, emotive rejection of it. While both are subconscious reactions, the one is not simply a mirror image of the other; they have a level of independence and their own complexities.

There is little or no role in this model for objective reason, which requires knowledge to feed it. Given that publics know very little about the climate and have been soaked for decades in the fairy tale of the Catastrophe Narrative, this is hardly a surprise.

9.7.1.1 Cultural belief is a group phenomenon

As noted earlier, the religiosity scale used in this book is based upon self-assessment; it is used as the x-axis for many charts, including Figure 7. The series in that figure represent faith-endorsing responses to questions about religiously-orientated values, but only the trendline

* While support for a culture always falls as reality constraints become stronger, in the case of climate catastrophism, due to Pascalian assent among the irreligious, for the weaker constraint trends (VWC, WC, MC) it is still equal to or higher than unconstrained support (any trend).

for the SA series maps closely to religiosity as expressed in this self-assessment. In nations at the far right-hand side of the figure, essentially the whole population claims to be religious, but the trendlines for the other series end there at *y*-axis values that are significantly lower, and sometimes dramatically so.

So, even if we know from self-assessment that a particular individual is religious, we can only hazard a guess about where they might sit on the chart in respect of each series. There's a very high chance, for instance, that they will give faith-endorsing responses to the questions for the SA and WC series, but a low chance they'll do so for the FC question. However, one individual may give faith-endorsing responses to all questions, including FC, and another perhaps only to the question for SA.

This in turn tells us that attitudes at the individual level are neither universal for believers, nor what actually matters for the phenomenon of cultural belief. It is a distribution of attitudes across the group that allows cultural entities to operate most effectively. Given their existence is owed to sustained cultural group selection (Section 3.1), this should not be a surprise. Success for the culture trumps individual traits, and a range of attitudes linked by a common theme delivers both adaptability and the best group reach, while maintaining cultural coherence (it may even be the case that an attitude which, due to external events, has become embarrassing and so could harm the culture, is prevented from doing so by jettisoning those adherents who express it). Nevertheless, it is not intuitive that two professed adherents of a strong culture may have differing, and possibly widely differing, attitudes about culturally relevant issues.

Religion is acknowledged as a belief system – a matter of faith rather than rationality – by both believers and sceptics, so self-assessment will be consistent for both. This makes self-assessment of belief (or disbelief) possible, allowing us to construct a religiosity scale covering many nations. However, climate catastrophism is not treated as a belief system, so no such scale can exist, and this is why we have to use the lens of religiosity to probe attitudes. This adds complexity, but it is nevertheless clear that attitudes are distributed in a similar manner as for religion.

With self-assessment impossible, it's tricky to even define 'belief' in climate catastrophism at the individual level, notwithstanding my convenient labels of allied and core belief for signature attitudes. However, at the far left-hand side of Figure 5, only about 1 in 4 of national populations is religious,* so the effects of religiosity are relatively limited and allied belief is very low. In contrast, the numbers of those who believe in climate catastrophism directly, rather than through religious faith, in other words the core believers manifest in the FC series, will be at their highest here, despite such ardent believers being minorities embedded in sceptical populations. But even at this high point, there is a very wide range of attitudes to climate change as revealed by the various series. Just as for religion, and even if we could magically remove the complication of Pascalian assent, we cannot predict where an individual believer in climate catastrophism might sit on the chart in respect of all these series. As for religion, what's critical to the functioning of climate catastrophism is not what attitudes are held by any particular believer, and indeed these could be very different between any two of them anyhow, but the distribution of attitudes across the whole group of believers.

To emphasise how unintuitive this group nature is, core belief, as represented by the FC trends in Figure 5 (climate catastrophism), and Figure 7 (religion), is extremely important to the respective cultural entities, but in both cases it is only expressed by a small minority of their total believers.

9.7.2 But what if?

Notwithstanding the considerable weight of evidence and theory behind the conclusions stated here, it is healthy to ask some challenging questions:

- What if my explanations are hogwash?
- What if there are outliers unduly influencing the data?
- What if a third variable drives both attitudes and religiosity?

These challenges can be more than adequately answered. However, doing so must wait until the end of the next chapter, which introduces more data series and cultural features, and also looks at

* Sweden has the lowest religiosity of all the nations I use, at about 24%.

why nobody has previously investigated the possibility of climate catastrophism as a new culture in its own right. With all this in place, answering the questions becomes more straightforward.

9.7.3 A final note – ironies

It strikes me that the cultural saga of religion and climate catastrophism is not short of ironies. For instance, in religious societies, the worst effects of the fairy tale of climate catastrophism are only prevented by belief in an older fairy tale! This is because the reality constraint inherent in policy decisions prompts sharp innate scepticism, which defends the older culture and hence society generally. Meanwhile, people in secular nations don't buy climate catastrophism at face value (their allied belief is very low), but many more of them accommodate the culture's zealous and misdirected policies, at least in part because they inherently trust science (Pascalian assent)!

Chapter 10

THE FULL MODEL, A DISMAL FAILURE, AND 'WHAT IF?'

In Chapter 8, I presented a set of very robust results, which, as I explained in Chapter 9, demonstrate the existence of a cultural entity, climate catastrophism. Section 10.1 presents and explains further results, reflecting the much wider range of cultural responses that surveys prompt in practice, all of which support the same conclusion. The expanded model that results is broad enough to explain the responses from every international (non-US) survey on attitudes to climate change taken between 2015 and early 2020, and is thus highly significant.* However, the section is in essence about nuance and detail, and doesn't add a lot to the more fundamental story of climate catastrophism or its social impacts, as traced thus far. Some readers might therefore wish to skip to Section 10.2, or possibly even to Section 10.4.

Section 10.2 briefly summarises the data series featured in this book, which is helpful context for what follows. Section 10.3 outlines the long-founded extensive literature on how international attitudes to climate change are formed, and considers the dismal failure of researchers in the field to identify the importance of national religiosity as a 'social predictor' of those attitudes. Section 10.4 answers the 'But what if?' questions raised towards the end of Chapter 9, before Section 10.5 speculates about the longevity of the model.

* Providing the questions do not also probe attitudes unrelated to climate-change. For instance, a question about preferences for addressing climate change through state or invidividual action would not address attitudes to climate change itself. Surveys subject to serious pre-selection bias are also unpredictable, for instance this UN one: www.undp.org/publications/peoples-climate-vote. After about mid-2020, the Covid pandemic may have significantly impacted survey responses too.

10.1 Further cultural measurements

As their name implies, the Basic series depicted in Chapter 8 cover only a subset of the expressions of international (non-US) attitudes to climate change. In particular, they only take into account the most-endorsing responses to survey questions. However, publics express a much wider range of attitudes to climate change: their responses may be less endorsing, or resistive. In addition, the survey questions eliciting responses are more varied than those considered so far. There are also several other series covering different question framings.

As in the Basic series, many of these additional patterns of response have a straightforward relationship with national religiosities: a robust linear correlation or anti-correlation. For others, the relationship is more complex, and in some cases is affected by secondary factors not directly related to national religiosity. However, all of them still fit into a single, straightforward model based upon cultural causation, albeit more complex than the one represented in Figure 5.

The additional features and/or relationships with religiosity that we need to consider are set out below. Firstly, there are three auxiliary series related to climate-change most-endorsing responses:

- *Lifted series:* responses are 'lifted' upwards on the *y*-axis, while retaining the gradients that would be expected from their framings.
- *Extremely weakly framed series:* responses to extremely weakly-framed questions, where the modality of responses between unconstrained or reality-constrained types, is lost. These series are non-linear.
- *Mixed-mode series:* responses to questions that mix both unconstrained and reality-constrained elements within the same question, so that the modality of responses is mixed. Most of these series are non-linear.

In addition, for responses to the weaker strength survey questions (either unconstrained or reality-constrained), a secondary variable causes some systemic variation around the response trends. This feature is examined in Section 10.1.5.

Finally, I look at series for responses that are not climate-change most-endorsing:

- *Equivocal series:* responses to unconstrained questions that are not most-endorsing of climate change, but are still endorsing to a lesser degree. These series may be non-linear.
- *Resistive series:* responses to unconstrained questions that are resistive to climate change (in practice, resistive to the Catastrophe Narrative).

10.1.1 The full model for most-endorsing attitudes

Figure 8 represents the full model for climate-change most-endorsing attitudes. It shows, on a single chart, the three auxiliary series and the secondary variable from the list above, along with the Basic series from Figure 5.* In principle, the most-endorsing attitudes revealed by every international (non-US) survey taken between 2015 and early 2020† should match one of the series in this figure.

Climate-change 'equivocal' and 'resistive' attitudes are briefly explained below, and some examples are charted. However, as we shall see, they don't add much that can't be deduced from the most-endorsing attitudes, apart from useful confirmation. Resistive attitudes, in particular, are more or less a mirror of the most-endorsing attitudes. Hence Figure 8 essentially represents the full model of cultural causation, although it formally includes only the climate-change most-endorsing responses.

As noted for Figure 5, the MC trend (dashed) is intuited, and the FC trend is estimated from WC (with limited spot-confining through some data measurements). All survey questions, results and sources for the series in Figure 8 are listed in Appendix C. Each of the additional data series and features, over and above Figure 5, are explained in a subsection below.

In the rest of this book, where I refer to the robustness of trends for attitudes to climate change as plotted against national religiosity, I use the following terminology:

---

* See p. 147.

† See also footnote on p. 179.

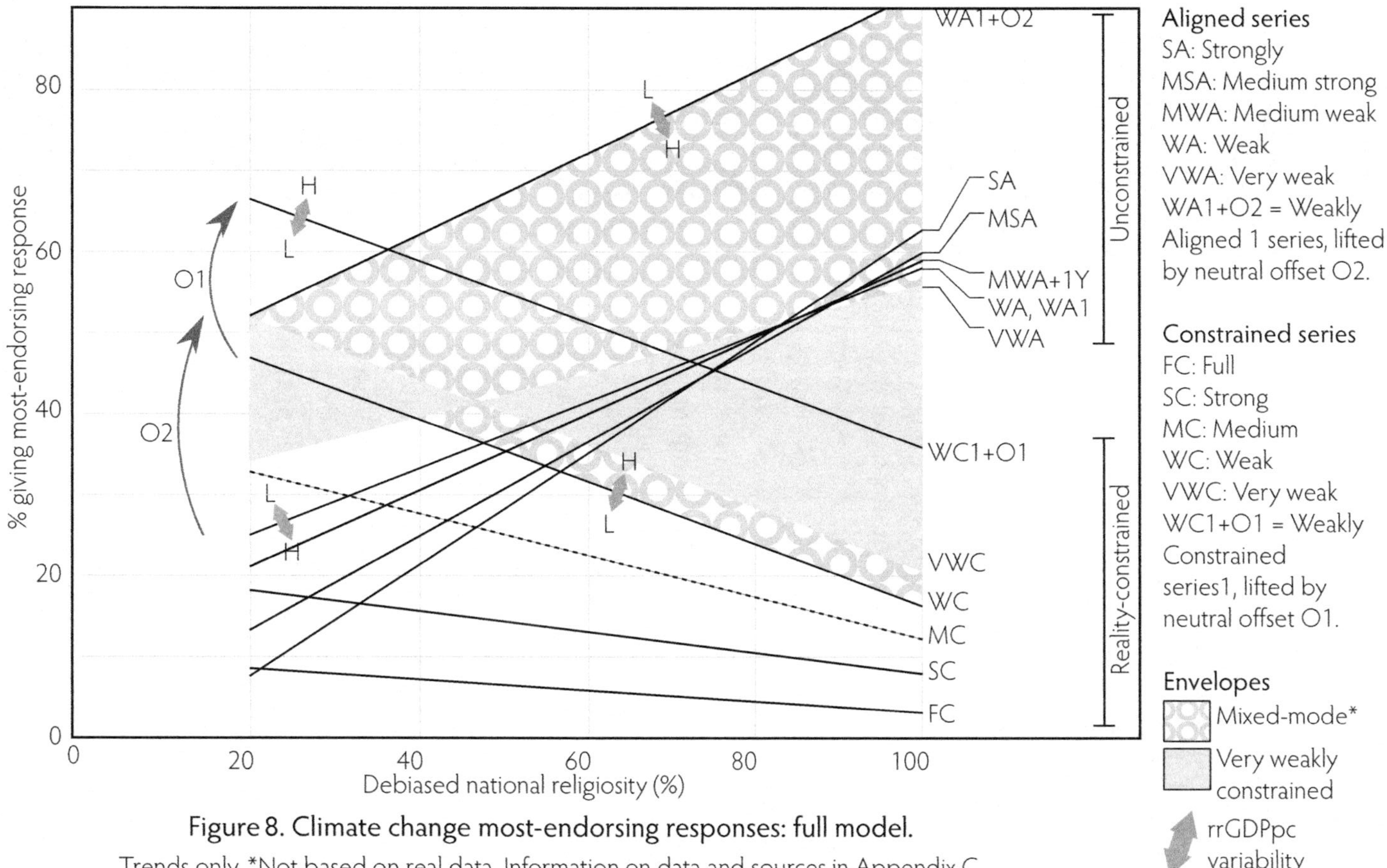

Figure 8. Climate change most-endorsing responses: full model.

Trends only. *Not based on real data. Information on data and sources in Appendix C.

- Where the correlation R^2 is greater than about 0.36 ($R > 0.6$), I use the term 'robust'
- Where R^2 is less than about 0.1 ($R < \sim 0.32$), I treat these trends as 'not significant' or 'culturally neutral'.
- Between these two values, I refer to trends in the attitudes that are 'modest'.

So, for instance, for the Basic series from Chapter 8, the correlations are robust, while, as we shall see later, for the 'equivocal' series they are modest. In both cases, this is true for all the series, and whether the correlations are positive or negative.

10.1.2 Lifted series

The climate-change endorsing responses to some survey questions look like those typically prompted by reality-constrained or unconstrained questions, except they are also offset to higher *y*-axis values; in other words, support is generally higher across all nations. In principle, such lifted series can be broken down into two (or possibly more) components, comprising at least:

- a culturally modest or neutral offset that does the lifting
- a 'normal' unconstrained or reality-constrained series (of any strength)

An extra element within the question framing is causing the offset. However, as we will see, the separate components might not be identifiable in practice.

In principle, there could be a lifted series corresponding to any one of the trendlines in Figure 8; in fact, there could be a range of them, each having different offsets. Even the boundaries of the 'envelopes' for extremely weakly framed or mixed-mode questions, as discussed later in this chapter, could be lifted.

Two such series are shown in Figure 8. WC1+O1 is a reality-constrained series featuring an offset, O1. WA1+O2 is an unconstrained series featuring an offset, O2. While lifted up on the *y*-axis, these series have approximately the same cultural gradient as their counterparts, WA1[252] and WC (WC1 alone isn't plottable in practice, but all weakly-constrained trends should be about the same).

We will first look at the WA1+O2 series, for which the separate components can be identified. The question asked was: 'How serious a problem, if at all, do you think climate change is?' The published results combined two separate climate-change endorsing answers: 'Extremely serious' and 'Very serious'. Lacking more granular data, I couldn't be sure, but surmised that 'Extremely serious' – the more emotive option – attracted a cultural endorsement and was therefore causing the robust positive correlation with religiosity. Meanwhile, 'Very serious' was, in these circumstances (see Section 10.1.6), attracting only an equivocal response, so causing little or no corresponding cultural gradient. Fortunately, I was eventually able to obtain the detailed results breakdown, which showed my assumption was correct. The 'Extremely serious' series was a good match with other weakly-aligned series, while the 'Very serious' series has no statistically significant trend (it is culturally neutral), yet produces an average offset of about +30 units on the *y*-axis.

It might be counterintuitive that 'Very serious' fails to prompt a cultural response. However, many things are labelled 'Very serious', but most people pay little attention to them. In contrast, the word 'extremely' crosses the threshold to produce a cultural response, although the effect is weak, because there is no explicit emotional or personal context (as there is, for instance, in the SA series).

It is fortunate that the WA1+O2 series is so cleanly separable into its two components, but in practice it's not always possible to do so:

- The components may be more ambiguous, implying that further splits are necessary to resolve the most basic elements.
- The data might not be available (as was originally the case for me with WA1+O2).
- The data might not be amenable to breaking down at all.

An example of a series that resists a reductive analysis is WC1+O1, prompted by the question: 'Which countries, if any, do you think have had the most negative impact on global warming and climate change?', and reflects the share of people who chose 'China' from a list of five possible countries. Given this is a single-word answer, it is very hard to demonstrate conclusively, but I suggest that the pattern of responses stems from two distinct social psychological elements: a modest or neutral offset, and a robust cultural trend.

The act of comparing the total emissions of countries is a reality constraint, but only a weak one, as the consideration of relative emissions is not particularly emotive. However, there is some cultural engagement from both committed believers and the innately sceptical, because the question is essentially seeking to apportion blame for global warming, and is thus, to some extent, a Catastrophe Narrative endorsing one. We would therefore expect the pattern of responses across nations to be an anti-correlation with religiosity, of about the same gradient as the WC series.

However, the question is also a technical one, which doesn't explicitly support or reject climate catastrophism. It is therefore likely that, across nations, many people – and far more than in the Basic WC series – simply give an answer that they know to be correct: 'China'.* Some actual knowledge is required to do this, but China's huge population and vast industrial capacity are widely recognised, and these facts are not culturally contested. The confidence in this 'common knowledge' answer, which can be given without any commitment to a cultural position, is probably what gives the culturally neutral 'lift' on the *y*-axis. However, I don't know how to prove this. And why is the lift up the *y*-axis about 20 percentage points, and not, say, 10 or 35? I don't know.

The survey question texts and correlation parameters for these two 'lifted' series are summarised in Table 12. See Appendix E for some other lifted series, including one that represents the priority expressed for climate change in the economic recovery post Covid.

Sometimes – currently rarely, I think – what appears to be a culturally neutral lift may instead come from a question wording that simultaneously encourages Pascalian assent among the irreligious and allied belief among the religious. For an example, see Online-appendix D. It's also worth noting that when the cause of a lift is 'common knowledge', this may not always be correct; sometimes what publics deem to be commonly known is actually wrong.

* Responses for the other obvious answer, 'the US', also anti-correlate, although weakly so, because Taiwan and Hong Kong in particular are less willing to blame the US for emissions.

Table 12. Lifted series: questions and correlation parameters.

Question	Reponse	Series	N	R^2	p
How serious a problem, if at all, do you think climate change is?	'Very' or 'Extremely'	WA1+O2	37	0.59	4×10^{-8}
	'Very'	No trend	37	0.008	6×10^{-1}
Which countries, if any, do you think have had the most negative impact on global warming and climate change?	'China'	WC1+O1	22	0.72	2×10^{-7}

Appendix D provides a full-data graphic of the WA1+O2 composite, and some more detail. Links to the survey source, and pointers to the original data for both series, can be found in Appendix B.

10.1.3 Extremely weakly framed questions

When the survey question framing strength is extremely weak – for unconstrained questions, even weaker than very weakly aligned, and for reality-constrained questions, even weaker than very weakly constrained – the climate-change most-endorsing responses from publics are non-linear against national religiosities. This is because the framings aren't emotive or emphatic or reality-charged enough to invoke cultural responses. Given that publics have no significant domain knowledge to guide them, their responses (in fact, not only climate-change endorsing, but *any* responses) lose modality and simply drift between the available two *cultural* options.

As a result, the responses have a non-linear pattern, but all fall into a single envelope defined by those framings that are still just strong enough to produce a linear result. This envelope is indicated in Figure 8 by the grey cones between the very weakly aligned (VWA) and the very weakly constrained (VWC) trendlines; responses can appear anywhere in these areas, and somewhat beyond them due to the normal scatter of data around a trendline. They probably reflect a host of lesser local influences for each nation.

In practice, identifying exactly where trends dissolve isn't easy, and there'll be a small ambiguous zone where trends are not robust, but may still be modest (like the equivocal response options to more strongly framed unconstrained questions).

In Appendix F, I explain the pattern of responses to an example extremely weakly-framed question, in a way that should give a more intuitive feel for how this effect occurs. Appendix F also explores an exception, whereby a small minority of nations may 'escape' the grey cones, and always in the same direction; I don't currently have a good explanation for this exception.

10.1.4 Mixed-mode questions

Mixed-mode survey questions are those that contain both unconstrained and reality-constrained elements within the same text. Publics react to both elements, but in different ratios, depending upon secondary factors. As a result, climate-change most-endorsing responses to mixed-mode questions typically have a non-linear pattern. They should, in principle, all fall within the boundaries of an envelope that is defined by the patterns we'd expect if each of the elements were presented as a separate question, with its own framing type and strength. However, if we use the trendlines of these separate elements to represent such boundaries, some data points will inevitably stray beyond them due to the normal scatter around a trend. In some circumstances, mixed-mode responses may retain a linear trend, but still within the same envelope; see Section 10.1.4.2.

10.1.4.1 Non-linear mixed-mode series

In Figure 8, the area filled with grey rings, between the WA1+O2 and WC lines, represents the envelope for a non-linear series, as described above (in this case, not based upon real data). A mixed-mode question with elements that match the framing type and strength of the two questions that prompted these series will fill this envelope. In principle, such a mixed-mode envelope can exist between any two lines in this figure, lifted or otherwise, where one anti-correlates with national religiosity, and the other correlates.

In practice there may not be enough data points to determine the envelope boundaries by observation, and it may also be impossible to deduce them from theoretical considerations, even where the different text elements in the question are explicit.

The (Pew 2019) question generating the example series depicted in Figure 9 consists of an explicitly unconstrained element mixed with an explicit comparative list that provides a reality constraint. The former comes from nominating climate-change as a 'major', 'minor' or 'no' threat, while the latter comes from the fact that one of these three options has to be chosen for each entry in a list of nine (global) threats, of which climate change is just one. There isn't enough data to fill the envelope, even if repeated instances of the survey are used to try and give better coverage. The presumed linear boundaries of the envelope as shown in the figure are highly speculative; they are drawn here as the WA and WC trendlines from Figure 8, with arbitrary offsets added to both in a manner that seems to fit. However, it's very challenging to tie the particular strengths and offset values to the question text elements.

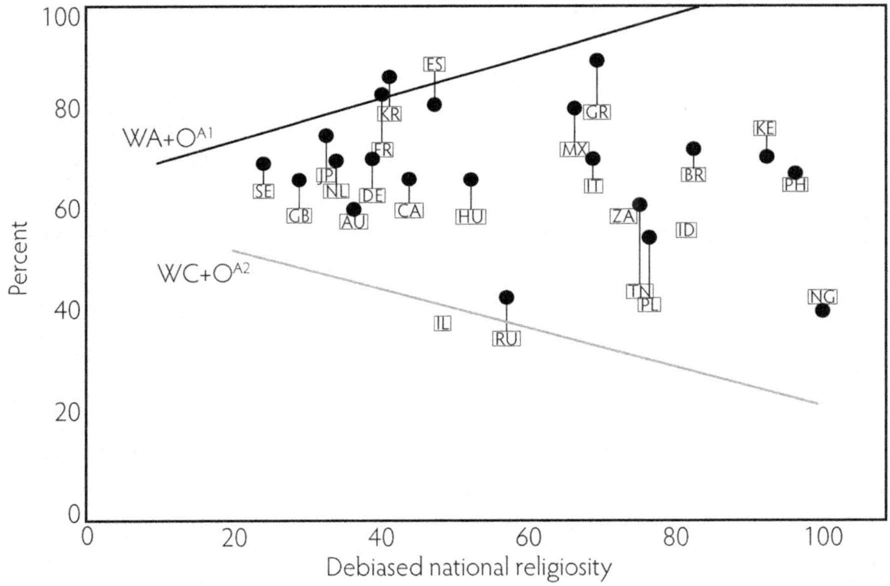

Figure 9. The MM1 series.

MM1 is a mixed-mode question with speculative boundaries. See main text for question. The original chart can be found on the 'Extra' sheet of the Excel-Ref. Country codes represent the 2017 survey data and black circles are the equivalent 2018 data. Where no circle is shown, the two years overlap. A list of country codes can be seen in Appendix L. $O^{A1} \sim 49$, $O^{A2} \sim 4$.

Even when the bounds of the envelope seem clearer, some mixed-mode questions are still hard to analyse. This is because subtle differences in language can dilute the question framing, or introduce constraints in respondents' minds. When this happens, it can be very difficult to understand exactly what aspects of the question text caused the pattern of responses. An example is shown in Figure 10; its boundaries appear to be a VWA-type trend with offset, and a VWC-type trend. Recall that both of these trends are *estimates* (in Figure 8) anyhow, although they are very close to the real WA and WC respectively.

The question that generated this series is: 'Do you think global climate change is harming people around the world now, will harm people in the next few years, will not harm people for many years or will never harm people?' Splitting global harm into time buckets is a *technical* task, for which publics are completely unprepared.

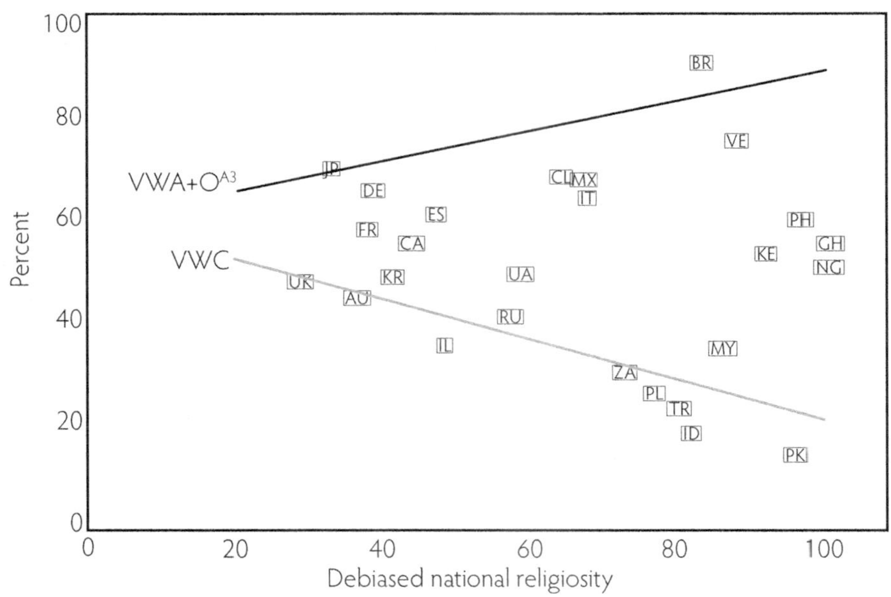

Figure 10. The MM2 series.

MM2 is a mixed-mode question with less speculative boundaries. Refer to main text for question text. The original chart can be found on the 'Extra' sheet of the Excel-Ref, series MM2. A list of country codes can be seen in Appendix L. O^{A3} ~ 34.

This probably dilutes the cultural framing enough to result in a very weakly aligned trend. The act of considering such exact timescales also introduces an element of reality constraint; for most of the public it's unlikely this could be done without considering natural disasters (climate related or otherwise) and other harms witnessed during their lives, which will contextualise return times and prompt consideration of the threshold for 'harm' in regard to all such events. Without the timescale element (or if this was vaguer), I suggest that 'harm' would be interpreted much more emotively.

However, this explanation is still some way from being able to confirm a hard linkage between particular text elements and the candidate VWA/VWC trends. It is very challenging to explain the precise offset of the VWA series too, according to the principles in Section 10.1.2. Because the question is sharply focused on the timescale element, it doesn't have any breakdown of 'harm' (e.g. 'extreme', 'serious' or 'moderate'), which makes for a very low threshold of what 'harm' might actually consist of; it is probably this that produces the culturally neutral 'lift'.

Notwithstanding these ambiguities, some mixed-mode series tell us more about the alliance between religiosity and climate-catastrophism. Consider the MM3 series (Figure 11), which is based on 67 national responses to a mixed-mode question: 'Do you think that climate change is a very serious threat, a somewhat serious threat, or not a threat at all to the people in this country in the next 20 years? If you do not know, please just say so.'

The responses almost fill the expected envelope, and its boundaries are clear. In addition, their shape and position correspond well to expectation. 'Threat' is an emotive word, but because there is no category for an 'extreme' threat, the climate-change most-endorsing response will likely include a culturally neutral element – producing an uplift – from those who think that the threat is 'very serious', and a cultural response (of at least medium-strong alignment) from those who think the threat is 'extremely serious' but are forced to choose 'very serious' as the strongest response on offer.* As a result, the upper boundary is set by the MSA trend with an added uplift

* See Section 10.1.2 for a similar situation, where the separate elements are measurable.

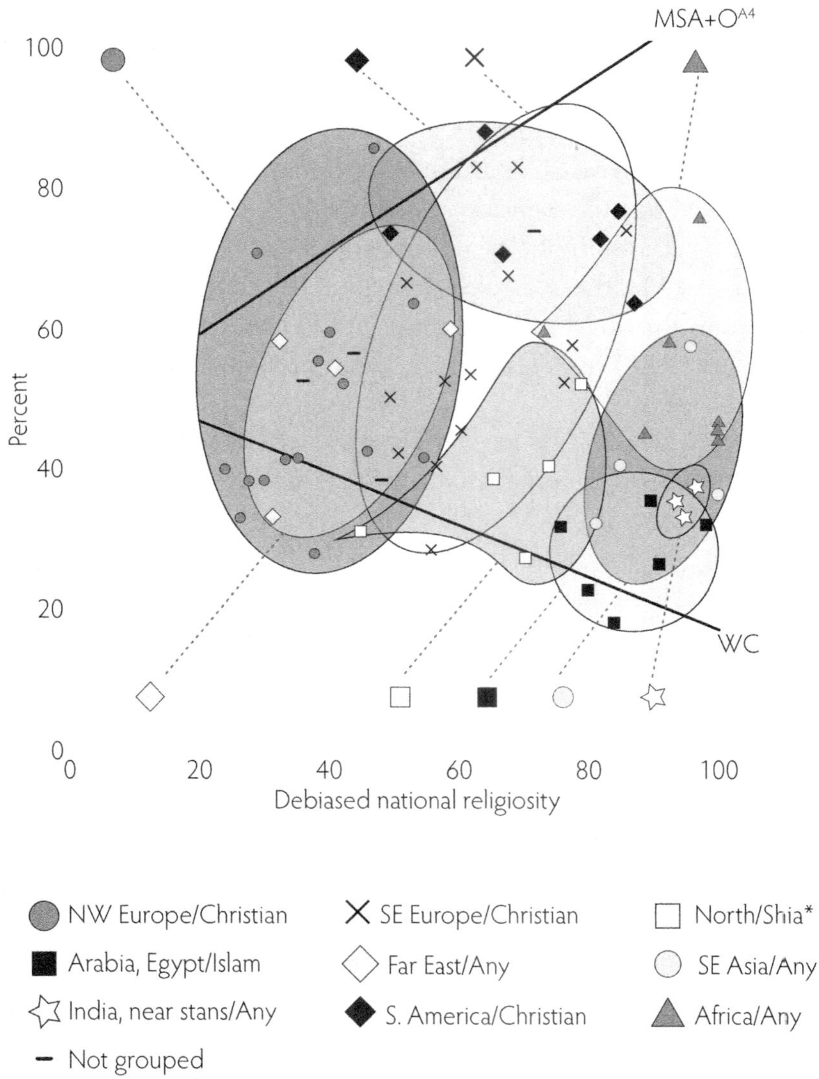

Figure 11. The MM3 series.

MM3 is a mixed-mode series with well-defined boundaries, as well as religio-regional groupings. *The Eurasian arm of Islam, including countries such as Iran and Turkey. $O^{A4} \sim 50$. The original chart can be found on the 'Extra' sheet of the Excel-Ref.

(although, again, its value is not determinable). The lower boundary is defined by the WC series because, similar to the MM2 series in Figure 9, the timescale element of the question imposes a reality constraint (but only a weak one).

Figure 11 also reveals that there is secondary systemic structure in the responses to mixed-mode questions. It is clear that countries with similar levels of religiosity and in geographical proximity to one another – I refer to these as 'religio-regional' groups – respond in similar ways to the question. Religio-regional groups are considered in much more detail in Section 10.1.5, but are ringed and encoded on Figure 11.

At the high religiosity end of the *x*-axis, where the mixed-mode envelope boundaries diverge much more, the two elements of the question produce different effects in different religio-regional groups, despite their similar levels of religiosity. So, for example, in the South America/Christian group, relatively more of the respective publics react to the unconstrained element of the question, and give climate-change most-endorsing responses. In the Arabia, Egypt/Islam group, meanwhile, the predominant reaction of publics is to the reality-constrained component of the question, and so fewer climate-change most-endorsing responses are recorded. The net effect is for the groups to be teased apart vertically. This effect does not occur for the less religious nations, which sit where the envelope boundaries are much closer together.

The effect suggests that, in principle, groups higher on the y-axis have a somewhat stronger alliance with climate-catastrophism than groups lower down, despite their similar levels of religiosity. Why this should be remains obscure – presumably secondary factors come into play when there is tension between the two types of question element. However, mixed-mode questions are giving us useful information that cannot be gleaned from the responses to unconstrained or reality-constrained questions alone, because these tend to prompt responses that are undiluted allied belief or innate scepticism.

Also, I don't know why the far top-right of the MM3 envelope is unoccupied. Even when a series is 'lifted', it is unlikely that the most-endorsing responses would reach 100%, but there are no nations in the 80–90%+ range either.

Assuming common cultural reactions, we expect similar distributions of national responses for different mixed-mode questions about climate change, even though they target different sub-issues. Despite the sparseness of the data, we can indeed see hints of the same teasing out of religio-regional groups along the *y*-axis at higher religiosities, in both the MM1 and MM2 series (Figures 9 and 10). However, there is only partial conformance to the pattern of the rich dataset of Figure 11. This is probably because the boundaries defining the envelopes in each of the three series are different: if, say, the upper boundary of one was a strongly framed trend and the upper boundary of the other was a weakly framed trend, the distribution of nations would be unlikely to match well, although there might be some resemblance. This would also be the case if the first boundary had a very high 'lift', but for the other it was very low or zero.

See Appendix C.i.d for links to the surveys that generated the MM1, MM2 and MM3 series.

10.1.4.2 Linear mixed-mode series

Some mixed-mode series do not produce the expected non-linear spread within an envelope. Instead, they retain a linear trend, which is situated somewhere between the envelope boundaries. This appears to occur where the reality constraint is presented as a list of competing issues, and respondents are asked to rate each in terms of a level of threat or worry (this rating forming the unconstrained element). This is different to the case in MM1 above, where a separate question covers each of the nine global threats to be assessed.

It seems that providing visibility of the whole list in a single action changes how people respond; it's as though possessing all the possibilities enables them to consider more deeply the trade-off between their cultural alignment and the reality constraint. As a result, their responses converge on a compromise that still reflects their religiosity to some degree, rather than drifting across the entire envelope due to secondary factors. I think this is the main characteristic of the examples LMM and LMM1 in the 'PostCovid' sheet of the Excel-Ref.*

* My examples are recent, but I believe post-Covid effects are not behind the phenomenon.

Linearity might also be preserved when the reality constraint in a mixed-mode question is extremely mild. For instance, any question about action on climate change, or the priority it should be given, even if it is written in a manner that manages to avoid an explicit reality constraint, will tend to have an unspoken constraint. Even if this constraint is very mild, action on climate change, whether at a personal or organisational level, must to some extent impact other priorities within people's minds. See endnotes for an example series.[253]

10.1.4.3 Implicit constraints can create mixed-mode responses

The reality constraint element that results in a mixed-mode series might not be within the actual survey question. There can be an implicit constraint, which might come, for instance, from the position of the question within a list. A question that is asked immediately before one that would otherwise be unconstrained may contain a strong reality constraint that lingers in people's minds, resulting in the mixed-mode responses.*

10.1.5 Systemic variability about trends: GDP per capita

In Chapter 8, I noted that the data for the WC series is quite widely dispersed around the trendline.† Some familiarity with the GDP of nations led me to suspect that part of this variance was not random, but systemic, and probably due to a secondary variable that might be GDP-related. The fact that the same kind of variance seemed to occur on other (weaker reality-constrained or unconstrained) series, strengthened this possibility. I already had a version of the WC chart on which I'd encoded religio-regional groups (these are explained below); simply by adding GDP data to this chart as well, the nature of the secondary variable became clear. It is indeed GDP-related. The rest of this section explains the systemic variance, starting with the definition of religio-regional groups.

There is a kind of geographical religiosity gradient in the world, created by irreligious practice spreading slowly outwards from north-west Europe into other countries. As a result, one can split much of

* For a possible example, see the series Y5 in the PostCovid sheet of the Excel-Ref. There are other possible explanations for the form of this series though; see associated text in the Excel-Ref. There is no data to distinguish these possibilities.

† See Figure 4 on page 139.

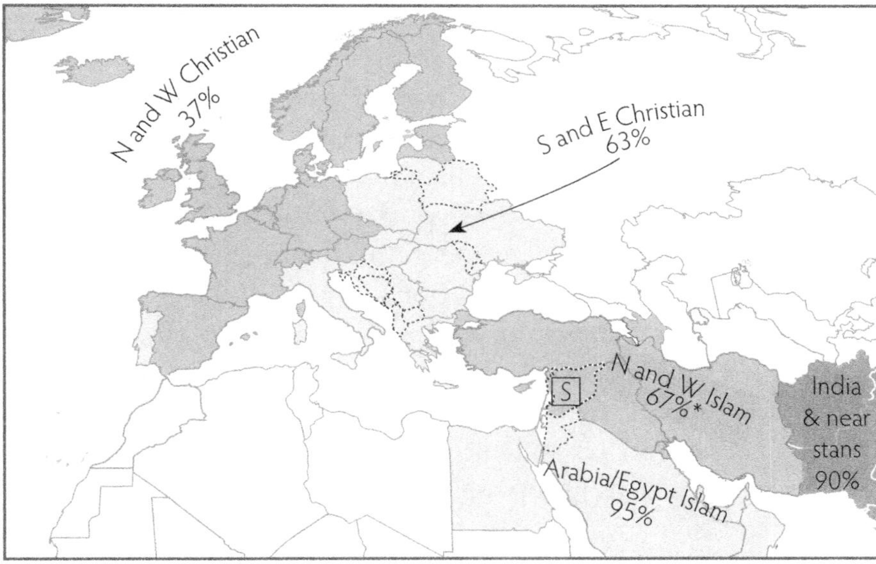

Figure 12. Religio-regional groups in Eurasia.
S, Syria. *If Azerbaijan, on the fringes of the region, is excluded, the percentage rises to 72%. Countries in white are not evaluated. Countries with dotted borders have also not been evaluated but are assumed to be within the religio-regional grouping indicated.

the world into geographical regions, the group of nations within each having similar levels of religiosity, no matter which faith this is owed to (although most groups are a single faith). Hence, according to the data presented in Chapter 8, these groups will also have similar reactions to climate catastrophism. I term the groups 'religio-regional', given their reactions hold across both the geographical region and across different main faiths (if indeed there is more than one within a group).

Figure 12 depicts 5 religio-regional groups, covering 42 nations, mainly in Europe and the Middle-East. The groups are shown with varying greyscale shadings simply to make them easier to pick out. Boundaries are empirically determined; countries that are geographically within a group, but have not actually been evaluated, are indicated by dotted borders. It's unlikely that war-ravaged Syria (indicated by letter A) could be evaluated anyhow. The percentages represent the average national religiosity per group.*

While the GDP per capita of nations somewhat modifies their publics' main reactions to climate catastrophism, it turns out that this effect for each nation is determined by their GDP per capita value *relative to the average for their religio-regional group*, and *not* to their absolute GDP per capita value. I term this their religio-regional GDP per capita (rrGDPpc). Religio-regional groups are also depicted in the chart in Online-appendix A. As an example, Estonia is ranked 40th in the world,[254] and being a poorer member of the 'NW Europe Christian' group, is well below the trendline. However, Turkey is ranked 52nd, i.e. poorer still but, being relatively well-off within the 'North and Shia Islam' group, is well above the trendline.

As Figure 8 indicates with the fat arrows that straddle some trendlines, for responses to weakly framed unconstrained questions (positive gradient), nations that rank highly for rrGDPpc are typically close to the trend-line or *below* it, while low-ranking ones are close to the trend-line or *above* it. The opposite is seen for responses to weakly framed reality-constrained questions (negative gradient): high-ranking nations are close to the trend-line or *above* it, while low-ranking ones are close to the trend-line or *below* it.

Figure 13 shows the full data for the WA+O2 series (which we first met in Figure 8), revealing the rrGDPpc variance around the trend. In the top panel, filled shapes represent a high rrGDPpc figure (better/low rank number). In the lower panel, hollow data-point shapes represent a low rrGDPpc figure (worse/high rank number). The normal rrGDPpc pattern suggests that the hollow shapes should be above or at least near the trendline, while the filled shapes should be below or near the trendline.

The data points represented as pale grey crosses are nations that don't have near religio-regional peers to compare with. In practice,

* The N&W Islam group is fully inside the religiosity spread of S&E Christian. However, a neater subdivision is perhaps possible if Azerbaijan were to be allocated to the latter group, despite being Islamic (which would change the group name); Azerbaijan has a much lower religiosity than the other N&W Islam nations. For now, I have grouped it with other nations of the same faith; the same GDP divider seems appropriate in both groupings anyhow (see chart WCrr in the 'Main Trends' sheet of the Excel-Ref), and unless more data suggests otherwise, the allocation is arbitrary.

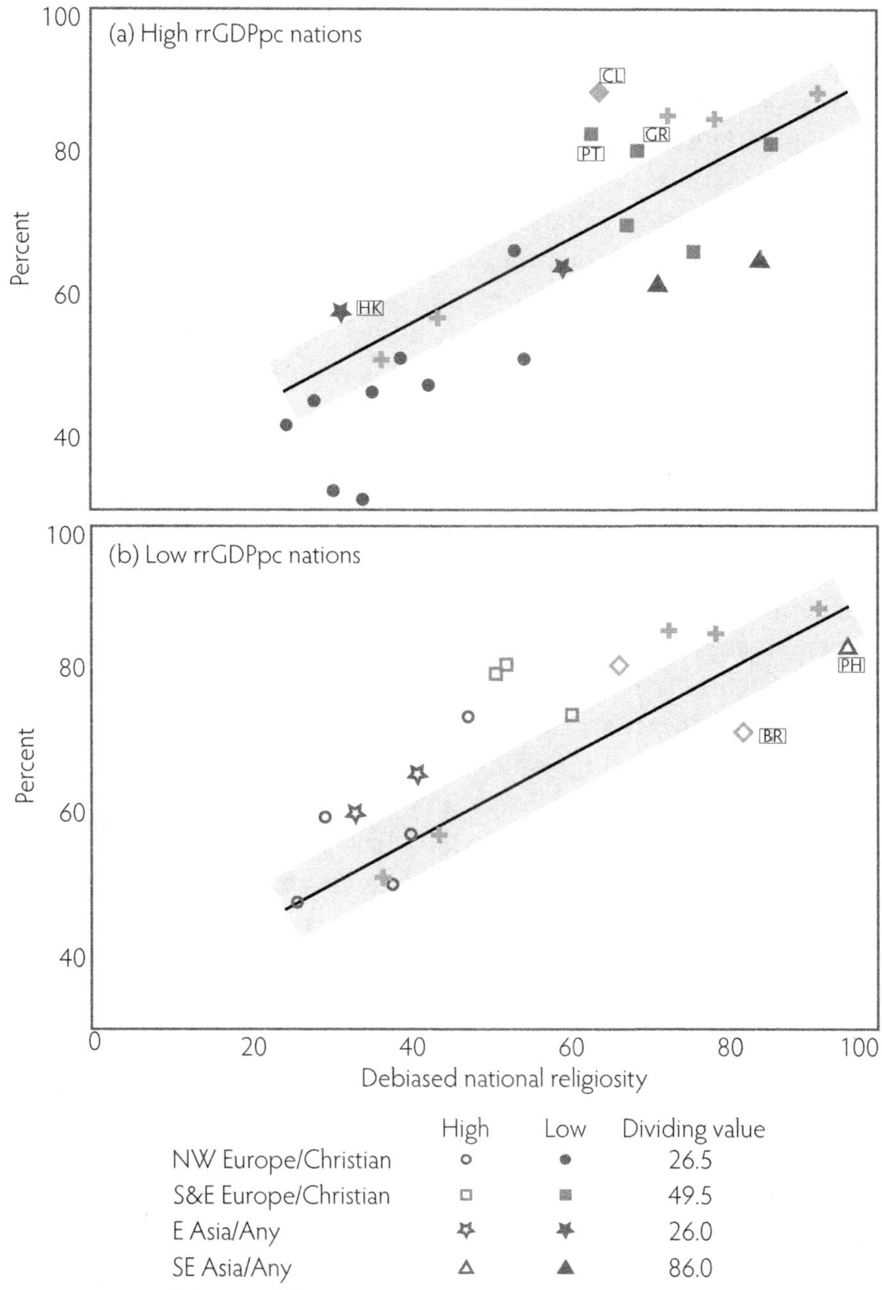

Figure 13. Effect of religio-regional GDP on WA+O2 series.

Grey shaded area defines 'near' the trendline. Crosses are points with insufficient peers.

some of these have peers within other data series, so overall we can approximate a map such as Figure 12. However, within any single series we can't be sure of the conformance of unpeered nations to the rule of rrGDPpc variance.

A few nations in Figure 13 do *not* conform to the same pattern – these have country codes against them. There's probably just too much measurement noise for full conformance. 27 nations conform, 5 nations do not. Five nations don't have peers (although three of them *must* conform, because they are very near the trendline: Australia, Canada and Kenya). So without the unpeered nations, 84% conform. In the worst case – if all the unpeered data points are wrong – conformance with the pattern is still 81%.

Figure 14 shows the rrGDPpc variance around the trend for the WC series (which we first met in Figure 4). However, for a reality-constrained series having a negative correlation with religiosity, it is upside-down with respect to Figure 13; that is, the hollow shapes are generally above or at least near the trendline, while the filled shapes are below or near it. Just as in Figure 13, the data points represented as pale grey crosses are nations that don't have near religio-regional peers to compare with. Nations that do *not* conform to expectation are again highlighted with country codes. 28 nations conform, 2 nations do not. Seven nations don't have peers (though of the latter Morocco *must* conform, because it's very near the trendline). So without the unpeered nations, 93% conform. The worst case conformance with unpeered is 78%, and more likely ~87% (say 3 of the 6 unknowns conforming). Both Figure 13 and Figure 14 can be found in the Excel-Ref on sheet 'WA1+O2 and WC37'.*

The systemic rrGDPpc variance about trendlines might, in principle, occur for all the series, but is only visible in certain circumstances, such as when there is less noise and/or larger data samples, and probably when the emotive responses are less emphatic. This would explain why weakly aligned and weakly constrained series exhibit the effect. The SC series has minimal and noisy data, which

* Figure 14 depicts a variant of the WC trend with data from 37 'Renewables' nations (this number also matching Figure 13). For the original 47 nation data of the WC series ($R = -0.76$, $p = 4.1 \times 10^{-10}$), see charts WCrf, WCrr, and WCrr1 in the Excel-Ref (which each have different extra information encoded) on sheet 'main trends'.

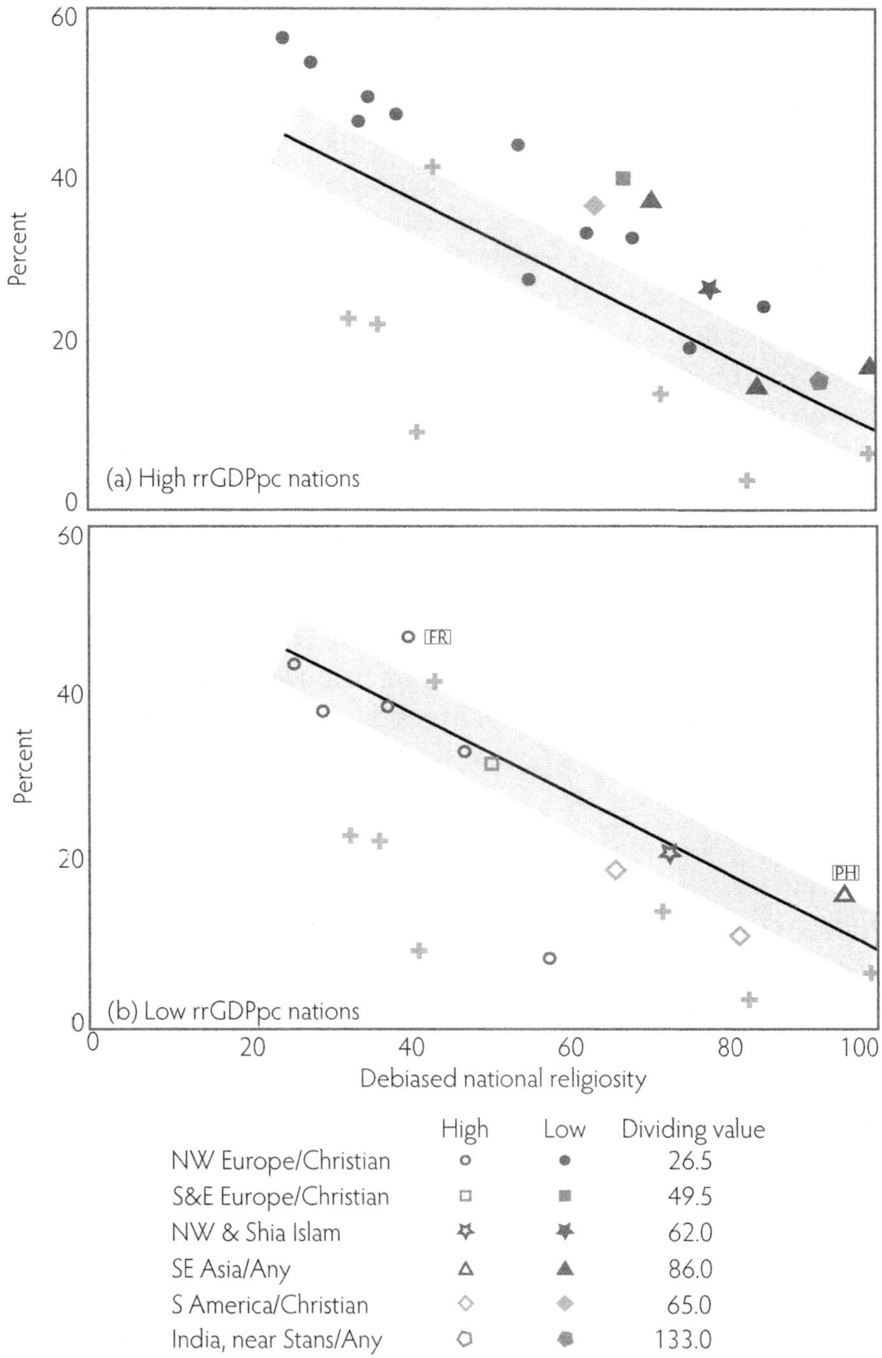

Figure 14. Effect of religio-regional GDP on WC series.

probably blots it out. Data for the very emotive and personal question prompting the SA series is very tight to trend, probably reducing the effect.

Having first observed the effect in the weak reality-constrained series, which anti-correlated with religiosity, I assumed this was because a lack of wealth represented an 'extra' reality constraint. So, those individual nations subject to more financial pressures relative to their religio-regional peer-group norm were still less accepting of the 'alien' culture of climate catastrophism, and even less willing to emotively invest in it as well as their religious faith (or alternatively switch horses), and vice versa for the wealthier group members, producing a scatter around the main trend. I haven't abandoned this idea but, as we see from Figure 13, the same kind of variance also occurs, the other way up, on the weakly-aligned series that *correlate* with religiosity. This suggests that *more* wealth buffers or lessens cultural behaviours: it reduces allied belief (in unconstrained questions), and it reduces innate scepticism (in reality constrained ones). Equivalently, therefore, less wealth should intensify cultural behaviours. In other words, my previous understanding may simply have been just a part of the whole story.

To reiterate:

- The effect is driven, not by absolute GDP per capita, but by GDP relative to the religio-regional peer-group norm.
- Straight religiosity is still by far the main determinant of attitudes to climate change at the national level.

This area needs further investigation. For example it should be possible to control for GDP.[255]

10.1.6 Equivocal responses

So far, we have only considered most-endorsing responses to climate change surveys; for example, saying that climate change is 'Extremely serious' or will cause 'A great deal' of personal impact. In Section 10.1.7, I look at resistive responses, but first I will consider equivocal ones: those that are less endorsing of the Catastrophe Narrative; for example, saying that the issue is 'Somewhat serious', or will cause 'A fair amount' of impact.

For equivocal responses to unconstrained survey questions about climate change, the robust correlation with national religiosity seen for the most-endorsing responses is typically replaced by either a modest one, or no correlation at all (i.e. a culturally neutral effect). This is because, for any given strength, equivocal responses to unconstrained questions are given by those who are less culturally engaged. Hence, we'd expect a wider spread of data and so less correlation across our *x*-axis of national religiosities (which is purely cultural).

Some equivocal responses can even anti-correlate with national religiosities, although they will remain modest. This is because cultural responses are ultimately a reflection of an alignment to (or against) the Catastrophe Narrative. So, for instance, someone who responds to an unconstrained question about the seriousness of climate change by suggesting it is 'Somewhat serious' is actually offering up a (mildly) emotive rejection of the climate catastrophism: a certain global catastrophe cannot be only 'somewhat' serious! This technically climate-change endorsing response is therefore actually an emotive rejection of Catastrophe Narrative, so may well anti-correlate, although to a modest extent only.

Equivocal responses represent the unstable ground between emotive acceptance and rejection of the Catastrophe Narrative. This means that they are highly sensitive to the precise question text and all the response options available. For instance, when survey participants are asked about the seriousness of climate change and are offered both 'Extremely serious' and 'Very serious' as possible responses, the latter might harvest a much more equivocal result than if 'Very serious' was the climate-change most-endorsing option on offer. In addition, responses will change over time, as the influence of the Catastrophe Narrative grows, in a kind of 'grade inflation'. So in the past, 'Very serious' may well have implied a higher level of cultural commitment than in recent times; as Catastrophe Narrative variants such as 'climate emergency' or 'climate crisis' proliferate, 'extremely serious' has to be expressed in order to stay loyal to the cultural narrative (again, assuming this response option or something equivalent is actually available), while 'Very serious' falls to the status of a lesser commitment. See Section 9.5 for further discussion of changes in response patterns over long timescales.

Given the sensitivity to subtle differences in the wording of survey questions, which are very hard to pin down, it is unwise to use equivocal series as social predictors of international attitudes to climate change. Nevertheless, some still perform better than most of the predictors in the current literature, the poor performance of which is outlined in Section 10.3.

Equivocal series aren't the main event in explaining climate catastrophism, so I don't catalogue them in this book. Nor do I let them clutter up the main model in Figure 8. However, for a neutral example, see the dashed equivocal series in Figure 16.* For an example of a modest anti-correlation with national religiosities, see the dashed equivocal series in Figure 34.† It's worth noting that there are no equivocal responses to reality-constrained questions. The nomination of climate change (or a climate-change issue) as one of the X most important issues in a larger list of Y concerns is a digital pick: in or out. So, there isn't an equivocal option, unless unconstrained levels are added to each pick. This is done in some surveys, but it then becomes a mixed-mode question (see Section 10.1.4).

10.1.7 Resistive responses

Climate-change resistive responses to unconstrained questions anticorrelate with national religiosity. To be climate-change resistive is also to resist the Catastrophe Narrative, which the innately sceptical will do. Given the default innate scepticism of the irreligious, many more people within irreligious nations will give resistive responses. In religious nations, where allied belief is high, many fewer people will give resistive responses. The pattern of the climate-change resistive responses is therefore largely a mirror image of the endorsing responses, although resistive series are generally lower on the *y*-axis (less overall support). This is also true at different framing strengths; each of the most-endorsing series in Figure 5 has a mirror-image most-resistive series. There are also, in principle, less-resistive mirror images of the less-endorsing series discussed in Section 10.1.6, although when we enter equivocal territory from either the endors-

* See p. 213.
† See p. 321.

ing or resistive poles, the nature of the series (correlating, anti-correlating, neutral) becomes far less predictable anyhow.

In principle, there are resistive opposites to the reality-constrained questions too. However, because of the nature of the 'X out of Y issues' reality constraints used here (as noted in Section 10.1.6 above), unless there happens to be a lot of 'Don't Know' or 'No answer' responses, the climate-change resistive ones will even more closely mirror the endorsing ones.

Because resistive series are essentially mirror images of their endorsing equivalents, plotting them all on a graph won't tell us anything more about the character of climate catastrophism. The series for which R = −0.71 in Figure 16,* is a solitary example shown in the main text, and there are further examples in Appendix B.† Similarly, I don't add any resistive series to the Figure 8 model, because it would only make for more clutter, yet add little information.

The polarisation of climate-change most-endorsing and most-resistive attitudes, and the mirror-image gradients that result (when plotted on a cultural axis), together represent an important clue as to their origin. However, in the literature, they are rarely plotted in this manner, but instead tend to be combined on a Likert or similar additive scale, before comparing to a social predictor. As a result, the polarisation is unlikely to be seen. As noted in Section 10.3.3, this practice may have contributed to national religiosity being missed as a key predictor of international public attitudes to climate change, or indeed why the inherently cultural nature of these attitudes has not been appreciated.

10.1.8 An SC series for Europe only; more polarisation?

This section does not present a specific feature of the cultural causation model, but instead discusses some insights from a survey conducted across Europe only. Covering 24 European countries, the SCe series represents the climate-change most-endorsing responses to a strongly constrained question, taken in 2019, three years after the data for the SC trend depicted in Figure 5.[256] The trend is steeper.[257] Climate change rose in prominence in public discourse over the

* See p. 213.
† See p. 320.

three-year gap, especially in the most irreligious European nations, and attitudes might, in any case, have been more polarised in Europe than worldwide to start with. A global dataset could therefore reduce the gradient with religiosity again, and perhaps the average level of support somewhat too. See chart 'SCe' in the 'Extra' sheet of the Excel-Ref.

These trends will all vary with time; my speculation is that a whole-world SC series for 2019 would sit between the MC and SC trends in Figure 5. Indeed, removing two outliers – Sweden and Denmark – from the dataset moves the trend to this position. These two countries may be anomalous because of the advent of the School Strike movement and the arrival of Greta Thunberg on the scene, which would have pushed up support for climate catastrophism in the intervening years. Support may in time return to its previous levels, because events can cause priorities to fall as well as rise. For example, the Covid pandemic may well have pushed all of the reality-constrained trends down vertically again. The results of the same survey, repeated in 2021, appears to confirm this idea; see chart 'SCe1' in the 'PostCovid' sheet of the Excel-Ref.

10.2 Summary of data-series in this book

This book presents 20 primary linear series, all of which are plots against national religiosities (on the *x*-axis). 17 represent international (non-US) attitudes to climate change. Two represent international activism on climate change. One represents international policy on climate-change (the deployment of wind and solar energy installations across nations). Sixteen of the twenty are the results from independent questions, and for four of these, there are companion series representing related responses – lifted and resistive and so on. Table 13 summarises their statistically significant correlations with national religiosity.

Four non-linear series, which reflect envelopes of data effectively bounded by linear trends, are also presented as primary support for the cultural causation model.

Full details of all these series, along with references to where they are first mentioned in this book and sourced in the accompanying

Table 13. Primary series correlations with religiosity.

R^2 range	Number of series
0.33–0.50	5
0.50–0.69	10
0.69–0.87	5

Excel-Ref, are provided in Table 23 (Appendix G). Online-appendix C provides a description of the main climate-change attitudes as revealed by all of these series, as a set of text-based 'rules', with a brief explanation for each rule based upon the cultural causation model.

A further 13 series are generated from data taken a long time (>1 year) after the onset of Covid; 10 of these are linear and 2 non-linear. I haven't fully evaluated how the pandemic affects attitudes to climate change, but these series still appear to be largely consistent with pre-Covid patterns, despite a couple of minor unresolved issues. Hence, they provide further support for the cultural causation model. Full details of these series, along with references to where they are first mentioned in this book (for those that are) and sourced in the accompanying Excel-Ref, are provided in Table 23 (Appendix G). Table 14 shows their statistically significant correlations with national religiosity.

All these figures are worth bearing in mind when we examine the existing literature on factors that govern attitudes to climate change.

10.3 The failure to identify climate catastrophism

As far as I can tell, the social sciences are blind to the presence of a culture of climate catastrophism. This failure is probably the result of widespread bias engendered by the culture itself; for instance, the

Table 14. Post-Covid series correlations with religiosity.

R^2 range	Number of series
0.32–0.50	7
0.50–0.69	3
0.69–0.74	1

false notion that certain imminent global catastrophe is backed by the hard sciences, or a fear in some academics – subconscious or otherwise – of being portrayed as anti-science or a denialist if they suggest that the dominant narrative in the public domain might be more about culture than science.

Given most social scientists don't study the nature of public attitudes to climate change and what motivates them, we might charitably suggest that many in the field are simply 'going with the flow'. However, there is an entire sub-discipline devoted exclusively to public attitudes on climate change, and it has produced an extensive literature on 'social predictors' that might explain the underlying motivations. This being the case, the failure to find the culture of climate catastrophism is hard to view as anything but extraordinary.

10.3.1 National religiosity as a predictor of attitudes

An R^2 coefficient gives the proportion of the full range of the dependent variable (for most series here, an attitude to climate change) determined by the independent variable (national religiosity). So, for instance, an R^2 of 0.6 or 60%, means that national religiosity predicts a particular attitude to this extent; in other words, it is a 'social predictor' for attitudes to climate change (and also, for other series presented in this book, for activism and climate policy commitment). The values for the primary linear series presented here range from 0.33 for the SC series, to 0.87 for the SA series, and fall into the ranges noted in Tables 13 and 14 (see Appendix G for detail).

Expectations for how 'good' R^2 should be depend upon the field of study. While benchmarks may be high in engineering, for example, lower ones may still be considered meaningful in, say, medicine and social psychology. Notwithstanding that the data are measured at the national level (which likely reduces overall variation[258]), the results presented in this book are *far* better than anything previously reported in the relevant literature (outlined below), and even the least robust of my results beat almost all predictors reported to date at both individual *and* national level. The most robust results are extraordinarily good; as good as a physical relationship, such as the one between shoe size and height, which is frequently cited in statistics training modules. In fact, 10 of the 20 primary linear series summarised in Table 23 have R^2 values equal to or surpassing the value described as 'impressive' in a typical training presentation.[259] Religiosity is therefore clearly an outstanding predictor of international attitudes to climate change.*

10.3.2 Missing the wood for the trees

As noted above, the 'relevant literature' examines 'social predictors' for international (non-US) public attitudes to climate change. Over the last fifteen years or so, this has taken in a broad range of predictors, but results thus far have been disappointing, to say the least. The power ascribed to each individual predictor is typically low (<20%) or very low (<10%). Despite reasonable claims of greater power for multiple predictors used together, the results remain poor, and the field is still characterised by disagreement over which ones are truly important. Latterly, the focus has shifted to more complex models and/or statistical techniques.

To give an idea of the size and range of the literature, the 2019 meta-study by Ruiz and colleagues reviews 33 predictors from 64 studies, in an attempt to provide a more coherent 'big picture'.[260] The authors don't actually rule out any of these as having importance; rather, they construct fiendishly complex diagrams emphasising that all have a significant role, and suggest that most interact with each other in a highly convoluted manner.

One such diagram, covering just the basic positive or negative influences upon attitudes, is shown in my Figure 15a. The equivalent explanatory model from this book is essentially Figure 8,† which is considerably more straightforward, being based on a single predictor. Moreover, unlike the Ruiz diagram, more hard information about particular attitudes in particular nations is clear in my model depiction. Recasting Figure 8 in the same form as the Ruiz diagram, to give Figure 15b, emphasises the marked difference between the approach in this book and that in the general literature.

I think this Ruiz et al. diagram is a fair representation of the gen-

* Although not in the US – the unique situation of which is examined in Chapter 11 – or nations where religion is suppressed, such as Vietnam or China. Vietnam and the US are shown to be exceptions in one of my *Climate Etc.* guest posts.

† Although it would really need the 'resistive' trends from Appendix B to be added to make the comparison more stringent.

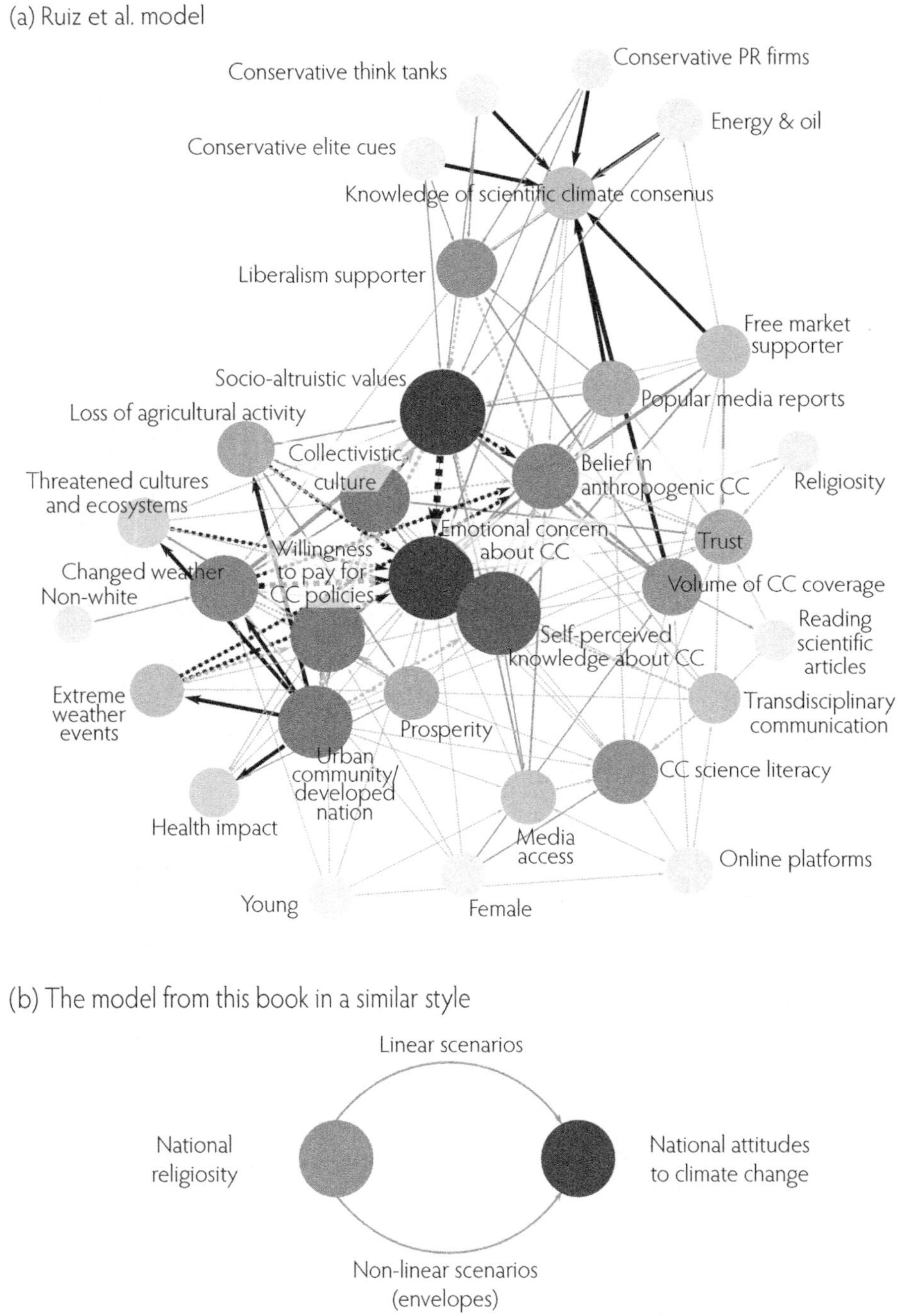

Figure 15. Influences on attitudes to climate change.

(a) Redrawn from Figure 3 of Ruiz et al.

eral perception of the climate change domain in the field, which is to say both highly complex and packed with different influencing and interacting variables. In my opinion, almost nothing of use can actually result from this view; not because most of the variables discussed are wrong, but because they represent lower-order effects.

To give a feel, some brief examples from the literature follow. Papers may assess predictors at the individual level, national level, or both. A study across 47 nations by Kvaløy et al.[261] (2012) found the top three positive predictors for the perceived seriousness of global warming from a list of 16 to be education (~21%), left political stance (~15% between maximally opposed stances), and 'God important' (~10%).* Across 14 western European nations, McCright et al.[262] (2015) found 'political ideology' and 'perceived understanding' as the best of 5 variables for predicting the perceived seriousness of climate change, scoring about 9% and 10% respectively over the full variable ranges. Interestingly, they found no significant similar relationship for political ideology in eleven former communist countries in Eastern Europe.

More recently, a study by Levi[263] (2021) featured 17 predictor candidates for a belief that climate change is human-caused. For national (country) level variables,[264] Levi states:

> Country-level conditions that predict higher levels of climate change belief are country-level environmental protection (by 7–11 percent), civil liberties (by 7 percent), exposure to climate-impacts (by 4–7 percent), and the number of domestic NGOs and climate scientists (both 1–2 percent).

The power of all the predictors described here is very modest, and it has not improved over time despite the increasing sophistication of more recent studies. Importantly, to a large extent, the predictors are also inconsistent between studies. It may be claimed that both of these problems are to some extent attributable to the differences between widely separated regions. However, in a study conducted within a single region (Europe, plus Israel), which one would think

* 4 of the 16 predictors are political stances, and 3 are related to religiosity. I recast their results to express as a percentage of the full range of attitudes the paper finds across national publics; they were not reported in this manner.

would limit predictor diversity, Poortinga et al.[265] (2019) put as much stress on the differences of predictor types and their variance across nations as upon the fact that 'some of the associations are remarkably consistent across the 23 countries'.

They say:

> ...both socio-political and demographic factors are needed to understand public perceptions of climate change...Even if the direction of the associations were to a large extent consistent, the sizes of the effects were not. This shows the importance of cross-cultural research, and the need to validate results in multiple countries and cultural contexts before assuming certain effects are universal.

The vast gulf between the existing literature and the results reported here suggest that researchers in the field have simply failed to see the wood for the trees. They haven't failed to imagine that religious affiliations might have some influence upon views about climate change; they have overlooked the possibility that attitudes to climate change might be overwhelmingly cultural *in their own right*. And they have done so despite the obvious presence of a highly emotive Catastrophe Narrative that contradicts *mainstream* science and has been ubiquitous within the public domain for decades. As noted in the following section, this reality changes how the search for social predictors should be undertaken.

10.3.3 The failure to identify religiosity as a predictor

There are many reasons why social scientists searching for influences on international public attitudes to climate change might have overlooked the key social predictor of national religiosity. Online-appendix A sets out eleven likely contributing factors, but it's worth looking at a few here.

In general, it's a lot easier to miss something if you were never really looking for it. Many studies don't even consider religiosity as a possibility (and many others only do so in only a limited manner). The Ruiz et al. study mentioned above shows that 21 other candidate predictors for international (non-US) attitudes to climate change have been studied more often than religiosity. Secondly, the context is hard to get right if one doesn't believe one might be investigating a potential global cultural phenomenon in the first place. For instance, because religiosity levels are usually similar across regions (irreligiosity has spread slowly outwards from north-west Europe, see Section 10.1.5), studies that only look at one or maybe two adjacent regions will find much less variance of attitudes, and will probably form an impression that religiosity only has a limited role. If you start with an assumption that a culture is involved, you will, as a matter of course, look for global data to support the hypothesis, and will naturally consider religion as a prime candidate.

Ideally, several nations from all world regions should be sampled. When this is done, all the main faiths are likely to be encompassed, so the religiosity scale employed must work across them all. As noted in Section 8.2, the scale used in this book is simple and straightforward, being derived from self-assessment polls, which helps to ensure consistency across faiths. However, as also noted in that section, and briefly expanded upon in Online-appendix Avii, most studies that actually consider religiosity use behavioural proxies (e.g. the number of visits to church), either on a standalone basis or incorporated into more complex scales. I suspect this approach gives less consistent results, because the behaviours are themselves faith dependent. If so, this will again tend to obscure or diminish the role of religiosity.

Yet another issue regarding the treatment of religiosity is that even where an appropriate range of nations is targeted, some surveys don't examine results on a per-nation basis, but instead combine individuals from all nations into religious/irreligious categories or, say, low, medium and high religiosity. However, because cultures are a group phenomenon, as the data presented here confirms (see Section 9.7.1.1), aggregation may blur the characteristics that help to reveal the cultural nature of the responses. Although it's a secondary effect, cultural subgroups that are tightly connected by geography may not support exactly the same range of attitudes when they are set within very different contexts. So, for example, religious people in a nation that is very religious overall may not express quite the same attitude set as religious people in a nation that is overwhelmingly irreligious; cultures will to some extent adjust to their local circumstances. We can intuit this from Figure 8.[266] This doesn't mean that religiosity is a poor predictor of attitudes to climate change, only that, when measuring a cultural phenomenon, it is more revealing to assess data at the level of subgroups rather than individuals. This is much less likely to happen, however, if researchers don't think they are looking for a group phenomenon in the first place; if they aggregate responses in the above manner, information will be lost.

Further issues are revealed in Figure 16, which shows international attitudes on the seriousness of climate change – from an unconstrained question, 'How serious a problem, if at all, do you think climate-change is?'[267] Different response types are plotted against national religiosity, and the trends are revealing. The most emotive response available, 'Extremely serious', correlates with religiosity (R = 0.70). The aggregation of the least emotive responses – ranging from 'Somewhat serious' to 'Not very serious at all' – anti-correlates (R = –0.70). In the middle, 'Very serious' is uncorrelated (R = 0.09).

Charting the survey results in this way can provide important clues to the presence of a cultural entity in the climate domain. However, it is common in the predictor literature to aggregate the response types, and to work only with the resulting synthesis. This is typically done via a Likert or similar type of additive scale – allocating, say, 1 for 'Not serious at all', 2 for 'Not very serious', and so on up to 5 for 'Extremely serious'. The aggregate score would then be used to represent the survey responses. This approach essentially causes serious information loss (even if the results are charted to get a visual feel, which isn't always the case): firstly of the fact that attitudes are strongly polarised, and secondly of the nature of the balance point between opposing poles; in other words, the particular response that represents the neutral position between the poles.*

As discussed in Section 10.1.6, that culturally neutral line is not determined by whether responses express concern about climate change or not (in which case we might expect it to be the 'Not very serious' response), but whether responses are emotively aligned to Catastrophe Narrative or not. Hence, three responses that all express some concern about climate change ('Extremely/Very/Somewhat

* Note: I have aggregated three of the responses myself in this case, without any weightings, but first checked that in this case there isn't significant information loss.

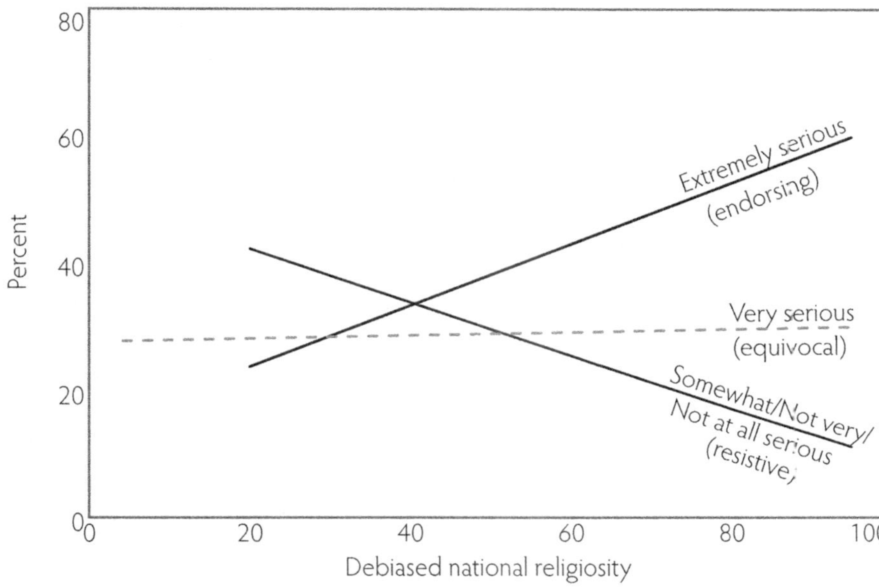

Figure 16. Endorsing, equivocal and resistive responses.
How serious a problem to you think climate change is?

serious') generate positive, no, and negative trendlines, emphasising that the responses are emotive, not rational. If these responses are aggregated, the existence of strong polarisation and emotive responses, both fingerprints of cultural origin, will be overlooked.

This is a barrier to developing an understanding of what is going on. That said, it is possible that aggregation might not completely obscure the importance of national religiosity as a social predictor. If used for the survey in Figure 16, a graph of the aggregated responses would still have a positive gradient (although a very shallow one) and a robust R^2 value. However, if 'Very serious' was the climate-change most-endorsing option on offer, as noted in Section 10.1.6, all the responses would redistribute somewhat. In particular, 'Very serious' would probably gain a positive trend, although not a particularly robust one. This being the case, aggregation might not only distort the visual impression, it might also seriously weaken the net predictive power of the data, thus obscuring the role of national religiosity altogether. In principle, a significant anti-correlation might result – in other words, the relationship that emerged between religiosity and most-endorsing responses would be reversed.

Once more then, the Likert-style aggregation could obscure the role of culture in general and national religiosity in particular, or at least cause confusion about its nature.

While any one reason alone may not prevent the detection of national religiosity as a potent predictor, in most studies in the literature, several of them are in play at the same time. Indeed, a number of surveys don't offer emotive response options along the lines of 'Extremely serious' for their unconstrained questions.* Others avoid emotive options entirely, believing them to be biased. However, to see cultural belief (or rejection), you need to ask questions that are biased towards the prime cultural narrative (for climate, Catastrophe Narrative). If respondents aren't provoked by emotive options, their responses will be equivocal or neutral. This doesn't mean they aren't brimming with emotion about the topic; it's just that it will be missed unless the language of surveys allows its expression.

A further factor that probably contributes to national religiosity being overlooked as a robust predictor is that many survey questions from mainstream pollsters mix unconstrained and reality-constrained elements. There's no reason this couldn't occur in academic studies too, especially given that some use mainstream survey results, as I do extensively in this book. As we have seen in Section 10.1.4, responses to such mixed-mode questions are mostly non-linear and look chaotic when plotted against national religiosity (or likely, any other cultural axis). However, they occupy an 'envelope', bounded by the linear trends that would occur for each element type alone. But no-one is likely to guess that the chaotic data obeys such a rule, unless they already had a theory of culture, along with some established and robust linear results. A similar effect occurs with responses to extremely weakly framed questions, although I believe these are less likely to be used as the subject of predictor studies.

* 'The seriousness of climate change' is only weakly-framed, as it doesn't directly invoke the personal or concerns about harm.

Finally, many studies concentrate on a single attitude, although they may test it against a long list of candidate predictors. So, they may not even get to the 'first base' of the fundamental difference in responses to reality-constrained and unconstrained questions, let alone framing-strength effects. This is particularly the case if their results are diluted and/or obscured by some or all of the issues set out in this section; the 'big picture' described in Section 9.2 and the explanations summarised in Section 9.7 will be missed. Indeed, they *have been missed*.

Online-appendix A sets out further reasons why national religiosity was almost entirely missed as a predictor of international attitudes to climate change. It's likely that this wouldn't have happened if there had been an acknowledgement that public attitudes to climate change might be cultural in their own right. For researchers in the field to have examined so many possible factors without ever hitting on the correct one indicates a pervading bias. However, there is one partial exception in the literature.

10.3.4 Not entirely missed! Lo and Chow (2015)

A lone bright point I can find in the literature is a paper by Lo and Chow.[268] This surveys attitudes to climate change across 33 nations, and reveals a good correlation and anti-correlation between the most-endorsing attitudes and a single predictor, namely national GDP per capita (GDPpc). This is ultimately because, due to long-term developmental issues, GDPpc has a good anti-correlation with national religiosity, although there are some notable exceptions.[255] Lo and Chow's most robust findings consist of a good correlation ($R = +0.61$) between the 'perceived importance' of climate change and GDPpc, and an anti-correlation ($R = -0.62$) for the 'perceived danger' of climate change. In straightforward terms, this says that wealthy countries do not see climate change as being particularly dangerous but (apparently perversely) think it is very important. Poorer countries see it as dangerous but, equally perversely, think it unimportant.

Lo and Chow make the perspicacious suggestion that the contrasting results might be something to do with 'identity conflicts', possibly guilt, which they say would be more prevalent in wealthy nations, and also a feeling of protection from wealth, although they also acknowledge issues with this hypothesis. They state:

> However, these explanations cannot fully explain the positive relationship between climate change concern and national wealth found elsewhere. If wealth accumulation contributes to low perception of danger, it should also result in reluctance to express concern. The expected coherence is not found in the following analysis. The observation that wealthier societies downplay the risk of climate change requires further explanation.

However, in light of the explanations in Chapter 9, these results are easy enough to understand. Lo and Chow's 'perception of importance' is a reality-constrained question framing, and their 'perceived danger' variable is an unconstrained question framing. The shift from reality-constrained to unconstrained is what causes these climate-change most-endorsing responses to flip from a correlation with national religiosity to an anti-correlation. It is this change in framing that is really driving their findings, although note that because GDPpc anti-correlates with national religiosity, their results are 'upside down' with respect to the ones reported here. Moreover, because their 'perceived danger' question includes an 'Extremely dangerous' response option, it will certainly ensure (perhaps fortunately, if it wasn't anticipated) that strong cultural reactions are prompted.

Hence Lo and Chow's results are similar to mine, if somewhat less robust, but they have still missed the primary driving factor. It is not, as they conjecture, mere guilt, but something more fundamental; it is *cultural identity*, as invested in both religious faith and climate catastrophism itself. Had they used national religiosity as their predictor instead of GDPpc,[269] they would almost certainly have improved these findings.[270]

There is a great deal about Lo and Chow's paper that seems, from my perspective, to be 'on the money'. In this respect, their work seems unique, but unfortunately, for such a powerful and interesting result, it has received little attention.

10.3.5 Predictor comparisons

Figure 17 shows the contrasting performances of national religiosity and a variety of other predictors from the literature. The 20 primary linear series in the book are shown alongside 20 series from seven prominent academic papers.*

Some of the papers in the literature include a US measurement. Some evaluate predictors at the individual level, others evaluate at the country level or do both. Rather than evaluate single social predictors, researchers in the field generally use at least straightforward multivariate analysis and, especially for the more recent publications, complex models/techniques. The papers selected represent all such approaches. All the associated 'p' values reported are below 0.05, many well below.

As noted above, the literature on social predictors for attitudes to climate change is extensive, and so this is not a comprehensive evaluation. However, these are the best predictors I can find from a range of older and oft-referenced papers, together with more recent ones featuring the latest sophisticaed statistical approaches. Confidence that my selection is reasonably representative is increased by the fact that Hornsey et al. (from which I have drawn the most predictors) is a meta study for 'belief in climate change' spanning 56 countries (including the US) and 171 studies.

I can't find any values approaching those of Lo and Chow, apart from Tranter's T1 series (in some nations but not others – its range represents applicability to different nations). Many predictors considered notable in the field have R^2 (or equivalent) below 5%, although in a multi-predictor model the emphasis is not on any individual value.† It's also the case that some of these predictors are themselves high-level/complex social attitudes, such as 'New ecological paradigm' or 'Class' (engaged, pessimistic, indifferent, doubtful about climate change), which themselves will rest upon a mix of simpler social values and/or demographics.

Although I always evaluate religiosity as a predictor at the national level, the results in the literature, whether at individual or national level, are clearly far weaker.

* Full details can be found in Appendix G for the former, and Appendix K for the latter.

† Indeed, some models don't report easily comparable metrics for individual predictors.

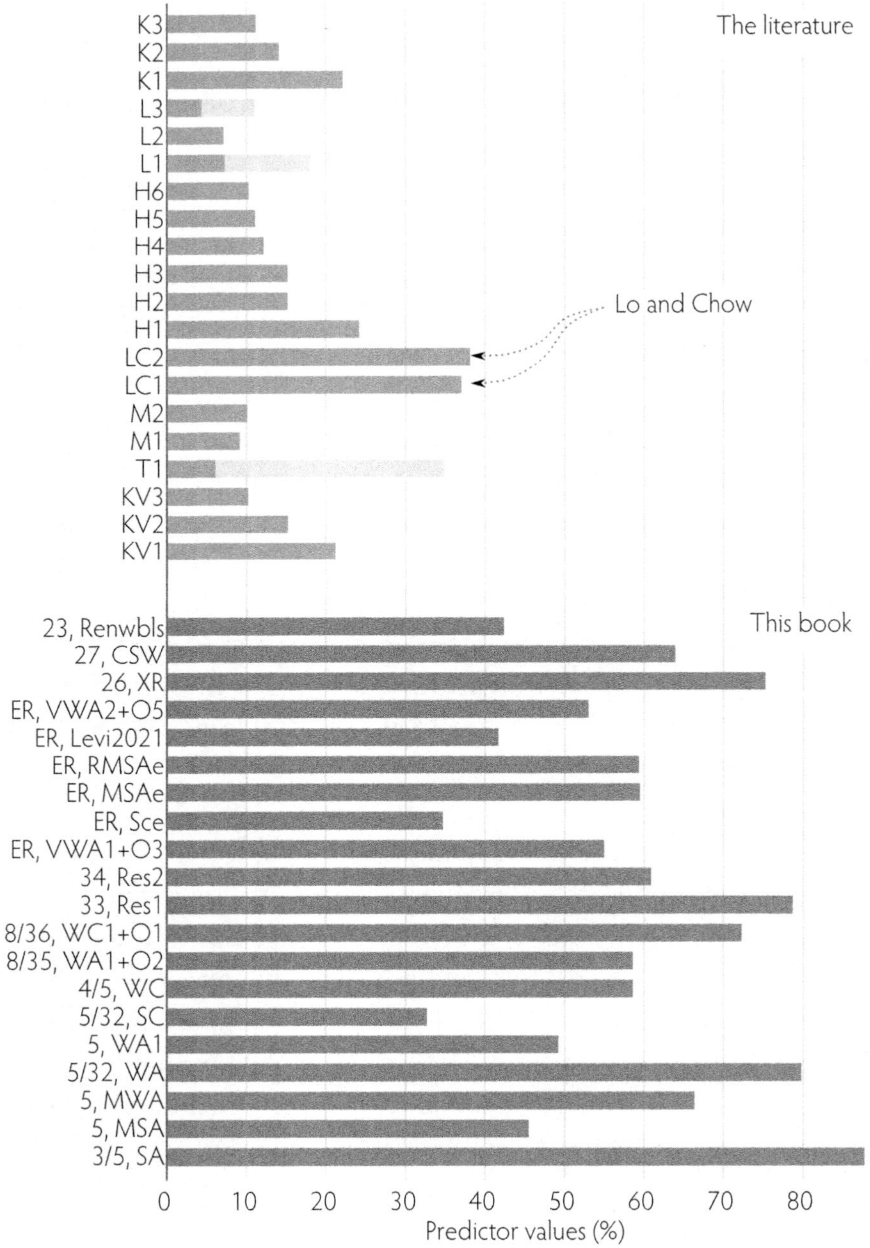

Figure 17. Social predictor values in this book and the literature.

Literature code formats are: paper, predictor code. KV, Kvaløy; T, Tranter; M, McCright; LC, Lo & Chow; H, Hornsey; L, Levi, K = Kácha. Appendix K has details. Codes for this book are figure number then series ID. Appendix G has details.

10.4 But what if...

10.4.1 ...my explanations are hogwash?

Even disregarding all of the explanations as set out in this chapter and Chapter 9, the data presented in this book leaves the social predictor literature obsolete. And, given the nature of religiosity and its strong (dual) relationship with attitudes to climate change across nations, I believe it will be very hard for alternative explanations to avoid a significant role for cultural mechanisms.

10.4.2 ...there are outliers unduly influencing the data?

On two or three of the weaker series, especially where the number of data points is low, some undue influence from outliers is possible. However, all the p values are reported too, and none are close to exceeding 0.05 (0.02 is the closest, and this is exceptionally high), which increases confidence generally. In addition, the great advantage of working at national level is that many independent attitude surveys can be employed as part of the analysis. These all fit well into a single coherent framework, which reduces greatly the likelihood that any particular series is misleading, and helps to validate the underlying theoretical explanation, namely of cultural causation.

10.4.3 ...a different variable drives attitudes?

As noted in Section 10.3.4, with some exceptions, national GDP per capita (GDPpc) has a good anti-correlation with religiosity, and could therefore be a candidate for an 'underlying variable' that drives attitudes to climate change. However, firstly, while the relationship between GDPpc and religion is thought to be mutually interactive (they drive each other), it is not fully understood, although I believe the majority of researchers suggest that GDPpc is more driven by religiosity than vice versa. If this is correct, it cannot be a candidate underlying variable. Secondly, replacing national religiosity with GDPpc generally degrades R^2 values.* Nor can GDPpc directly

* Doing this for the SA and WA series causes them to lose linearity altogether. This is because the normal correlation between religiosity and GDPpc breaks down both in oil-rich nations (because of their sudden acquisition of wealth) and Singapore (a city state). Without them, the correlation remains almost as strong as with national religiosity. The fact that climate attitudes are predicted by national religiosity even in such places strengthens the case for cultural causation.

explain the framework set out in this book, into which all the series presented here fit.

Cultural causation and the relationship between national religiosity and climate catastrophism *can* explain the framework. Moreover, it is able to explain the very different nature of the historic (2005–09) unconstrained series;* apart from elements such as the rise of China, the GDPpc ranking of nations did not radically change from 2005–09 to 2015–20, so attitudes mostly dependent on this ranking would not be expected to change either. Additionally, the situation in the US† also points to a cultural cause rather than GDPpc.

So, our conclusion is that national religiosity largely drives GDPpc and attitudes to climate change at the national level. GDPpc is not the 'prime mover'.‡ I don't believe there are other viable candidates.

10.5 Will the cultural model hold indefinitely?

The principal feature of the model (as usual, excluding the US) is the alliance between religion and climate catastrophism, which produces high levels of support across much of the world, but which is shallow, and therefore collapses in the presence of reality constraints. This arrangement is currently advantageous to both cultures: climate catastrophism gets an enormous boost from the support of faith leaders and their flocks, and in return the aged and generally declining religions get a shot of cultural relevance by blending climate catastrophism into their spiritual pronouncements.

However, in theory this cosy relationship cannot last indefinitely. Cultures do not take kindly, so to speak, to those who obtain group benefits without paying their dues. This is essentially what religion is doing right now; it is benefiting from the emotive power of climate catastrophism, but its flocks have not surrendered their deeper cultural identity and remain largely uncommitted to the new culture's priorities (the reality-constrained trends are very low across religious

* See Section 9.5.

† See Chapter 11.

‡ And even if GDPpc drives religiosity more than vice versa, as some believe, there is no reason it should cause the change of attitudes to climate change over time that we see. Nor should it cause the radically different most-endorsing responses to unconstrained or reality-constrained questions. Even in this case, GDPpc is only a driver once removed; it would not cause the observed effects without the purely cultural intermediary of religiosity

nations). The increased presence enjoyed by climate catastrophism is of little use if it doesn't eventually translate into more tangible benefits. At some point it will therefore demand complete loyalty. It will try to wrench social control away from religious authority, and impose its worldview - expensive Net Zero policies included - on (often poorer) religious nations just as it has on Western secular ones.

It is not easy to predict what will happen. Will the religions crack, and be consumed by the new culture, or at least fall under its control? Will they finally give up their age-old primacy in society? Or will their leaderships quietly delete their grand statements of support for the 'climate crisis', recognise the existential threat that catastrophism represents, and redirect their flocks onto the path of resistance. My personal bet is on the latter; the religions are great survivors. Either way, the relationship is going to undergo dramatic change. I have no idea whether this will happen within a decade or several, but it must surely come eventually, unless a major geo-political or cultural shock of some kind intervenes and changes the relationship in ways we cannot predict.

Chapter 11

THE USA: SAME RULES, UNIQUE FACTORS

We now need to turn our attention to the USA, which is unique because of the strong political polarisation of the public along tribal (cultural) political lines – Rep/Cons versus Dem/Libs – on many issues, including climate change. However, the theory of cultural causation means that the same 'rules' regarding cultural entities and their interactions should hold everywhere, so we should be able to demonstrate this within the US, despite the complicating factor of there being more cultural entities in play.

As noted in Chapter 9, social psychologist Dan Kahan has suggested that US public attitudes to climate change are not based on 'what people know' about this issue, but on 'who they are' – in other words, their cultural identity. I agree with this, but reject Kahan's assumption that this cultural identity is exclusively about political 'tribe'. To reiterate from Section 9.2: '…cultural identities in the US actually come from a four-way interaction: between the two political tribes, religion and climate catastrophism; which is to say, four different cultures. In other countries, without the highly polarised politics, to a first approximation there remains just religion and the culture of climate catastrophism, which, as we have seen, are together sufficient to entirely explain the patterns in Figure 5.'

Much more effort appears to have gone into public surveys and academic studies on US attitudes to climate change than anywhere else. As a result, the polarised debate there has a high profile, which has tended to encourage confusion about which lessons learned are likely to be applicable to other nations, and which are not. So, the inference that US attitudes are mainly about cultural identity *is* applicable elsewhere; as noted earlier the measurements in Chapter 8 confirm that Kahan's rule is universally true. But the lesson of strong public polarisation about climate change along left-right political lines is largely inapplicable elsewhere.[271]

We now have to adapt our methodology to a situation where attitudes to climate change are indeed strongly polarised along left-right political lines, as numerous US studies and surveys have shown over many years. Beneath that polarisation, essentially the result of two opposed cultural entities, we should still see the effects of religion and climate catastrophism, as already measured in the Rest of the World (RoW). In turn, these effects should provide us with a fuller explanation of what's happening in the US, and we should in some formal sense be able to relate this to what's happening in the RoW. These are the challenges for this chapter.

11.1 Attitudes to climate change in the US

We will start by charting attitudes to climate change in the US, in a manner as close as possible to the way we mapped attitudes in the RoW. An immediate issue arises in that very few surveys collect US attitudes to climate change as related to levels of religiosity, so we must use a different cultural variable as our *x*-axis. Given the strong polarisation along 'tribal' political lines, an axis based on party affiliation should work just as well, because 'tribal' is essentially another word for 'cultural'.

Polls on climate change attitudes normally collect information about political affiliation too, and from this data it is straightforward to find a modern (2018–20) set of survey questions that approximately matches our previously defined (Section 8.4.4) framing buckets. The surveys derive from eight independent sources. Their questions and measured responses are shown in Table 15 and 16, along with links to the original surveys.

There are a few differences to the surveys for the RoW in Chapter 8. Among the unconstrained questions, I have a medium-aligned (MA) question; recall, for the RoW measurements, I was unable to find one with this strength, so I used a medium-weak aligned (MWA) and a medium-strong aligned (MSA) one instead. Conversely, for

Table 15. Unconstrained survey questions for the US.

Strength	Question*	Response
Strong	How worried are you that you or your family will be *personally harmed* by climate change in the next 50 years?[275]	Very worried
Medium	Climate change is best described as...[276]	*A crisis*
Weak+	How *worried*† are you that the *United States will be harmed* by climate change in the next 50 years?	Very worried
Weak−	How *concerned*† are you about climate change?[277]	Very concerned
Weaker aligned trend with neutral offset (2)	On the subject of climate change do you think...?	The world's climate is changing as *a result of human activity*

* Italics indicates emotive cue. †'Worry' is a slightly stronger cue than 'concern'.

the US, I have a 'weak aligned minus' (WA-) and 'weak aligned plus' (WA+) series, which are averaged to get a weak aligned (WA) one. I also have a real MC series for the US; this was interpolated for the Chapter 8 measurements. I have found no fully-constrained data for the US, so the FC series has been estimated from a question that is 'nearly' fully constrained (NFC on the graph), which measures climate change as a single priority from 19 national issues, but also conflates it with other environmental concern, lessening the constraint. Finally, each of the tables includes a question that generates a lifted series (see Section 10.1.2), although for clarity I have not shown these when I graph the series.*

Figure 18 plots the proportion of climate-change most-endorsing responses from these unconstrained (Figure 18a) and reality-constrained (Figure 18b) questions against a three-point political x-axis (Democrat, Independent, and Republican). Ignore the grey sections to the left, for now. The trendlines aren't statistically valid because the corresponding series have only three data points each (all of which

* See sheet 'The US' in the Excel-Ref for a comprehensive chart.

Table 16. Constrained survey questions for the US.

Strength	Question	Select	Comment
Almost full	What is the most important issue facing the nation today?[278]	Environment/ climate as 1 from list of 19	Environment dilutes, but very little at this strength, where climate dominates. Long list, national issues.
Strong	Which 2 issues are the most important to your vote?[279]	Climate as 1 of 2 from list of 14	Long list, national issues.
Medium	Which 3 of the following issues should be top priority for US foreign policy in the next 5 years?[280]	Climate change as 1 of 3 from list of 12	Long list, pitched between global and national in scope.
Weak	Which of the following issues are very important to you personally?[281]	Climate change from list of 5.	Also, have 3 other options to allocate, including 'somewhat'.
Weaker constrained trend with neutral offset (1)	'Federal government is not doing enough to doing too little to...[282]	'Reduce the effects of climate change', from list of 5 environmental only issues	Single topic list and can pick all. Very weak constraint, and invokes some 'care for environment' across all cultural positions.

are marked), but they are useful signposts that help us to understand what's happening.

The first point to note is that the unconstrained responses and the reality-constrained responses are grouped in the same kind of patterns as we saw when interrogating the cultural entities of climate catastrophism (Chapter 8), and religion (Section 9.6), across many nations outside the US. I believe this is the signature of cultural responses; the reality-constrained series lose gradient and sink down the *y*-axis with increasing strength; the unconstrained series pivot about a common point. The pivot point is very low on the *y*-axis, just to the left of the Republican line. Noise and a dearth of datapoints mean the series don't all cross over at quite the same point, but they are close. So, it seems we are indeed looking at cultural responses;

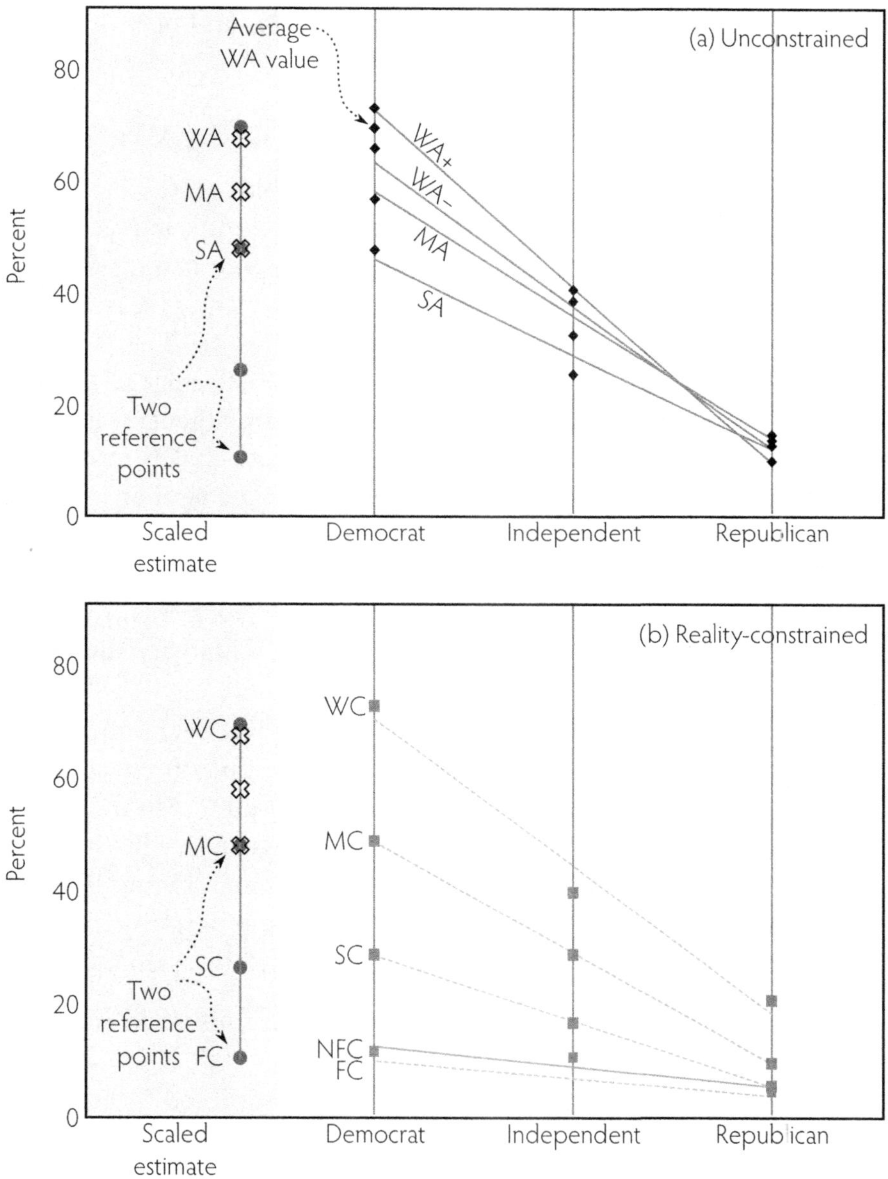

Figure 18. Climate-change attitudes and political tribes in the US.

Top: Unconstrained questions, bottom, reality-constrained. Constraints: WC, weak; MC, medium; SC, strong; NFC, nearly full; FC, full. Alignment: WA, weak; MA, medium; SA, strong. Dotted line indicates an estimate (from NFC line). Shaded area contains scaled estimates (per the text).

but how do the four cultures involved contribute, and how does this situation relate to what happens in the RoW?

11.2 Connecting the US and the Rest of the World

Although we were forced to use different variables on the *x*-axes for the US (Figure 18) and the RoW (Figure 8), if the underlying cultural mechanics are indeed invariant, we should be able to link the two frameworks represented by these figures in a way that gives a deeper insight into the more complex interactions in the US.

We need to search for specific connections between the US and RoW scenarios, beyond the fact that both feature a generic cultural signature. With this in mind, it is useful to note that the religiosity of Democrat supporters, as measured on a scale similar to mine, is 51%.[272] So, if we think of the US political sides as quasi-nations, which happen to occupy the same geographical space, a comparison of the levels of climate-change most-endorsing responses for Democrats with those of people in nations of similar religiosity should be illuminating.

In response to unconstrained questions, between 45 and 75% of Democrats give most-endorsing answers, depending on the strength of the framing. This is much higher than the 25–40% seen in nations of about 50% religiosity in the RoW data (see Figure 8). This strongly suggests that something is making allied belief much higher than we would expect from religiosity alone. It therefore looks very much as though climate catastrophism has formed two separate alliances:

- with Democrat supporters primarily committed to religious culture;
- with Democrat supporters primarily committed to Dem/Lib 'tribal' culture.

While the two subpopulations – religious Democrats and tribal Democrats – may not be the same size, the 51% religiosity figure suggests that they are not very different. We also can't say anything for sure about the relative power of the two alliances to motivate most-endorsing responses. However, given both stem from major cultures that have developed their relationships with climate-catastrophism over decades, we will hypothesise that they are similar in this respect, and see where this takes us.

If the hypothesis is correct, then an unconstrained question on climate change, of any strength, should provoke similar proportions of both groups to express their allied belief by giving a most-endorsing response. That means that we might expect the proportion of US Democrats expressing allied belief to be a simple multiple of the proportion in a nation of equivalent religiosity. Naively, we would expect a doubling, but it is possible that, when both are operating within the same population, the cultures might hit some kind of absolute ceiling, resulting in a lower figure.

It is less clear what might happen when different strengths of reality-constraint kick in. In the RoW, allied belief falls away proportionately to innate scepticism, so in principle this effect should occur for both groups. However, we should also recall from Chapter 9 that the dramatic rise of innate scepticism among the religious is accompanied by apparently less scepticism among the irreligious, which in turn may stem from what I referred to as 'Pascalian assent'.* As a result, in secular countries, reality constraints actually provoke *higher* levels of climate-change most-endorsing responses. From the RoW data, it wasn't possible to determine whether this rise stems from a direct relationship of the irreligious with climate propositions, or whether it is something to do with the alliance of religion and climate catastrophism. If the latter is the case, we can also wonder if the same effect would be seen for the alliance of the Dem/Lib political tribe with climate culture.

All this means it is harder to say whether the level of reality-constrained responses should still be a simple multiple of the figure from a nation of equivalent religiosity.

Because the unconstrained case is straightforward, it allows us to test the hypothesis that the two cultures have equal powers to provoke allied belief. This should result in a simple scaling factor between its expression among US Democrats and among the population of a nation of equivalent religiosity. As we proceed, we can see whether the same scaling applies to the constrained responses too.

* See p. 155.

11.2.1 Democrat responses match the RoW pattern

The first step is to find the exact equivalence between the two frameworks. We don't have a precise translation between my religiosity scale, used for the RoW, and the US scale, on which the Democrats score 51%. For the rough comparison above, I just assumed the scales were equivalent, but this might not be the case. Fortunately, a happy coincidence in the data allows us to avoid any impact from a difference in the religiosity scales. In Figure 18, we can see that Democrats give almost exactly the same level of most-endorsing responses to the SA and MC questions: 47% and 48% on the *y*-axis respectively. And in Figure 8 for the RoW, there is also a place where the level of most-endorsing responses for SA and MC coincide. This is therefore an accurate common reference point. On Figure 8, it is at 46% religiosity on the *x*-axis. This is not too far from the 51% figure cited above, so it seems that the different religiosity scales are close, but not exact.

Next, we need to isolate the allied belief element from the total level of most-endorsing responses to unconstrained questions, since it is only this that is expected to scale between the two frameworks. Doing so is straightforward. We first note that most-endorsing responses come from two distinct groups: those who directly believe in climate catastrophism, and those who only give most-endorsing responses as a result of allied belief. The direct believers should approximate to the small number of responders in the FC series (for either framework); these individuals believe even when the strongest reality-constraint is applied. The unconstrained series, meanwhile, include responses from direct believers *and* allied believers. Hence quantifying allied belief is as simple as deducting the value of the FC series from the value of each unconstrained series.

Calculating the scaling factor is then straightforward too. Figure 19 shows how it can be done in visual fashion. It is, in essence, a simplified and modified version of Figure 8. The only significant difference arises because for the US there is an MA series, whereas for the RoW, we have only MSA and MWA series. To make the two frameworks directly comparable, I interpolate between the two to create a hypothetical MA series for the RoW. This slightly revised model is then overlaid with a transect at 46% religiosity. The transect

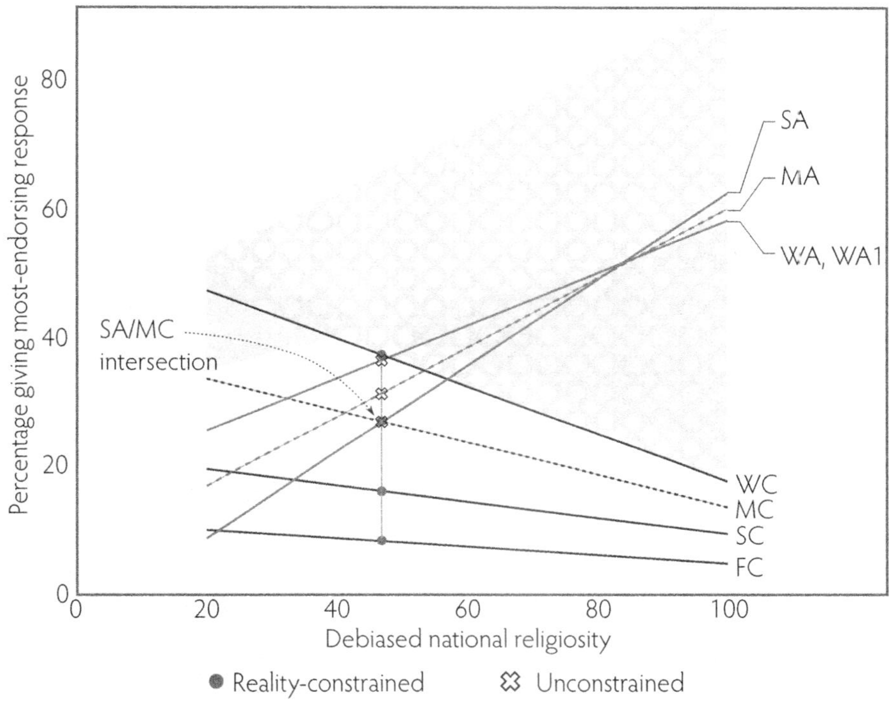

Figure 19. Mapping the US to the Rest of the World.

The full model, adjusted as per the text, with an overlaid transect of series, at 46% religiosity, where SA and MC intersect.

is marked with hollow crosses where it intersects the unconstrained trendlines, and solid circles where it intersects the reality-constrained ones. Both symbols therefore appear at the point where the SA and MC series cross. 'Ground level' for the resulting totem-pole of symbols is the FC series, representing the level of direct belief. The distances from the foot of the totem-pole to each hollow cross represent the levels of allied belief expressed with different framing strengths for our hypothetical nation of 46% religiosity.

To move to the US framework, we need first to shift the totem pole to the slightly different ground level of the US Democrat FC series, and then stretch it until the position of the SA/MC intersection matches the equivalent values in the US data.* Because of noise,

* Since the two points do not precisely coincide in the US data, I scale to their midpoint.

we cannot expect an exact match for the other series, but in fact they are very close. The grey shaded areas at the left of Figure 18 shows the rescaled totem-pole for comparison with the original US Democrat data. The correspondence is excellent, not only for the unconstrained series, but for the reality-constrained series too.

Table 17 repeats this exercise in mathematical terms. It shows the values for all the RoW series at the 46% transect, the new values when these have been rescaled (including the 'regrounding' to the value of the US FC series), and the very small differences between the latter and the equivalent US series.

The fact that the scaling is the same for all the series supports our hypothesis that the two cultures – religion and political tribe – have the same power to motivate allied belief. The value of the scaling factor – 1.97 – is very close to 2.* As noted above, since both cultures are operating in the same population, one might have expected to see some limit on their combined effect, and thus a lower figure. That there is 'no limit' is interesting; however, around 30% of Democrat supporters do not provide climate-change most-endorsing responses even to weakly-constrained or weakly-aligned propositions, which suggests a limit has not yet been reached.

That the same scaling factor also works for the reality-constrained series moves us closer to an explanation for the conundrum in the RoW data, namely why irreligious people exhibit *more* support for climate-change propositions in reality-constrained scenarios than for unconstrained ones. In Chapter 9, I noted that this must be due to cultural effects; the irreligious may be less sceptical because they are not defending strong cultural (religious) values, while their lack of allied belief encourages the 'mental insurance' of Pascalian assent in its stead, which works to the same end. Additionally, we were not able to determine whether this reduced scepticism was dependent on the alliance of religion with climate catastrophism, or stemmed only from a direct relationship of the irreligious with climate propositions. Since the effect is indeed seen among the irreligious, I made the provisional supposition that the latter was the case.

* In practice, as the position of the series in each framework comes ultimately from textual analysis, the resulting measurement error is likely much bigger than the 0.03 difference.

Table 17. Expectation and reality for US Democrat attitudes.

Series	Transect value	Scaled transect value*	US Democrats	Difference
Reality-constrained series				
WC	37.0	69.0	72.0	3.0
MC (interpolated)	26.0	47.5	48.0	0.5
SC	15.0	26.0	28.0	2.0
FC	7.0	10.0	10.0	0.0
Unconstrained series				
WA	36.0	67.0	68.5	1.5
MA (interpolated)	31.0	57.5	56.0	−1.5
SA	26.0	47.5	47.0	−0.5

*((Transect value − RoW FC transect value) × 1.97) + US Dem FC value. All figures rounded to nearest 0.5. There are minor rounding differences in the table.

If this supposition is correct, then, when we move to the US framework, the simple scaling behaviour seen for the unconstrained trends should break down, because we would not expect the presence of another culture – political tribe – to have any effect on factors that come only from a direct relationship between irreligious people and climate change propositions. That the scaling of factor of around 2 still applies, shows the supposition to be incorrect. However, I think this actually strengthens the case for Pascalian assent, by supplying a more obvious cultural driver; in other words, the Pascalian assent to climate catastrophism is being driven:

- in the RoW, by a single cultural alliance – with religion;
- in the US, by two alliances – with Dem/Lib tribalism and with religion.

Intuitively, this now makes sense. A second powerful cultural alliance pushing climate catastrophism could well result in twice as many people 'buying' mental insurance, and thus exhibiting Pascalian assent, even though they are not actual cultural believers. In essence, they are making a probability bargain, which they appear to base on their perception of how many others in society believe the proposition; a doubled level of allied belief in climate catastrophism in their peer group makes the 'offer' proportionally harder to resist.

11.2.2 Unequal alliances do not change the scaling factor

So far, we have considered the two cultural alliances operating in the US to be similar in all respects. In fact this may not be the case, although a little thought suggests that this possibility would not change the principles outlined above, or the expectation for the scaling factor between the US and the RoW. The only thing that would be different would be the number of cultural adherents associated with each alliance.

An IPSOS survey informs us that more Democrat supporters say their primary identity is about political tribe (38%) than say it is about religion (27%).[273,*] This doesn't tell us about absolute numbers of adherents associated with each alliance, but it implies there will be somewhat more 'tribally' motivated Democrats than religiously motivated ones. However, because both cultural alliances are operating in the same manner in respect of climate catastrophism, up to a limit, it does not matter what the split of religious and tribal Democrats is, because the gains in one alliance are offset by losses in the other. In these circumstances, the scaling factor will always come out at 2, so long as the asymmetry between the number of adherents in each alliance remains moderate.[†]

If the skew were more extreme, however, we'd expect to see something different. For example, if Dem/Lib tribalism completely eclipsed religiosity, the pattern of trends should look very similar to the high religiosity end of the RoW framework (in Figure 8), rather than being a close match to a scaled-up version of the pattern at medium religiosity. This scenario would also imply that religious Democrat supporters (who are likely less Dem/Lib 'tribal' overall), would have systemically different responses to the religiously unaffiliated. Surveys that cross-poll US attitudes to climate-change with

* Note that many of those who say their political leaning is their primary identity may still be religious, hence the difference between the religion figure here and the religiosity figure quoted at the start of Section 11.2.

† Although this would mean the assumption that the religiosity scales were equivalent at US 51% versus RoW 46% may not be valid; however the scales translated, religiosity would have less influence with Democrat supporters.

political *and* religious affiliation, are rare, and they do not cover a range of framings and strengths, which would be extremely useful for confirming the speculations in this chapter. However, data from one such survey shows that for a single (unconstrained, weakly-aligned and lifted) framing at least, 83% of religious Democrats give climate-change endorsing responses, almost the same as religiously unaffiliated ones (86%).[274]

11.2.3 It's different for Republicans

In Figure 18a, religiosity mildly increases from left to right on the *x*-axis, because there are more religious people, and more who are strongly so, among Republican supporters. Despite this, all of the unconstrained series drop sharply in this direction, to register minimal climate-change most-endorsing responses from these people. This is the opposite direction to the RoW, for which the unconstrained series all rise sharply, to register very high levels of climate-change most-endorsing responses in the most religious nations (Figure 8).

So, what is happening? We must again note our supposition that, no matter what the surface features, the same cultural mechanics will be operating beneath. We therefore have to find out which of those mechanics dominates here, and why; how the model of cultural causation fits this situation. Or alternatively, if it doesn't fit, what remains unexplained.

We start with the most studied and publicised fact of US attitudes to climate change, which is the very strong polarisation between Rep/Con resistive attitudes, and Dem/Lib supportive attitudes. From a cultural point of view, one can say that the Rep/Con political tribe is aggressively opposed to climate-change values simply because their Dem/Lib opponents embrace them, and vice versa; neither of these tribes actually has significant knowledge about climate change.

So, an expectation from the cultural model is that strong Rep/Con opposition to their cultural opponents will trigger their innate scepticism about climate change, which is to say about climate catastrophism. And once triggered in one cultural context, innate scepticism of climate catastrophism will remain in other cultural contexts too, and may even be reinforced; for the large majority of Republican supporters who are religious too, this innate scepticism will be fuelled by the need to protect their religious values as well as their Rep/Con values.

This will mean a collapse in allied belief among Republican supporters, which is exactly what we see; so the model does explain this observation. And given that innate scepticism is already active, reality-constrained questions will not produce very different results; those that are mildly constrained garner somewhat more climate-change most-endorsing responses, and those that are strongly constrained garner somewhat less, but the percentages are low for all the questions. One way of viewing this situation is that Rep/Con culture acts like a strong reality constraint, triggering innate scepticism of climate catastrophism in defence of religious values.

This collapse in allied belief means that an equivalence cannot be drawn between Republican responses to unconstrained questions and the same responses in the framework for the RoW, as we did for the Democrats. The fact that the cultural mechanics are the same in the US as for everywhere else does not mean that the response patterns for two arbitrary mixes of cultures will precisely match; only that both will exhibit typical cultural features. In theory, an equivalence could still be drawn for the reality-constrained series, but there is no tell-tale clue in this case that would boost our confidence as to how the patterns may be matched (as occurs for the Democrats above where SA and MC coincide). Added to which, Republican results are all very low on the *y*-axis, meaning that noise will be much more of an issue, perhaps to the point of overwhelming the cultural signals.

11.3 Summary for the US

Dan Kahan and others have established the cultural nature of attitudes to climate change in the US. The findings set out in this book indicate that their proposition of attitudes being motivated by political tribe is correct, but that it is only part of the full story.

It appears that a culture of climate catastrophism is a key driver of attitudes in its own right. Religiosity contributes too, but via a direct interaction with climate catastrophism, rather than as a secondary factor resulting from its asymmetric relationship with the Rep/Con and Dem/Lib publics. The details may not yet be fully elucidated, but it does seem clear that there are four cultures interacting within the US, all making significant contributions to the pattern of observed attitudes.

There are two interesting corollaries to set out. Firstly, the FC series for the US looks very similar to the equivalent series from Figure 8; it seems that core believers in climate catastrophism are everywhere rather few in number, never rising above 10% of populations even in the most favourable circumstances, and falling to as low as 3% when there are strong reality constraints. Secondly, a big uplift of allied belief in catastrophic climate change for the Dem/Libs appears, ironically, to be caused by the same factor that helps to reduce support for the Rep/Cons, namely religious belief.

Chapter 12

CLIMATE CATASTROPHISM AND POLICY: RENEWABLES

12.1 Motivations for renewables deployment

There have been many technical analyses of wind and solar energy, covering a range of issues, from efficiency and energy density, through subsidies and land usage, to maintenance, grid impacts, intermittency and more. While very useful, these analyses nevertheless share a common limitation, an aspect that they're unable to probe, although it often appears among the questions and speculations of the authors. This is the social motivation for renewables deployment, which is the subject of this chapter.

As a case in point, the first of a series of excellent renewable energy analyses at *Climate Etc*[283] by Russ Schussler, a former Vice-President of Transmission Planning at Georgia Transmission Corporation,[284] observes that a full cost-benefit analysis requires expertise beyond that of power systems specialists. If it is to take in externalities, such as the environmental impacts, it will involve climate scientists (to quantify the direct harms), economists (to cost them), and environmental scientists (to quantify the indirect harms, such as the impacts of climate change and policy measures on the environment). Input will also be required from a variety of specialisms to determine what is and may ultimately be possible, in political and societal terms: sociologists, politicians, government departments and intergovernmental agencies, for example.

However, the parent domain for renewables is climate change, which as we have seen has long been subverted by cultural influence. Climate catastrophism has triumphed here; its hi-jacking of the authority of science* has enabled it to corrupt the entire policy arena, including all the fields of expertise mentioned above. It has overridden rational discourse, replacing technical justifications for renewables with social (moral and subjective) ones, which technical authors are ill-equipped to analyse. Schussler acknowledges the powerful cultural presence:

> I hope that the voices of concerned utility experts are not drowned out by the noise of 'true believers' or disbelieved because of false accusations of self-interest.

Schussler's text is infused with references to the kind of cultural behaviours we have examined. 'True belief' is of course commonly expressed by ardent cultural adherents, while 'false accusations of self-interest' is a clear reference to demonisation in general, and the merchants of doubt meme in particular. Even his title, '*Myths* and realities of renewable energy' (emphasis mine), alludes to the presence of a cultural phenomenon.

Other authors also note the presence of an influential culture that subverts rational consideration of their technical concerns. For example, environmental scientist Vijay Jayaraj, like Schussler, uses the word 'myth' to allude to cultural motivation for renewables deployment.[285] Others go further, referring to renewables advocates as a cult, for example the article entitled 'Age of Un-reason: How fear and ignorance drives wind and solar worship cult'.[286]

Geologist, climate scientist and energy analyst David Archibald directly frames renewables motivation as 'religious' in nature, religions being the most familiar brand of cultural entities. In a brief treatment of renewables,† he employs religiously-orientated terms such as 'believers' and 'government encyclicals':[287]

> Briefly, the only reason solar and wind get a look-in is because solar panels and wind turbines are made using energy from coal at $0.04 per kWh and turn out power at $0.20 per kWh...You can't use solar and wind power to make solar and wind power equipment; as such they are neither renewable nor sustainable. And they certainly won't be replacing fossil fuels when the fossil fuels run out.

* See p. 295.

† The article is actually focused on hydrogen.

> Even some lefties are figuring this out and thus the documentary *Planet of the Humans*. So the global warming clerisy, headed by Alan Finkel in this country [Australia], needs to keep coming up with new content to satisfy their simple-minded believers...Vast sums are to be spent on hydrogen. The language of the Government encyclicals suggests that hydrogen is a new source of energy that just has to be tapped to guarantee a wonderful future.

Needless to say, Archibald views this 'religious' influence as overwhelmingly damaging, adding as part of his conclusion: 'Global warming doesn't build orphanages or hospitals. As a religion it doesn't do any good at all.'

All of these authors are highly critical of renewables on technical grounds. For instance, Schussler is convinced that the downsides involved are so severe as to be intolerable except in the most extreme circumstances:

> ...I believe most planners and utilities recognize that the overall impacts to society (unless needed to avert environmental disaster) would be extremely harmful in the net analysis.

In other words, the technical case is so inadequate, all these authors are sensing that something irrational – essentially cultural – is behind society's enthusiasm for renewable energy.

This is not to discount the many positive takes on the costs and benefits of renewables. However, even these (apart from fringe contributions[288]) generally agree that there will be financial pain and environmental compromise (industrialising entire landscapes is hardly eco-friendly). They just argue that these are necessary collateral costs in preventing global catastrophe.

In other words, Schussler's words 'unless needed' are the nub of the matter. Whether the pain is justified by the prospect of calamity is precisely the point that is culturally disputed. It may be that the extent to which any author believes the costs are justified is simply a function of how much they are gripped by cultural fear of climate catastrophe.

So this is the key question: has the renewables revolution come about largely due to pragmatic considerations that are consistent with mainstream science? After all, the things do actually produce power. Or instead, is it, as Schussler, Jayaraj and Archibald suspect, largely due to cultural motivation, pushed through regardless of the costs – fiscal or societal or environmental – and therefore solar panels and wind turbines are largely icons for the 'secular religion' of climate catastrophism? They are unable to answer this question, which requires cultural not technical investigation. However we should be able to resolve it here, because the hard social data on cultural attitudes to climate change, as described in Chapter 8, can be mapped to renewables commitment across nations.

12.2 Probing motivations for renewables deployment

If a particular domain is dominated by cultural attitudes, support for resulting policy measures (and therefore spending) should ultimately be rooted in those attitudes. This isn't to say that for the climate domain, for instance, members of the public will rush out in proportion to their national attitudes and purchase wind turbines with their credit cards; it is politicians and other authorities who ultimately define and enact policy on behalf of their publics.

So support for climate policy measures should be a function of belief in catastrophism among those political elites.* However, those attitudes, while enhanced, should still be in proportion to the beliefs of their corresponding publics. Similarly, public scepticism - mostly innate, rather than rational, because of a lack of domain expertise - will push back proportionately in the opposite direction, and will therefore be the main restriction on policy excess.

This should be the case whether or not nations are democracies. Where they are, cultures are adept at gaming the democratic process to ensure no awkward rational questions are asked, which might alert elites or public alike to policy problems; in the UK for instance, 'Net Zero by 2050' was nodded through Parliament with only cursory scrutiny and no meaningful opposition. Where nations are not democracies, it is typically even easier for policy to be dictated by cultural expression.

The parent domain for renewables is climate change; that is to

* At least outside the USA.

say, whether their deployment is rationally or culturally pursued, the justification for their deployment comes from the desire to mitigate or avoid global warming. We know, from the measurements in Chapter 8 and the explanations in Chapter 9, that public attitudes to climate change across nations reflect the culture of climate catastrophism. And additionally, from Chapter 5, we know that the primary narrative of this culture dominates the pronouncements of public authorities and elites on the topic. So, if our assumptions about culture governing policy are true, then the commitment to renewables across nations should be proportional to the (cultural) attitudes that each national public expresses about climate change; we should see a very strong correlation between these two phenomena.

However, as Figure 5 shows,* there are a range of different attitudes to climate change depending on the question asked. Which is relevant to our probe of renewables motivation?

All nations have limited budgets and competing demands for resources; these represent reality constraints upon renewables deployment. So, the correct Figure 5 trend to use is one that comes from the response to a reality-constrained question. However, because neither national publics nor the elites determining policy have a good technical understanding of renewables, the constraint can only be weak at best (if the downsides of deployment, as noted in Section 12.1, were better understood, the constraint would be stronger). In summary, the commitment to renewables across nations should correlate strongly with the responses to a weakly-constrained climate-change survey question, as posed within the corresponding countries. This is a testable hypothesis.

The weakly-constrained responses we will use for our test come from an extensive UN survey of policy priorities across many nations,[289] and represent the proportion of respondents who selected 'action on climate change' as one of their six priorities out of seventeen global issues. See the WC series in Figure 4† for these responses as plotted against national religiosity. For absolute clarity, we have already established that this series is cultural in nature, because there

* See p. 147.
† See p. 139.

is a robust anti-correlation with national religiosity ($R^2=0.57$), and because it fits the pattern of all the other attitude trends in our model of cultural causation. The breadth of cover in the UN survey is useful because it provides plenty of overlap with readily available renewables deployment data for nations. I select 40 nations that have the most significant wind turbine deployment,[290] and 40 nations that have the most significant solar power deployment,[291] and which in both cases are also covered by my religiosity scale. Thirty-five nations are common to both these datasets; they have deployed both types of renewables at scale.

The combined commitment for wind and solar should produce a more reliable result for our test, because the individual technologies may have features or idiosyncrasies that could make them buck the trend in some way, but which are more likely to average out over the combination of both. That said, we can start by examining each technology separately.

In practice, the correlation is unlikely to be zero or perfect; policy decisions are very rarely 100% free of cultural factors, but they will rarely be wholly cultural either. So, we can set reasonable thresholds for the test. I'd suggest that a correlation of $R<0.33$ ($R^2<0.11$) would be too weak to claim that the culture of climate catastrophism is a key motivator for the commitment of nations to renewable energy, even allowing for the generosity typically applied to social data (in which there are usually many things going on). But if the correlation is $R>0.66$ ($R^2>0.44$), then we can reasonably claim that the culture of climate catastrophism is indeed a key motivator, and considering that there's bound to be a fair amount of noise in this data too, almost certainly *the* main motivator. If the value lands in-between, then I guess we'd have to dig a little deeper and think about other variables.

12.3 Wind turbine commitment across nations

12.3.1 Plot preamble

It is well-known that GDP per capita across nations strongly anti-correlates with national religiosity.* Therefore, if implementing wind

* There are some exceptions, but these aren't an issue here. See Chart F1 in the Excel-Ref for confirmation.

power is motivated by a WC-type cultural attitude, which itself anticorrelates with religiosity (see Figure 5), then plotting national wind turbine capacity per-capita against national religiosity should yield a power-type function. It does; see Chart F2 in the Excel-Ref.

Power functions are more difficult to deal with or apply our test thresholds for R to (based on linear trends). However, normalising the wind turbine capacity per capita with respect to GDP per capita removes the long-term GDP-religiosity relationship* and so any unequal fulfilment of motivation due to higher GDP (the same national motivation for renewables purchases more wind turbines if the GDP is bigger). This resets our expectation back to a linear function with either national religiosity, or with any data having a strong relationship to this, such as the WC responses, which we now use for our x-axis below.

12.3.2 Plot for wind turbine commitment versus WC cultural attitude

Figure 20 shows the deployed wind turbine capacity per capita for each of 40 nations, normalised as described above (this is the 'wind turbine commitment') and plotted against the weakly-constrained (WC) series response data. There is clearly a robust correlation. So, the real-world infrastructure outcome from a weak constraint represented by wind turbine policy, aligns well with the cultural attitude of national publics to a weakly-constrained question on climate change. This strongly suggests that these two reactions to what is essentially the same framing scenario, are largely the same too, which is to say that they are both culturally motivated by climate catastrophism.

The correlation coefficient R is 0.64, which is just slightly under our test threshold for detecting a dominant cultural motivation. As noted above, for any single renewable technology a looser correlation is likely, because there will be idiosyncrasies regarding the policies of some individual nations, which appears to be the case here. For example, the Czech Republic has very little wind power despite a high vote for 'action on climate change' in the WC series. But Portugal, with a vote share that is only a little lower, has lots.

* I used Spain's GDP per capita as the arbitrary standard, so each capacity is divided by its own national GDPpc and multiplied by Spain's.

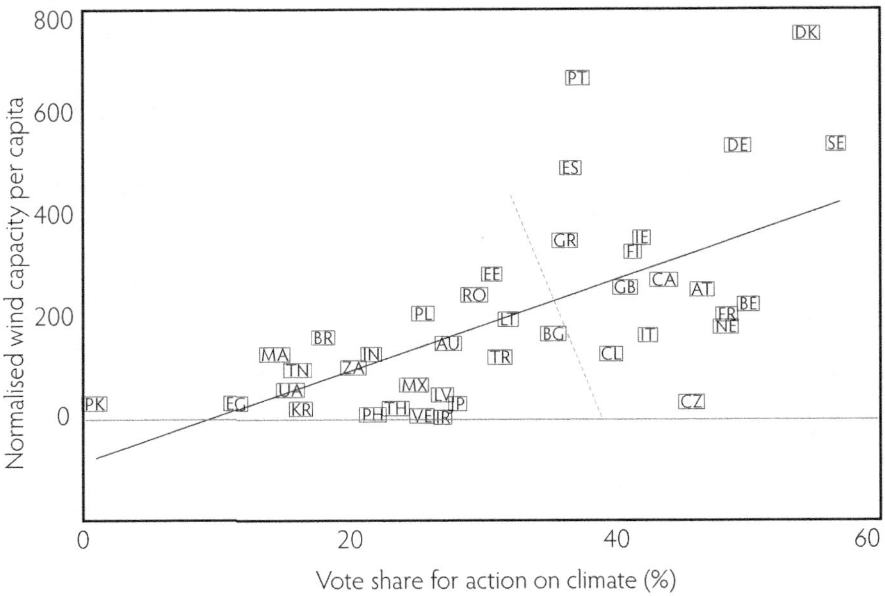

Figure 20. Wind power deployment is culturally driven.

Average wind turbine capacity over 2017 and 2018, normalised for GDP per capita, then divided by population. $R^2 = 0.41$, $R = 0.64$, $p = 9 \times 10^{-6}$.

In summary, it seems that as we move right along the *x*-axis and especially beyond the dashed line, individual nations don't necessarily have a higher deployment of wind turbines; they do *on average*, but there is more variability of deployment too. This is probably because a higher commitment to renewables is being expressed in deployment of other technologies. So, we need to add into the mix a similar analysis for solar deployment, which will better insulate us against the effects of such targeted policies for each technology.

12.4 Solar power commitment across nations

12.4.1 Plot preamble

The process for the solar data starts in the same way.* The power-relationship is made linear by normalising the solar capacity data with respect to GDP per capita. However, there's an extra issue for

* For wind, the power-relationship chart in the Excel-ref was F2. The equivalent chart for solar is F4.

solar power, which is that we need to adjust the capacity each nation has deployed according to its annual sunshine hours; the latter are systemically different across countries. Otherwise, the same spend (where spend corresponds to motivation), will produce different actual power per year in different nations. So, the deployed solar capacity is also normalised for annual sunshine duration.*

12.4.2 Plot for solar power commitment versus WC cultural attitude

Figure 21 shows the deployed solar power capacity per capita for each of 40 nations, normalised as described above (this is the 'solar power commitment') and plotted against the weakly-constrained (WC) series response data.

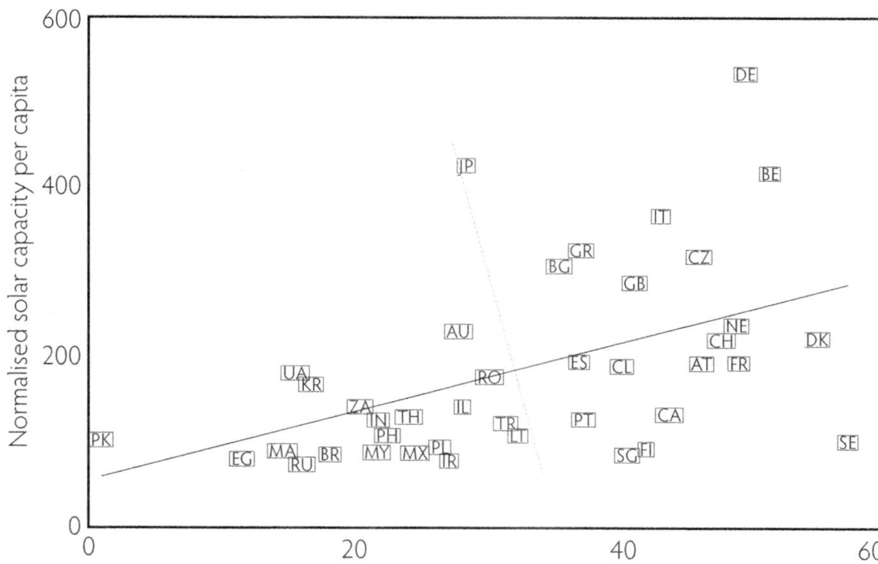

Figure 21. Solar power deployment is culturally driven.

Average solar capacity over 2016–18, normalised for sunshine hours and GDP per capita, then divided by population. $R^2 = 0.23$, $R = 0.48$, $p = 1.5 \times 10^{-3}$.

There is again a good correlation, although not quite so robust ($R = 0.48$). The appearance of the chart is very similar to Figure 20, with a greater range of expression in solar power going rightwards

* Again, using Spain as the abitrary standard.

on the x-axis, and especially after the dashed line. Similarly to the wind turbine case, this encouraging result suggests that the real-world infrastructure outcome from a weak constraint represented by solar power policy, aligns fairly well with the cultural attitude of national publics to a weakly-constrained question on climate change; in other words these are likely the same culturally motivated reaction to essentially the same framing scenario.

However, because the national idiosyncrasies are presumably different for solar and wind policy, in each case, different nations produce high or low outliers (on the y-axis). Indeed, much of the reduction in the R value appears to stem from Japan's extraordinarily high commitment to solar power, which is despite its middling vote share for 'action on climate change', while Sweden has a very modest solar power commitment despite a very high vote share.

However, like the wind turbine case above, this is only half the story. Combining the results for both the technologies will give us a clearer picture.

12.5 Combined wind and solar commitment

There are 35 nations common to the last two charts, having both significant wind and solar installations. Figure 22 shows the average renewables commitment across both technologies, plotted against the weakly-constrained (WC) series response data. The correlation has significantly improved, and at $R=0.73$ easily surpasses our test threshold.

The greater variance of the results towards the right of the graph has considerably reduced (except for Germany), confirming the expectation we noted above when looking at wind and solar in isolation. The commitment to renewables generally cannot be assessed by looking at any single technology in isolation, but can be when looking at several, although we are only covering the main two here.

So, the real-world infrastructure outcome, which we have hypothesised is a reaction to a weak constraint, correlates strongly with the attitudes of national publics to a weakly-constrained question on climate change, with an R value above our test threshold. These two reactions to what is essentially the same framing scenario,

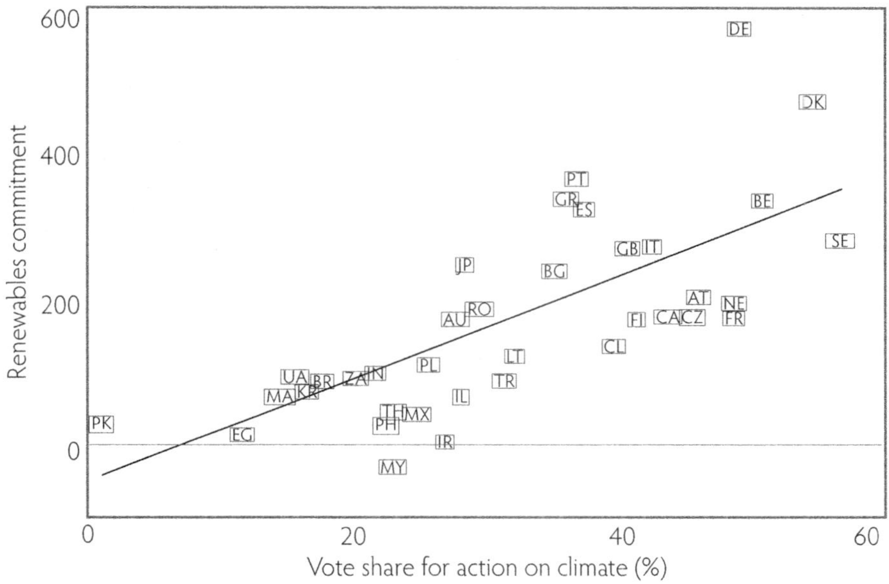

Figure 22. Renewables deployment is culturally driven.

Sum of renewables capacities normalised as per the previous two figures. $R^2 = 0.53$, $R = 0.73$, $p = 7.0 \times 10^{-7}$.

are the same too; they are both culturally motivated by climate catastrophism.

To put this a different way, the (cultural) attitudes of national publics to a weakly-framed survey question on climate change allow us to fairly accurately predict the commitment to renewables across nations. $R = 0.73$ gives an R^2 of 0.53; so, our predictor for any particular nation is 53%. This may not look particularly precise, but is unprecedented for a simple single-attitude prediction of a massive international spend in the climate domain, or perhaps in any domain.

Yet beware: we can't simply use any climate-change most-endorsing attitude for this prediction. For instance, the most-endorsing responses to an extremely weakly-framed question, or a mixed-mode one, will not have the same predictive power (and in most cases won't even produce a linear series). Moreover, using the responses to an unconstrained question would produce an 'upside down prediction'; as the following section reminds us with a chart, these anti-correlate with national religiosities. What makes the WC series a good predictor is not only that votes for 'action on climate change' are climate change endorsing, but that it represents responses to a question with a weak (but not too weak) constraint, the same constraint that renewables policies themselves present.

12.6 Renewables commitment versus religiosity

Chapter 8 used the 'lens' of religiosity to allow us to see that attitudes to climate change across many nations are cultural, the full explanation for which is in Chapter 9. In particular, climate-change most-endorsing responses to unconstrained survey questions correlate with national religiosities, whereas with reality-constrained questions, they anti-correlate. Given that the WC series responses we used above are in the latter category, we should expect the renewables commitment across nations to anti-correlate with national religiosities too. It does, as shown by the black series in Figure 23, which plots the rank (high number = high commitment) of renewables commitment on the *y*-axis, against national religiosity on the *x*-axis, using the same 35 nations as Figure 22.*

This robust anti-correlation with national religiosity, a purely cultural phenomenon, underscores the suggestion that the commitment to renewables across nations is culturally motivated. As a further reminder, the trend for the SA series (in grey), which consists of responses to an unconstrained question, is placed on the same chart. This is also a climate-change most-endorsing response, in this instance expressing concern about the personal impacts of climate change; however, it trends in the opposite direction to renewables commitment; i.e. a high concern about climate change impacts corresponds to a low national commitment to renewables, and vice versa. The contradiction between these two trends, which is somewhat counterintuitive, nevertheless fits the model of cultural causation set out in Chapter 9.

The data-points of nations for the black series in Figure 23 are quite widely dispersed about the trendline. In part, this is almost certainly because there is more noise when using religiosity as a proxy for

* For further insight, especially the use of rank rather than absolute value, see Appendix H.

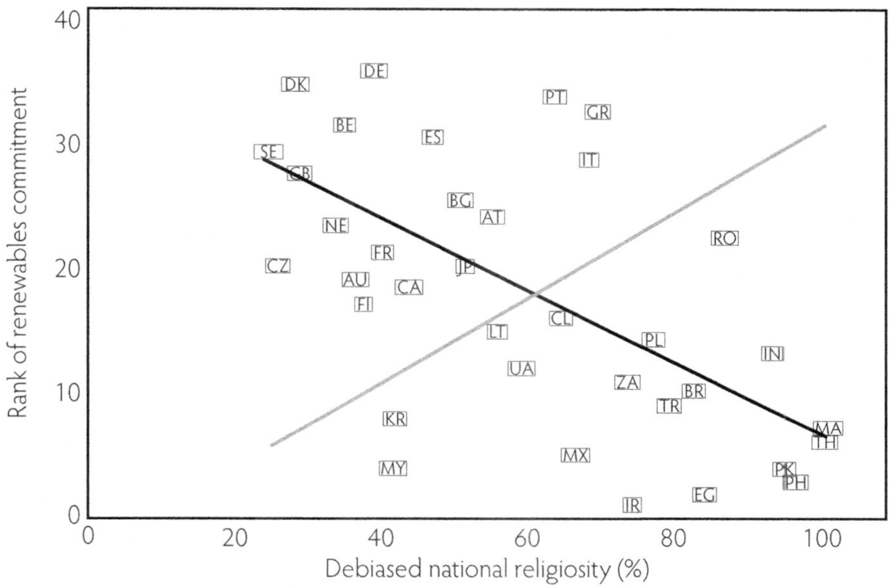

Figure 23. Rank of renewables commitment against religiosity.

With the SA series as contrast (trend only). Renewables commitment, N = 35, $R^2 = 0.42$, R = 0.65, $p = 2.5 \times 10^{-5}$.

the cultural attitude that motivates renewables commitment; so, we'd expect a somewhat reduced R value. Indeed this is the case, though R = 0.65 is still respectable. If we'd added similar analyses for other renewables technologies (biofuels and hydro, say*) into the mix our result would likely improve; however, the point is adequately made with the two main technologies of wind turbines and solar power.

A second reason for the wide dispersal of nations around the trend in this view against religiosity, is systemic variance due to religio-regional GDP per capita. This effect was explained in Section 10.1.5, and is visualised for renewables commitment in Appendix H; it occurs for attitudes not directly related to GDP, and so is not a function of the fact that renewables infrastructure happens to have a more direct connection.

* In fact, Norway is excluded from the above charts as a policy bias towards that country's abundant hydro-power attenuates motivation for solar and wind. If we'd done an analysis of hydro too, Norway could have been included, because commitment should average out over all technologies.

12.7 Electric vehicle commitment versus religiosity

We would expect other major Net Zero policies to result in patterns of commitment across nations similar to those seen for renewables. This is the case for instance regarding electric vehicles (EVs). If we skip the intermediate stages of the analysis used for renewables, and replicate only the very last chart, of commitment ranking versus national religiosity, we get the pattern shown in Figure 24.*

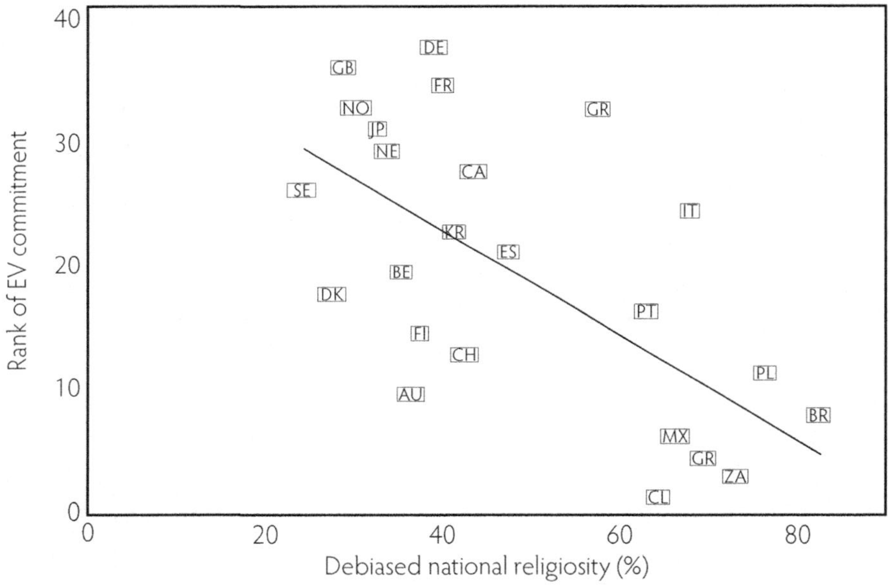

Figure 24. Rank of EV commitment against religiosity.

EV commitment is percentage BEVs and PHEVs (2021 data), normalised for GDPpc. N = 23, $R^2 = 0.46$, R = 0.68.

12.8 Conclusion for renewables motivation

The results for the renewables series are summarised in Table 18.† The weakly-constrained (WC) series is a leading predictor of the commitment to renewables across nations, because that commitment is culturally motivated; it is a product of climate catastrophism. So, although wind turbines and solar panels do actually produce some electricity, they are nevertheless essentially cultural icons; their elec-

* The outlier of India is excluded, but even if kept in, R^2 still scrapes into my 'robust' category. See Online-Appendix E for further details.

† Original charts and sources are in the Excel-Ref, sheet 'Relig DB and Renewables'.

Table 18. Parameters for renewables series.

Series	N	R	R^2	p
Wind turbine commitment against WC series responses (UN Poll vote share for 'action on climate change')	40	0.64	0.41	9.1×10^{-6}
Solar power commitment against WC series responses (UN Poll vote share for 'action on climate change')	40	0.48	0.24	1.5×10^{-3}
Combined renewables commitment (wind and solar) against WC series responses (UN Poll vote share for 'action on climate change')	35	0.73	0.53	7.9×10^{-7}
Rank of combined renewables commitment (wind and solar) against national religiosities	35	0.65	0.42	2.5×10^{-5}

tricity is a by-product. Chapter 15 looks at the balance of practical and cultural upsides and downsides for renewables.

Had I simply presented the correlation in Figure 22 without further detail, it would no doubt look intuitive that countries with a higher support for 'action on climate change' have correspondingly higher commitment to renewables, with no need to invoke cultural motivation. The several steps in this analysis help to tell us that this intuition is wrong and, for instance, that if instead we'd plotted renewables commitment against the most-endorsing responses from an unconstrained climate survey question, and relied on similar intuition, we would still be wondering why more commitment to renewables correlates with less concern about climate change; or for responses to a mixed-mode question, say, why there was perhaps no significant correlation at all. The true meaning of the results would have been missed.

Further insights can be gained by looking at the series in Figure 8. For instance, if knowledge of the technical and financial downsides of renewables made it past the cultural filter to become obvious to the public (and indeed public authority), this would represent an increase in the strength of the reality constraint, which in turn would shrink the motivation for renewables downwards, possibly to the point where deployment would essentially collapse.

12.8.1 Critical speculation is confirmed

As I noted at the start of this chapter, some commentators have discerned that the renewables debate is not entirely governed by rationality; it is subject to myths and religious-like thinking. This chapter shows that they are correct, and why. Michael Shellenberger,[292] the environmentalist who tirelessly advocates for nuclear power as a solution to our energy needs, characterises the irrational situation with respect to renewables as follows (and he's talking largely about solar and wind deployments):[293]

> For me the question now is, now that we know that renewables can't save the planet, are we going to keep letting them destroy it?

This is an insight consistent with the fact that the 'purpose' of cultural narratives is to gain emotive commitment, but only in order to hold the cultural group together; whether the resulting actions undermine or even reverse their stated purpose is irrelevant; this happens with cultures.

In the blurb for his book *Apocalypse Never,*[5] Shellenberger speaks about the character of, and underlying motivations for, modern environmentalism. These, he suggests, are behind the impetus to deploy renewables, and the strong resistance to nuclear power as a 'solution':

> What's really behind the rise of apocalyptic environmentalism? There are powerful financial interests. There are desires for status and power. But most of all there is a desire among supposedly secular people for transcendence. This spiritual impulse can be natural and healthy. But in preaching fear without love, and guilt without redemption, *the new religion* is failing to satisfy our deepest psychological and existential needs.

So Shellenberger rightly identifies the overall motivation as cultural; he uses the term 'religion' – as indeed do many others – simply because this is the example of a cultural entity with which people are most familiar. The social data in Chapters 8–10 confirms his view; the culture of climate catastrophism dominates public and public authority attitudes, and is behind the motivation for renewables deployment.

However, I believe Shellenberger has one thing wrong. Climate catastrophism does satisfy a deep psychological need, namely to be in a cultural group, and to develop an emotive commitment to it. The culture has become so pervasive for precisely this reason. Group membership inevitably leads to cultural behaviours, including all the irrationality surrounding renewables deployment – demonisation of opponents, irrational assessments of the facts, and all the rest of it. This is hard to explain unless climate catastrophism, in common with all other religions, is satisfying a deep-seated need.

12.9 Coda: An irony

It is a historical coincidence that atheism spread slowly outwards from the very cloudy countries of north-west Europe towards sunnier climes. As a result, and despite some important outliers, there is a reasonable linear correlation between the annual sunshine duration of countries and their national religiosities.[294]

This means that in the plot of solar power capacity per capita against religiosity,* we can substitute sunshine hours for religiosity, and still get a power-type function.† Normalising as before with respect to GDP per capita removes the simple effect of spending power (to get at motivation), and at the same stroke the long-term relationship of GDP with religiosity. This operation reveals the relative commitment to solar power for each of the nations, but now charted against their national sunshine hours. The results are shown Figure 25.

It reveals a considerable irony, namely that many countries with lower annual sunshine hours tend to deploy *more* solar capacity per capita than sunnier places. The overall effect is that much more solar power is installed in those geographies where it is least useful! This is a result of the cultural motivation behind renewables deployment, coupled with the historical coincidence mentioned above.‡

* Chart F4, as noted above.

† Chart F6 in the Excel-Ref.

‡ The increasing spread of nations right to left in the grey triangle, is the same phenomenon seen in Figure 21 for solar, and Figure 20 for wind, namely that an increasing cultural pref-

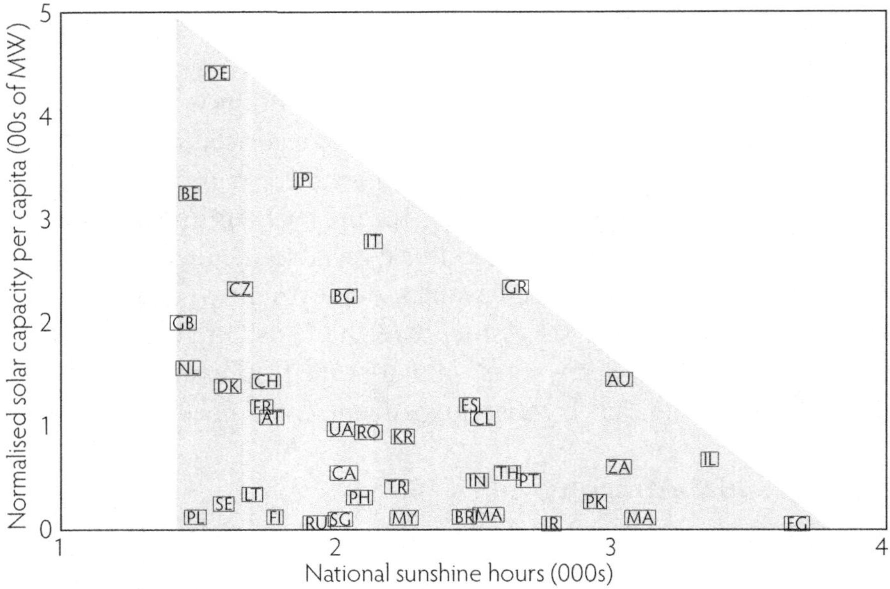

Figure 25. Solar installed where it is least useful.

Refer to main text for details of normalisation of capacity.

erence for renewables may nevertheless target specific technologies in particular nations, meaning overall a larger range of expression for any particular technology, in this case solar.

Chapter 13

CLIMATE CATASTROPHISM AND SOCIETY: ACTIVISM

Cultural attitudes stemming from climate catastrophism should determine the level of climate activism within nations, just as they do for renewables deployment. However, we again have to step carefully in order to avoid intuitive but incorrect assumptions about how these phenomena are related. To know what to expect, we need to assess which of the basic attitude trends is likely to be important for predicting levels of activism.

In Figure 5, the unconstrained responses show that allied belief (the SA trend particularly) is very low in largely secular nations, at the left-hand side of the chart. We can postulate that this widespread public rejection of the emotive Catastrophe Narrative will lead to a corresponding frustration for core believers in climate catastrophism. However, this shouldn't be the case in religious countries, where allied belief is very high.* This means that we should expect climate activism to be high in secular nations, where core believers in climate catastrophism will see themselves as defending their cherished values from large majorities of 'non-believers'; for cultures, as for armies, attack (aggressive proselytisation, civil disobedience) is a good form of defence. In religious nations, meanwhile, there is far less apparent need to rail against society, and there are also fewer core believers who might become activists (the FC trendline). In summary, the expectation is for high activism in secular nations, and low activism in religious nations.

There are issues with attempting to demonstrate that climate activism conforms to this pattern. Firstly, large international green organisations, such as Greenpeace and the World Wildlife Fund,

* Although coming from allied belief, such support is ephemeral.

existed before climate catastrophism took off, and they campaign on a variety of environmental issues. This makes it very hard to isolate the effect of climate catastrophism on their membership, number of offices, activities, or any other useful measure that might be a proxy for activism. Such large and long-lived organisations also tend to have strategic or financial reasons for maintaining offices in particular places, the mere presence of which is therefore not necessarily a reflection of local attitudes. In contrast, brand-new activist organisations spawned specifically to fight against climate-change catastrophe, such as Extinction Rebellion and School Strike for Climate, don't suffer from these issues. However, because they are still growing, their presence in any nation might in part reflect their expansion profile rather than national attitudes to climate change. Neither option seems ideal; I chose to analyse the newer groups.

13.1 Extinction Rebellion

Figure 26 shows Extinction Rebellion (XR) group presence per million of the population across 29 nations. No doubt because XR was founded in the UK, its presence there is huge relative to other places. So, to offset this issue and the related 'expansion' issue noted above, I plot the ranking of the number of XR groups against national religiosity (most groups = lowest rank), not the absolute number of XR groups per capita. Despite these issues, the data is more consistent than I expected. There is clearly a very robust correlation (R = 0.87) of the group ranking with national religiosities, which confirms the expectation of an anti-correlation for the absolute number of groups, this being a measure for activism.

As indicated by the grey horizontal stripe at the top of the chart, the measurement threshold of having at least one XR group becomes an issue for some of the most religious nations. A subset of these nations have very authoritative regimes that may not tolerate anti-social activities of the type XR encourage, although our model leads us not to expect much 'natural' support there anyhow.

Qatar 'luckily' supports one XR group, but also has a very low population, so ends up with an apparently very high ranking of groups per capita, considering its religiosity. Without that single group, it

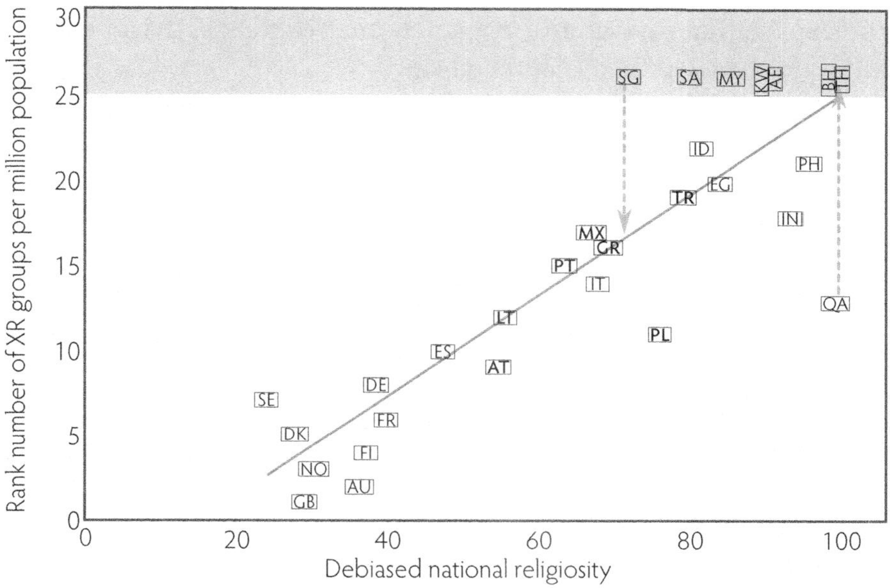

Figure 26. Extinction Rebellion as a cultural phenomenon.

Rank of number of XR groups per capita against debiased religiosity by country. Grey shading covers countries with no XR groups. $R^2 = 0.75$, $R = 0.87$, $P = 1.3 \times 10^{-9}$.

would sit next to Thailand and so roughly on the trendline. Singapore (which also has a low population) 'unluckily' has no groups at all. Were one to be formed, it too would sit almost on the trendline. These 'what if' scenarios aren't needed to demonstrate consistency with theoretical expectations, which look absolutely fine as the chart stands. But they are useful to make graphically clear the role of luck (and possibly factors such as stricter authority) in some of the datapoints in the top right-hand corner.

Appendix I explains why the particular nations in Figure 27 were used, and details a minor issue regarding the date at which the Extinction Rebellion data was scraped from web sources, relative to the onset of Covid.

13.2 Children's Strike Weekly

The Children's Strike Weekly (CSW) groups (part of School Strike for Climate), a protest initiated by Swedish teenager Greta Thunberg, are another climate-change activist presence that should conform to the expectation set out above, namely that climate activism should anti-correlate with national religiosities.

Figure 27 shows the rank of the number of CSW groups-per-capita plotted against religiosity, across 29 nations. Overall support for the CSW movement seems to be significantly higher than for XR, so there is not the same issue of smaller nations falling below the threshold of having a single group. That said, Bahrain and Qatar – marked with the grey circle – score a somewhat 'lucky' high ranking (low number) because they each have a single CSW group despite their low populations. Overall, the measurement once again confirms our expectation from cultural causation. Although the *y*-axis is a *ranking* and not the absolute values, an R value of ~0.8 is still very robust.

Appendix I spells out why the particular nations in Figure 27 were used, and details a minor issue regarding the date at which the CSW data was scraped from web sources, relative to the onset of Covid.

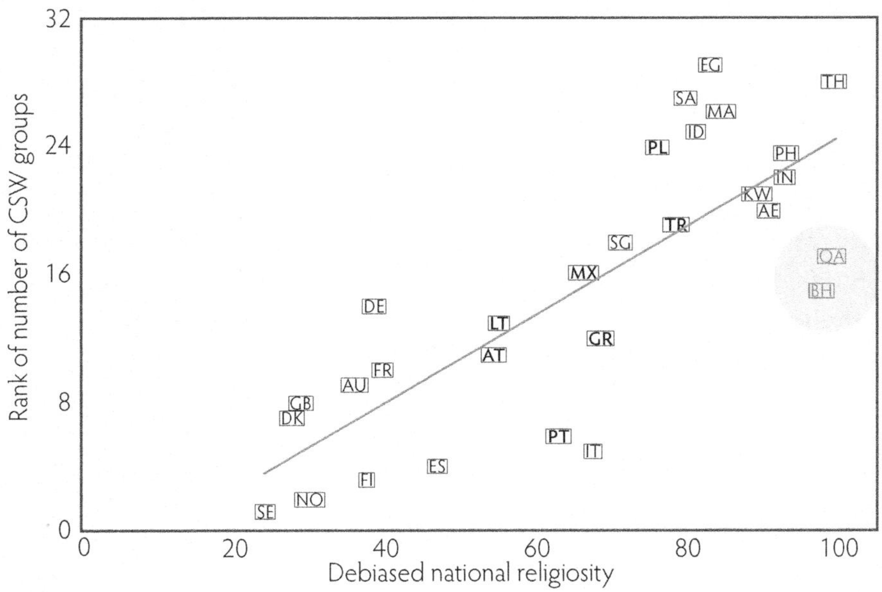

Figure 27. Children's Strike Weekly as a cultural phenomenon.

Rank of number of Childrens Strike Weekly groups against religiosity (high rank = low number). N = 29, $R^2 = 0.64$, R = 0.80, $p = 2.1 \times 10^{-7}$.

13.2.1 Are attitudes different for children?

Notwithstanding the robust confirmation of theoretical expectation, it's possible that Figure 27 may not be the most accurate reflection of the situation for CSW groups. This is because religion is receding in most Western nations, no doubt including those represented here. As a result, children are generally less religious than adults, but this 'religiosity gap' is almost certainly *different per nation*. Hence children will behave as though they're from a nation somewhat more to the left of the chart than their actual home nation. Moreover, the differing sizes of this religiosity gap mean that the left-right ordering of some nations may be different to the true ordering; the current debiased religiosity scale was derived purely from *adult* data.

While there are very few surveys involving children, projections from 'young versus older adult' data are a means to gain some insight into how significant the impact of child versus adult religiosity might be. And indeed, there's a reason to pursue this, because it may explain the apparently very high ranks (low numbers) for Italy and Portugal relative to their national religiosities, and to a lesser extent for Spain. There may also be a connection with the unexpected responses to extemely weakly-framed questions in Italy and Spain, as discussed in Appendix F and Section 10.1.3, which I speculated might be due to an exceptional influence of younger people in those nations.

Online-appendix B uses some 'young versus older adult' data to pursue this question of the child-adult religiosity gap. Overall, it appears that the impact on the Figure 27 results may not be too large. Although it could bring Spain much more into (the current) trend, there's only modest pull-in for Italy (and there isn't relevant data for Portugal). So, a different level of religiosity for children may not wholly explain Italy's position, unless a larger religiosity gap is still to emerge from children into (measurable) younger adults. However, the most important point here is that our overall result is likely to be robust to corrections for child religiosity, and may even improve.

Given they have authoritarian religious regimes, a possible suppression of children's strikes might contribute to the low ranking (high numbers) for Saudi Arabia and Egypt. Kuwait, which might have been expected to be in the same situation, is not anomalous anyway, although its position hangs upon only a single CSW group anyhow.

13.3 Aspirations for activism, and cultural hypocrisy

Given that strong cultural narratives have to be false in order to achieve their purpose (see Section 3.2.3.2), there is ultimately hypocrisy at the heart of all cultural attitudes, and the contradictions that result are sometimes so obvious it is hard to understand how they can pass unnoticed. A case in point is the stark difference between public aspirations to climate activism, and actual activism as measured above for Extinction Rebellion and the Children's Strike Weekly (CSW).

The aspirations to activism were measured in a Yale 2021 survey, which asked: 'How willing or unwilling are you to join a citizen's campaign to convince leaders in [country name] to take action to reduce climate change?' Twenty-six of the nations covered are also on my religiosity scale (and as usual I exclude the US). The aspiration is represented by combining those who responded 'Am participating' and those who indicated 'Definitely would'.* Standard public surveys would never pick up the tiny demographic of those who actually are serious activists, except possibly children taking part in CSW, but children are not typically included in surveys anyhow. The 'Am participating' probably indicates extremely 'low-cost' activism, such as clicking an on-line petition, or in religious countries endorsing a faith-based climate-orientated petition. 'Definitely would' participate, meanwhile, is even less of a burden if it never actually happens. Hence, although true activism would come with serious reality constraints (time and effort), these responses largely escape such downsides. So, we expect the trend for this series to look very similar to belief in climate catastrophism, which is to say allied belief.

The aspiration to activism is plotted in Figure 28, alongside

* The separate responses would still correlate with national religiosity, but with a lower gradient. 'Am participating' has low values, so that noise would be a problem. The form of the question text doesn't easily fit into the weak-to-strong framing system defined in Section 8.4, and it's possible that these responses, even plotted separately, reflect a composite series of some kind, maybe a linear mixed-mode series (see Section 10.1.4.2). This possibility is explored in the Excel-Ref; see the text associated with chart 'Y3' at the sheet 'PostCovid'.

actual activism levels, as represented by the ranking of the nation in terms of numbers of XR and CSW groups (the secondary *y*-axis has a higher rank-number for more groups per nation). Hence all the depicted trends are comparable. The trend of aspiration for activism is diametrically opposite to the two trends of actual activism. Cultural hypocrisy is clearly visible; we are seeing essentially 'virtue-signalling activism'.

Further cultural hypocrisy is revealed by plotting the public aspiration to activism, as above, against the public's desire to reduce fossil fuel usage. The same Yale 2021 survey includes the question 'Do you think that in the future [country name] should use more, less, or about the same amount of fossil fuels, like coal, oil, and gas, as it

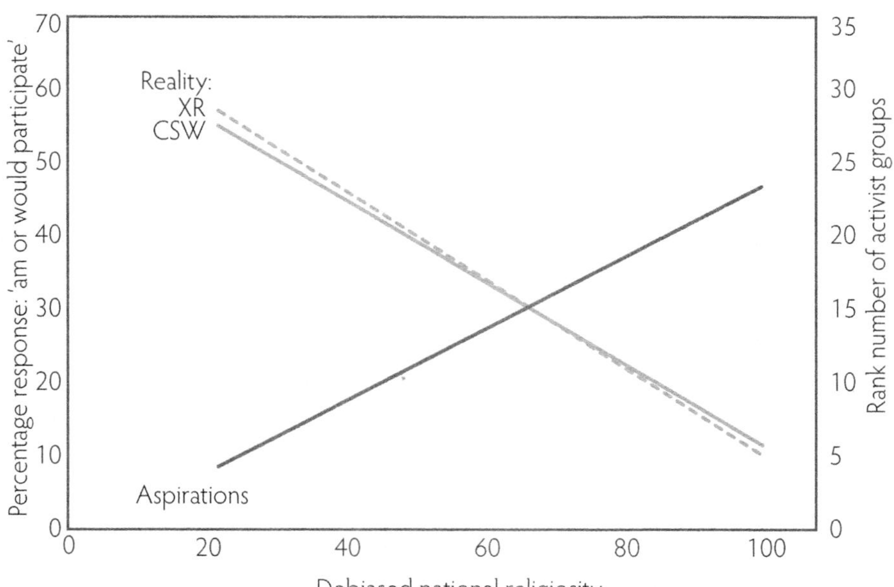

Figure 28. Activism: aspiration and reality.

Public aspiration to activism (dark line, left-hand scale, percentage responding 'am' or 'would' when asked if they would participate in a campaign to convince leaders to take action on climate change) and actual activist group membership (grey lines, right-hand scale, rank of number of CSW and XR groups), plotted against religiosity. The original chart is Y1 of the PostCovid sheet of the Excel-Ref.

does today?' Given that most climate change activism is ultimately about pressure to dramatically reduce fossil fuel usage, if any of these attitudes were rational, one might think that a plot of the response 'much less' would roughly track the trend for aspiration to activism. However, as Figure 29 shows, these two attitudes are in complete opposition. From a cultural perspective, this is not a surprise: cutting down on fossil fuel usage is a reality constraint, and one that bites at the personal as well as the communal level.

One can argue that, for the less-developed countries on the right-hand side of Figure 29, the constraint of using much less fossil fuel would bite all the harder, so perhaps this (grey) trend isn't only about culture. However, this wouldn't explain why, in complete contradiction, national aspirations to activism are so high in the same countries. Many other trends reported in this book tell exactly the same

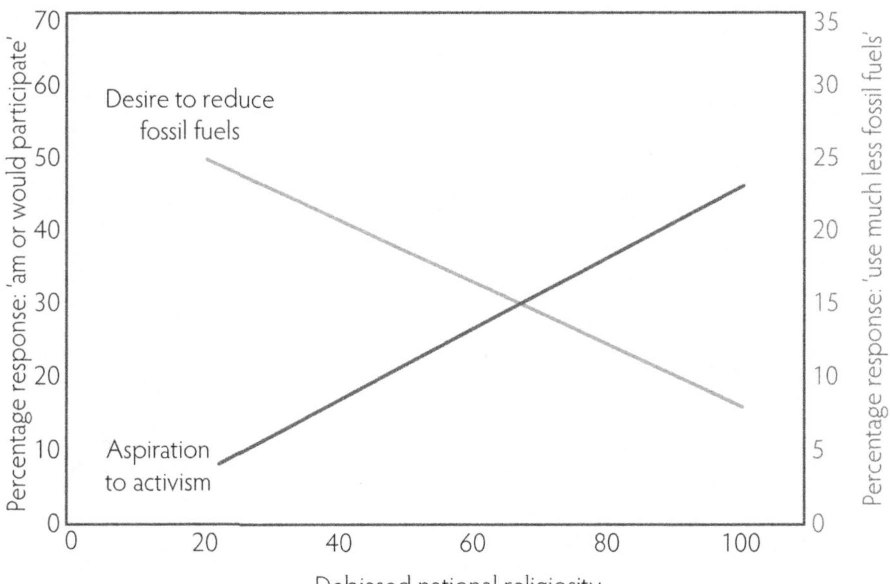

Figure 29. Aspiration to activism and desire to reduce fossil fuels.

Public aspiration to activism (dark line, left-hand scale, percentage responding 'am' or 'would' when asked if they would participate in a campaign to convince leaders to take action on climate change) and desire to reduce fossil fuel (grey line, right-hand scale, percentage responding that we should use 'much less' fossil fuel). The original chart is Y2 of the PostCovid sheet of the Excel-Ref.

cultural story, despite having no direct relationship with the fossil fuel resilience of nations.

It is worth adding that these cultural hypocrisies probably couldn't be resolved by feeding global publics further information. When asked about how much more climate information they needed, the responses correlate and anti-correlate with religiosity, like so many other series in this book (Figure 30). In other words, publics have an entrenched perception that climate information is cultural, and they will accept or reject it on that basis, whether it's right or not.

In irreligious nations, an innately sceptical majority probably feel they've been absolutely inundated with climate change information already. In religious nations, cultural narrative is a comfort to (allied) believers, despite its threat of doom (similarly to religious narratives that include the doom of Judgement Day). Both views are determined by instinctive reactions to information perceived as cultural.

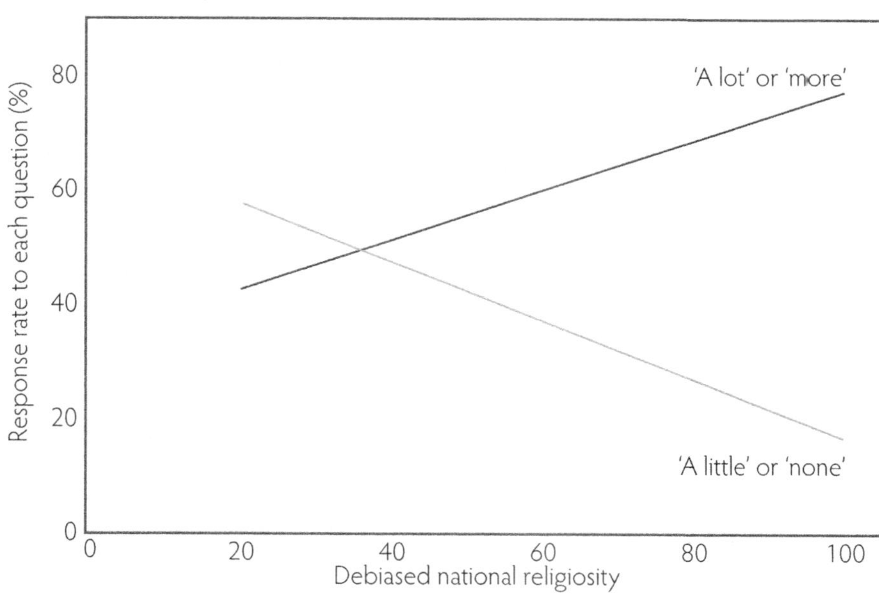

Figure 30. Climate information requirements and religiosity.

How much information publics feel they need, plotted against religiosity. For the black line, $R^2 = 0.74$, $p = 2 \times 10^{-8}$; for the grey line, $R^2 = 0.78$, $p = 2 \times 10^{-9}$. Original chart is Y4 of the PostCovid sheet of the Excel-Ref.

And this perception is mostly correct, in that masses of information about climate change or related policies, while having been created in good faith, is essentially cultural because it is based upon a starting assumption of imminent global catastrophe. As far as the public ear is concerned, any information not based on this assumption is probably lost in the noise.

13.4 Conclusion

The results for the series measuring the cultural motivation behind climate activism (XR and CSW), are summarised in the top two rows of Table 19.* For some people, it might be tempting to think that higher levels of climate activism in the more irreligious nations simply reflects greater objectivity. However, this also requires one to assume that groups of children 'striking' from their lessons are more rational than a big majority of adults within their respective nations, and likewise that the cult attire and antics of Extinction Rebellion represent a rationality apparently not present in the rest of the population. Even the name of the group is a cultural narrative variant; mainstream science does not say that climate change will cause human extinction.

Figure 5 tells us what's really going on: attitudes to climate change across nations are predominantly cultural, and, as this chapter has shown, the prevalence of both XR and CSW groups across nations conforms to the expectations from the cultural model presented in this book. Climate activism, like renewables commitment, is not rationally motivated, but driven by the culture of climate catastrophism. This fact ought to be discernible from the protests themselves, which are blatantly cultural in nature. And indeed, large swathes of the public do intuitively see this. Even parts of the media, normally so orthodox in its views on climate change, grasp that the XR Red Brigade and the bands of young children publicly chanting obscenities about mainstream politicians (see Section 4.4.1) are not behaving rationally.

* Note that the systemic secondary variance (of rrGDPpc) around trends for various series reported in this book, as described in Section 10.1.5, should not be seen in the above charts for climate activism; see Online-appendix Bii for details. All of the charts and original sources can be found in the Excel-Ref, see sheet 'XR & CSW'.

Table 19. Parameters for the climate activism series.

Series	N	R	R^2	p
Rank of Extinction Rebellion groups per capita	29	0.87	0.75	1.3×10^{-9}
Rank of Children's Strike Weekly groups per capita	29	0.80	0.64	2.1×10^{-7}
Public aspiration to activism	26	0.80	0.64	1.1×10^{-6}
Public desire to reduce fossil fuel usage	26	-0.69	0.47	9.9×10^{-5}
Publics need a lot or some more information about climate change*	25	0.86	0.74	2.0×10^{-8}
Publics need little or no more information about climate change*	25	0.89	0.78	2.0×10^{-9}

*Japan seems like an outlier here and so has been excluded. With Japan included, R^2 values reduce to 0.63 (row 4) and 0.70 (row5).

So, these movements are certainly *not* driven by reasoned considerations of science and policy. Nor does the activism in any way represent objective pressure from a large grass-roots majority. Rather, climate activism from XR and CSW groups is the high passion of a tiny minority of culturally convinced believers, whose approximate numbers across nations are represented at the very most by the FC trendline (and probably quite a lot less, as it is hard for surveys to truly represent the strongest reality constraints). Their belief is in certain, imminent, global climate catastrophe – a concept that contradicts mainstream climate science – and they express their considerable frustration because the societies in which they live largely don't accept their cultural beliefs.*

* As we have seen, most governments they protest at, and the UN elite too, largely welcome the protests, being culturally primed themselves.

Chapter 14

THE CHARACTERISTICS REVISITED

And so, after a long tour of the fundamentals of climate catastrophism – its origins, its critical features, a walk through its narrative population, measuring the culture, and predicting of some of its real-world impacts – we return to Professor Crusoe's scribbled list of likely cultural characteristics, which is repeated overleaf for convenience (Box 3).

A circumstantial fulfilment of many of these characteristics was provided back in Chapter 2, but a more complete treatment is now possible. Regarding the cultural consensus (Point 1), practically all of the world's top authorities, including presidents and prime ministers and the UN elite and religious leaders, propagate the Catastrophe Narrative (Chapter 5). It is also ubiquitous among lesser authorities, and has spread into businesses, schools, councils, charities and institutions of every sort. Consequently, there is a widespread public consensus on the narrative of certain, imminent global catastrophe (and the hope of salvation through the crash 'Net Zero' decarbonisation programme). However, as emphasised throughout this book, the consensus is simply wrong; it contradicts mainstream science as published by the IPCC. Section 3.2.6 explains why such cultural consensuses arise and shows that they are emotive, emergent, and not driven by top-down command.

The predicted hierarchy of cultural approval (Point 6), is seen across society. At one end of the scale, adherents of climate catastrophism frown on people who consume culturally disapproved foods, such as meat, or even milk, while applauding vegans and vegetarians. Cultural malefactors include oil and gas companies and min-

Box 3. Professor Crusoe's list

1. In large sectors of the public, there will be a common understanding of an existential threat and how society should respond to it (a cultural consensus).
2. The consensus will promote statements that are presented as all-explaining and/or indisputable.
3. The consensus will be actively policed, via status control – those criticising the consensus will be sidelined or downgraded – and emotive pressure – using fear and guilt to suppress dissent.
4. There will be uncritical acceptance, and possibly even adoration, of authority figures – and perhaps of 'prophets' too – who promote the cultural narrative.
5. Key information that makes the cultural entity and its narrative vulnerable to attack will be restricted to elite guardians.
6. Cultural adherents will place groups in society into an approval hierarchy, ranging from perpetrators of cultural misdeeds at one extreme, to their victims at the other. For a global culture, the groups could be large, incorporating whole ethnicities or nations or major demographic divisions.
7. Alongside the narrative of existential risk, there is also the prospect of salvation, rebirth and renewal, to be achieved through cultural conformance. This vision of hope will be propagated by the prophets in (4).
8. The visions of existential threat and salvation set out in the narrative will create unrealistic anxieties, fears, guilt, hopes and inspiration across society. This will cause immense bias – and a corresponding loss of objectivity – towards the culture in all areas of endeavour connected with it.
9. Double standards will prevail. Transgressions – legal, moral or ethical – of adherents will be overlooked, while those of critics will be harshly punished. This will create a feeling of threat, a shifting moral landscape and possibly large changes to the law (if the culture has been active long enough).
10. The narrative will be used to justify all sorts of social and infrastructural changes that benefit the culture, independent of their

true usefulness to mankind, and even if some are damaging.

11. There will be organisations that are particularly convinced of the narrative and, as a result, are particularly emotive in their cultural advocacy. They will police the internal ranks of the culture, and act as missionaries and recruiting sergeants (think Jesuits).
12. Many cultural adherents, especially those in the above advocacy groups, will self-identify with the culture, and will therefore be especially instinctive and emotional in their support, at the expense of reason.
13. False claims of conspiracy will be made against the culture (cultures emerge from subconscious processes).
14. Dissenters from cultural orthodoxy will be demonised, and possibly persecuted if the culture has penetrated society far enough.
15. Nevertheless, unless the culture completely dominates elites, expert opinion in the cultural domain will be highly polarised.
16. A large proportion of the public, possibly a majority, will remain unconvinced of the cultural narrative.
17. The culture will attempt to form coalitions with other cultures, both religious and secular.
18. Whole national leaderships and rafts of lesser authorities will have bought into the culture.
19. Huge resources will pour into activities that benefit and promote the culture, yet which do not help deliver the salvation it promises; they may even hinder it.
20. It will be claimed that there is no doubt about the existential threat. Nevertheless, the cultural narrative will slowly evolve.
21. There will be icons – visual reminders of the narrative. The evolution in (20) means that some will be de-emphasised or set aside for new ones, having lost their usefulness for some reason.
22. There will probably be positive elements to the culture (cultures are by no means all bad).
23. The societal effects are being caused by the cultural narrative, not any real-world phenomena.
24. Etcetera.

ing and construction enterprises, their employees and managers, the banks who fund them, and even those who have benefitted from their philanthropic contributions.[295] The Boomer generation are sometimes framed as climate criminals, both for their historic contribution to industrial society and their current (high) consumption,[296] while the young are lauded as saviours who will rescue the planet from the 'madness' of high greenhouse gas emissions. At the larger end of the scale, Germany is simultaneously favoured for its *Energiewende* and anti-nuclear stance,* and disfavoured for its historic emissions and still high industrial activity. Indeed, adherents denigrate many Western nations for climate-change 'crimes', while blessing China with cultural favour, an attitude that involves turning a blind eye to Beijing's extraordinary emissions and implausible decarbonisation plans. Worse, they also ignore the misdeeds of the Chinese Communist Party (perhaps inevitably so, since cultures have a soft spot for absolutism, perhaps casting envious eyes at its freedom of action). At the largest scale, as discussed briefly in the next chapter, the industrialised 'Global North' (including 'The West') is seen by adherents of climate catastrophism as an oppressor of the 'Global South'.

Climate catastrophism has conjured up numerous visions of 'extreme existential risk' (Point 7), often with 'the planet' or 'all life' at stake. Although the examples I collected focus on catastrophe, there are also some expressions of the corresponding salvation, even of a better world beyond the crisis. For instance, Paul Krugman's quote in Chapter 5† deploys climate catastrophe and salvation as a stick and carrot in relation to the US 2016 election. The CN-Archive also contains 'visionary' quotes from those who have taken up the role of prophets for climate catastrophism, such as Al Gore and Greta Thunberg. Thunberg's role was explored extensively in Chapter 4.

In Chapter 5 we looked at the many narrative variants that have, over decades, inculcated anxieties, fears, guilt and hopes in publics around the world (Point 8), including in children. All are linked to the dominant 'umbrella narrative' of certain imminent global catastrophe, as propagated by virtually all public authority sources. The

* Although as 2023 dawned, both of these policies are faltering in the face of harsh realities.

† See p. 65.

measurements in Chapters 8–10 show that the Catastrophe Narrative has done its work; across nations, the majority attitudes to climate change are determined culturally – which is to say emotively. For instance, publics respond to unconstrained survey questions in proportion to their emotive alignment with the Catastrophe Narrative (either with concern/belief, or with resistance if there is innate scepticism, the ratio of these depending upon national religiosity).

Extreme advocates for climate catastrophism, of the kind predicted in Point 11, appear in the form of organisations such as Extinction Rebellion and School Strike for Climate. Their cultural motivation is revealed by the fact that their disposition across nations is directly related to national religiosities (see Chapter 13). Additionally, Chapter 4 explored in detail the capture of children by the culture of climate catastrophism, their roles as prophets and proselytisers for the culture, and the psychological damage that many suffer due to emotive pressure.

Self-identification with the culture of climate catastrophism (Point 12) is very hard to measure directly. The surveys used in this book can only measure the results of climate catastrophism's impact on publics as a whole. However, there is ample evidence that the culture is indeed engendering such intense emotional behaviours in many people.

Joe Duggan's 'Is this is how you feel' project,[297] is a useful source for such evidence. It supports action on global warming, and has showcased climate and environmental scientists' responses to the question: 'How do you feel about climate change?' The idea is that the feelings revealed would, overall, bring home the seriousness of the issue to readers, and so inspire action. This may well be the case, but the feelings expressed in many letters are highly emotive, to say the least. Moreover, such high emotion is not usually considered compatible with scientific objectivity. Two example letters are provided below.

The first[298] is from Anthony Richardson, a professor at the University of Queensland who specialises in climate change impacts and marine ecology. Note the use in the last line of the engaging anxiety for children variant of the Catastrophe Narrative:

> How climate change makes me feel.
>
> I feel a maelstrom of emotions
>
> I am exasperated. Exasperated no one is listening.
>
> I am frustrated. Frustrated we are not solving the problem.
>
> I am anxious. Anxious that we start acting now.
>
> I am perplexed. Perplexed that the urgency is not appreciated.
>
> I am dumbfounded. Dumbfounded by our inaction.
>
> I am distressed. Distressed we are changing our planet.
>
> I am upset. Upset for what our inaction will mean for all life.
>
> I am annoyed. Annoyed with the media's portrayal of the science.
>
> I am angry. Angry that vested interests bias the debate.
>
> I am infuriated. Infuriated we are destroying our planet.
>
> But most of all I am apprehensive. Apprehensive about our children's future.

The second letter[299] is from Peter B. DeMenocal, a dean at Columbia University, who specialises in geochemistry and paleoclimate studies. Note the use of the terminal metaphor variant:

> I'd like to start with an analogy that best explains how I feel. Imagine how a medical doctor feels having to inform their patient, an old, life-long friend, of a dire but treatable diagnosis. The friend angrily disregards what you have to say, for a variety of very human reasons, and you watch helplessly as the pain and illness unfold over the rest of their shortened life.
>
> There is a similar closeness between climate scientists and the planet. There's a sense of wonder and respect. Nations and economies don't like uncertainty. Climate change destabilizes the institutions we've built over centuries of stable climate and sea level.
>
> Returning to our patient, I feel frustrated that my friend won't listen. But I hope they will listen to other doctors and come accept the diagnosis. I hope that, for very human reasons, the patient will see this as a positive, life-affirming choice.
>
> I hope that we see ourselves as the patient.

In Chapter 6 we explored how the term 'denialism' – a misframing – has enabled demonisation (Point 14) to take place on a large scale, particularly in the climate-change domain. Legitimate questioners, the innately sceptical, and even some highly qualified climate scientists, are frequently handed the 'black-spot'* of the denier label.

The continuing doubts of large proportions of publics (Point 16) is clear in the measurements in Chapter 8. This can largely be explained by innate scepticism, the nature and origin of which was comprehensively covered in Chapter 7. As the explanations in Chapter 9 make clear, the attitudes of national publics to climate change only make sense if they stem from cultural motivations, which include cultural disbelief (innate scepticism) as well as cultural belief. Bulk innate scepticism of climate catastrophism is active by default in national publics, but is disabled for many of the religious because of allied belief, which in turn is established by the supportive position of religious leaderships on climate change. However, reality constraints reactivate it for the religious, and heighten it for everyone in proportion to the strength of the constraint.

An alliance of the culture of climate catastrophism (Point 17) is revealed in its close relationship with Lib/Dem supporters in the US, as noted in Chapter 11.

The wasteful expenditure of resources on climate catastrophism (Point 19) is revealed in Chapter 12, which shows that the commitment to renewable energy across nations strongly conforms to a *cultural* pattern, and has nothing to do with their climates or climate exposures, or any other scientific or technological considerations. Strikingly, the greatest national commitments to solar power are in some of the cloudiest nations, *where it is least useful.* On average, there is far less commitment to solar power across sunnier nations (Figure 25).

Professor Crusoe would likely guess that the physical science behind the consensus mentioned in Point 1 must be immature and subject to significant uncertainty. Otherwise, a strong reality constraint would prevent the emotive memes that power cultural enti-

* A sign of death for a betrayer in the pirate community, as used in the book *Treasure Island*; a metaphor for out-group demonisation and cancellation.

ties from developing and obtaining a grip on society. In addition, the professor knows that if a potent new culture is not actively opposed, it can change people's behaviour in the most fundamental of ways, as suggested by Point 9, potentially hollowing-out the law and even shifting the very moral ground upon which society is based. This important issue is explored below.

14.1 Morals and the law

As we saw in Chapter 3, perceptions of what is morally right or wrong are rooted in cultural identity. In a society with several cultures operating, we can imagine a scenario in which the relationships between them are approximately in equilibrium, resulting in a stable moral landscape. If a potent new culture arises, not only will its adherents introduce new values into society, it will also disrupt the entire equilibrium, altering existing cultural alliances and antagonisms and creating new ones. The overall effect is to shift perceptions of what is moral.

The wider the scope of the new culture and the deeper its social penetration, the more the moral landscape will shift. Behaviours that were once considered acceptable may become offensive, and vice versa. Some people will find this profoundly unsettling. Others may find themselves edged out of social circles or organisations in which they once felt at home, their (unchanged) values no longer considered morally acceptable. Some may be badly treated, perhaps even by those they once trusted. Others will be inspired by the new culture, embracing its moral values to become converts or allies; this may bring them higher status as the culture flourishes.

With its new set of values, a potent new culture will inevitably come into conflict with the system that is the guardian of existing moral codes: the law. If it is to achieve permanence in society, it must overcome this obstacle; it must change statutes in its favour, both those that integrate and encapsulate the moral values of the old culture, and quite possibly some that are based on rationality too.

This is no easy task; the law is designed to be hard to change. This is especially the case for its core principles, which must be protected from fads and cultural wrong turns and the whims of individual power-brokers. Overcoming this inertia is therefore a formidable challenge; one that will take a long time to achieve.* Nevertheless, the law retains sufficient flexibility to accommodate social evolution,† and this provides an invading culture with some opportunities.

As we saw in Chapter 3, cultures are emergent phenomena that are capable of galvanising populations through subconscious means. Their adherents can be inspired to challenge the law, as can allies, and even some who are theoretically unaligned but are caught up in events. Although the role of elites is often disproportionate, cultures – even those viewed by history as repugnant[300] – are not usually driven top-down; the behaviours of all these people are therefore mostly self-motivated. The cultural assault on the law will therefore come from all levels of society, from the humblest individuals to the most influential. A vast social order is mobilised, a kind of moral tidal wave directed at the legal system.‡ This is hard to resist, giving the culture a fighting chance of success.

The degree of motivation will vary with the strength of an individual's attachment to the culture. At one end of the scale are those who merely turn a blind eye to the culture's activities. In the middle, there is shallow support from cultural allies, while at the other end there is strong support from those who are culturally convinced. It is ultimately a tectonic shift in the moral landscape that inspires all of these people.

Because members of social institutions are not separate from wider society, but are embedded within it, they can be as culturally biased as anyone else. This means that cultural influence can creep into – and sooner or later dominate – any institution.[§,301] For the institution of the law, any legal professional – in the broadest sense: politicians (law-makers), civil servants, judges, barristers, and the police – might be affected. They all stand upon the moral landscape;

* Unless a society is already damaged and weakened by other events.

† Depending on the country; religious dictatorships, for instance, tend to be very inflexible.

‡ Although the wave is subconsciously coordinated overall, there may be some patches of conscious coordination, for example where organisations freeride on cultural movements, or where due to noble-cause corruption, groups of particularly ardent adherents conspire to subvert the law.

§ For institutions, such as the law, that are designed to be hard to change, this would typically take longer.

as this shifts beneath them, maintaining an objective stance becomes difficult.

The nature of the assault on the law can also vary, from head-on challenges – lobbying politicians for new statutes or attempts to change the interpretation of existing ones through the courts – to subversion of the legal system – circumventing and demonising laws, preventing them from being upheld,[302] recruiting officials who can be proselytised and undermining those who cannot. Cultures will often present the changes they seek as modest and benign social evolution, rather than fundamental moral transformations that might create more problems than they solve; liberal democracies in particular are vulnerable to such representations.

A cultural assault is indiscriminate, at any time potentially impacting any relevant statutes and regulations, any officials, and any jurisdiction. This includes associated fields such as legal academia, and the assault may be international too if the culture has a presence in more than one country. Culturally relevant statutes and regulations are not only those that integrate and encapsulate the moral values of the old culture, but also some that are based on rationality. So the new culture will attack both, attempting not only to establish its moral authority over other cultures, but also to impose its arbitrary perception of reality. The latter case can lead to the ridiculous spectacle of cultures using the law to try and 'settle' scientific questions,[303] a notorious example being Philippe Sands, the legal academic who proposed using the courts as a way to end scientific discussion on climate change:[304,305]

> One of the most important things an international court could do – in my view it's probably the single most important thing – is to settle the scientific dispute.

Even more audacious are occasional attempts to simply overrule science. A recent example was when the Scottish National Party, captured by extreme trans rights culture and in contradiction of biological fact, attempted to legislate that men could become women (and vice versa) purely on the basis of a simple self-declaration.[306]

So, when a new culture is on a moral collision course with the legal system, this will manifest itself, as Professor Crusoe knows, as direct assaults on the law and on sceptics of the culture:

- There may be calls for that which is both accepted and lawful, but not aligned to the new moral compass, to be outlawed.
- More extreme commentators may call for the very system upon which the law rests to be changed or abandoned (for instance a revolution, the abandonment of democracy, or a major schism within a religious system/society).
- There may be calls for organisations or individuals who question the new culture – and/or its aspirations regarding the law – to be silenced, and perhaps socially 'cancelled'; this discourages people from resisting changes to the law.

And also as a variety of subversive activities:

- Adherents may bend or break existing laws that don't align with the new moral compass; others, working within legal system, will turn a blind eye.
- There may be extreme clemency for those who are successfully prosecuted. They may even earn an informal 'badge of honour' for their sacrifice, or a reward behind the scenes.
- There may be demands that organisations or individuals whose interests are counter to the new culture be penalised in some way, generally via new interpretations of existing law (but failing this, possibly via the introduction of new laws to this end).

Subversion – frequent bending or breaking of a law – may produce a de facto reality that will ease the way to actual legislative changes.

All these activities affect public and private law, along with the rules and regulations that govern conduct in all kinds of organisations. While the upper half of the list is mainly about pressuring for changes to the law, and the lower half is mainly about biased enforcement of existing law, the same cultural motivation is behind both. Both sets of actions can occur concurrently, and may sometimes be enacted by the same people.

Because, as noted in Chapter 3, cultures are polarising, there will always be some sceptics who resist these activities. Most of these people are likely to be innately sceptical, which is to say that their ultimate motivation is an instinctive reaction to the cultural invasion. However, there will be a minority of rational sceptics too, for instance legal professionals who see cultural bias creeping into the making and enforcing of law, and who will fight this tendency on principle, rather than as a matter of either belief or disbelief in the relevant cultural narratives. All those who resist will do so by a variety of activities, which will include attempts to stem the cultural advance through the diligent application of current law.

This resistance, which may be spirited, means that not everything will go the new culture's way. Self-righteous adherents will fall foul of the law if, for example, they go beyond legal and peaceful protest to achieve their ends. They may also do so where they adopt behaviours that are sanctioned by the new culture, but which are seen as immoral by the older one.

If a culture has made significant progress – even gaining influence in key institutions – but has managed to do so 'below the radar', the sudden exposure caused by, for example, achieving some limited statutory changes, can lead to a backlash. This can include a clampdown on its activities, and the launch of legal suits against organisations working on its behalf.[307] The exposure of cultural hypocrisy, along with fines, sanctions, and possibly the sight of zealots being carted off to prison, can then break its cultural 'spell'. From then on, it will be on the defensive, and the public may ultimately turn against it, preventing it from becoming dominant in society.

However, unless the new culture is completely crushed, it will not cease its activities. Cultures can be relentless; if one statute is resistant to change, another will be targeted; if one organisation or sector of society proves impervious, there are many more that may prove more vulnerable. This means that change will normally come about slowly and iteratively. But if pressure on the law can be sustained for long enough, statutes may eventually be reshaped. Any such changes may become a precedent that brings further advantage; in other words, success may breed more success. And precedent can be social as well as legal. For instance, a failure to forcefully police protests (in favour of the culture) can become a habit, as can a refusal to protect the rights of dissenters.

Eventually, the culture may be able to change enough laws to deliver a lasting dominance in society. Alternatively, its fairy-tale cultural narratives might at some point lead it into a head-on crash with reality, resulting in a major reversal of its fortunes.

All the points in the bullet list above occur in the climate domain, confirming the rising wave of climate catastrophism against the law. Box 4 lists examples, in the same order.* We can imagine each one occurring as a warning light appearing on a dashboard. For any single light, it is possible to debate whether the activity is justified in a particular context – but that isn't the main point. The fact that the whole dashboard is lit up like a Christmas tree, with multiple examples for each activity type, strongly suggests that we are dealing with a culture, and hence ultimately a fairy-tale narrative that cannot be true; one which defies reality. This accords with other evidence for the climate domain, such as the measurements of social data in this book. Hence the pressure that these activities cumulatively bring to bear upon the law should be resisted, whether or not an opposing position can be properly or fully articulated (there may not be sufficient data to do this).

Our scenario of a potent new culture in a previously stable society makes the principles easier to grasp, and also matches the rise of climate catastrophism, the main topic of this book. However, it is worth noting that all cultures that have sufficient influence, potentially at any stage of their development, will work to undermine or neutralise laws that harm or limit them, while creating and/or defending laws that promote their interests.

14.2 We need to acknowledge: it's a culture!

Professor Crusoe can predict an extraordinary amount about the climate-change movement, simply from knowing that it is a cultural entity. This therefore has to be the most important single fact one could possibly know about it.

If this fact isn't grasped more widely, especially by the relevant

* This list is perfectly adequate to demonstrate the point. Hence it hasn't been updated for any further examples occurring after 2015, when it was assembled.

Box 4. Law undermined by climate catastrophism

1. *Outlawing and banning things*

   - Fossil-fuels. From a coalition of hundreds of scientists, advocates and environmental groups to President Obama: 'We call on you to make our nation the first to commit to keeping all of its remaining, unleased public fossil fuels in the ground, thereby challenging other nations to do the same.'[339]
   - Beef. From the '*Sydney Morning Herald*': 'Some say cows are killing the earth. So do we need to ban beef?'[340]

2. *Democracy on hold or removed*

   - Scientist James Lovelock recommended putting democracy on hold while tackling climate change.[341]
   - Mayer Hillman, senior fellow emeritus at Britain's Policy Studies Institute, has recommended overriding democracy to better fight climate change.[342]

3. *Silencing dissent*

   - A British academic demands an international court declare climate sceptics wrong, once and for all.'[343] (This is another angle on the Professor Sands case from above).

4. *A blind eye*

   - There is a long record of the law being trampled to enable large-scale renewable energy deployment, in the name of saving the planet. See the efforts of Irish campaigner Pat Swords in this area.[344]

5. *Extreme clemency*

   - Senior scientist and head of the Pacific Institute, Peter Gleick, was caught committing and then admitted to, the federal felony of wire fraud (creating a false identity in order to steal private confidential documents over the Internet). This was done to advance the cause of climate change against what he considered a climate 'denier' organisation. He faced virtually no sanction,[345] and the criminal act represents more a badge of honour than a stigma in the climate community.

6. *Penalise conflicting interests*

   - Trying fossil fuel CEOs: James Hansen, a leading climate scientist, has demanded that chief executives of large fossil fuel companies be tried for crimes 'against humanity and nature'.[346] These efforts fizzled out after a few years, but begin again from time to time.[347]
   - The RICO 20 case: A US senator proposed using racketeering laws against climate change skeptics and fossil fuel companies.[348] His demand was supported in a public letter to President Obama by 20 climate scientists. This effort also fizzled out.

7. *Thwarting codes of conduct*

   - US Environmental Protection Agency officials colluded with environmental activists to use so-called 'sue and settle' tactics and to prevent the public being involved in development of environmental regulations.[349]
   - Funding bias in climate science. Analysis of US Federal budget documents has suggested a strong bias in favour of projects supporting the idea of human-induced climate change.[350]

social sciences, everything we know about cultures will have to be learned again, in a climate-change context, the hard way: climate catastrophism is already having a profound impact on our society – our morals, our laws, our way of life, and even our physical infrastructure. If we fail to recognise its cultural nature, we will not be able to prevent it extending and tightening its grip on society. We won't be able to put a stop to the distortion of science by emotion, still less will we be able to ensure that policy measures are rational and net beneficial.

Some might suggest that the downsides of catastrophism are justified as a means to an end, namely the environmental benefit. However, it is crucial to weigh the fact that cultural entities do not want* to solve 'the problem' that their narratives so emotively trumpet; they use up money and resources on a grand scale, but only for their own benefit. In other words, cultural entities are sustained by the continuous expression of high emotion about 'the problem', which a solution would kill. They may even work against their ostensible goals; cultural hypocrisy can do that. No one should want such a situation, whether they are very climate-change concerned or climate-change sceptical.

* The language here should not be taken as implying that cultures are sentient or agential.

Chapter 15

HISTORICAL COMPARISONS AND SOCIAL IMPACTS

15.1 Cultural entities and their shaping of society

In this book, I have explained how cultural entities work, and described the behaviours and attitudes they produce, with particular reference to social conflicts about science issues. We have seen that innate scepticism can be either apt or inapt, but that these are manifestations of the same underlying behaviour. We have seen how innate scepticism can be bypassed by cultural belief. We have also seen that most accusations of 'denialism' are out-group demonisation, that conspiracy and dishonesty are not the problem, and that rising cultural entities often use and abuse children. Appendix I.i summarises the 'rules' for cultural entities.

I have also shown that climate catastrophism is a cultural entity in its own right, with a ubiquitous presence that shapes public attitudes in the domain, although in the US, political tribes heavily influence views too. In essence, what matters everywhere is 'cultural identity'. We have seen that most support for action on global warming, as well as most resistance, is irrational, and that the culture interacts in a consistent way with all the main religious faiths. Finally, we have seen that the physics of the ocean-atmosphere system, and the findings of those who study it, are no longer relevant to the policy process, which instead has become an expression of the penetration of the culture of climate catastrophism into societies around the world. Appendix J.iii summarises the social psychology of the climate domain.

15.1.1 Living with cultural entities

To many readers, the phenomena measured and explained in this book will not be grasped intuitively. But nothing could be more normal for humans than to interact with cultures; we have experienced thousands of them – mostly religions, and, in more recent times, secular ones, such as Communism and Fascism, as well. This relationship may be confined to a relatively narrow social domain, but sometimes it is experienced more widely. For instance, religion once dominated all aspects of our lives, and for many people it still does.

Although some can be harmful, cultural entities have, overall, been a huge advantage in our distant past (Section 3.3). This is exactly why we're so susceptible to them, and why their simple narratives so easily bypass our rationality. However, in more modern times, the situation is less clear. On the one hand, there has now been an enormous accumulation of knowledge and the development of powerful technologies, which cultural entities are sometimes able to wield. However, they will do so irrationally, potentially causing huge damage to society and/or the environment. On the other hand, we now also have a variety of rational institutions – the law, democracy, science – that act to limit cultural excess. There is a constant war between rational institutions and the culture, with the former trying to tame the latter, and the latter trying to undermine the former. Climate catastrophism is an excellent and well-documented example of this struggle.

15.2 Living with climate catastrophism

So what will climate catastrophism bring in the future? Emergent systems are unpredictable, but because the same underlying mechanics are shared by all cultural entities, probable developments can be characterised in broad terms, with perhaps some tentative likelihoods placed on each.

We begin this process by looking at historic cultures. These provide the only detailed and realistic, indeed real, models that we have for how cultural entities express themselves in practice. We are interested in those aspects of historic cultures that had major impacts on society, so that we can consider how similar developments in climate catastrophism might affect us today, taking into account the differences in narratives and the different constraints that are now in operation. Many historical cultures have had destructive phases or tendencies; some have been harmful overall. Given this experience, we focus more upon these negative aspects.

Before we delve into history, it is important to note that such comparisons require great care. The context is important. While the underlying mechanics of all cultures are the same, allowing valid comparisons to be made, their surface expressions can be very different, as can the societies in which they operate. In particular, there will be differences in:

- the core narrative for each culture, which dictates who and what is considered to be either 'good' or 'bad'.
- the societal constraints on the culture.

These differences matter as much to outcomes as the common mechanisms that drive cultural expression. So while it is perfectly valid to observe that a contemporary culture demonises out-groupers, as all historic cultures have done, the differences in the narratives and the societal constraints mean that those targeted, how aggressively, and the outcomes, may be very different – a specifically racist narrative might have appalling consequences, as might a lack of societal constraints. In other words, it is not safe to make simplistic assumptions about who will be targeted for prejudice by one culture, or the level of intensity involved, from the example of who was targeted in a historic culture.

15.3 Features of historical cultures

15.3.1 Structural icons

Major cultural entities tend to produce large physical constructions, reflecting some aspect of a cultural narrative. Such structural icons, some of them very large indeed, can be important in the mechanics of the culture, helping in the process of binding adherents. However, they can also soak up extraordinary amounts of money and other resources, for little practical gain. They may even cause significant social damage. However, given that cultural entities are 'blind' – sustained by instinctive behaviours – such practical downsides may not prevent the projects from going ahead.

As suggested in Chapter 3,* structural icons can be of net benefit to their societies even when they lack a practical purpose – as in the Egyptian Pyramids – but they have a greater chance of being so when their cultural purpose is combined with a practical purpose, as with the irrigation systems built by early Sumerian city-states. Mediaeval cathedrals appear on the surface to be all about culture, yet one can argue that they are not wholly so. They stimulated the economy, inspired artisanship and artistry, acted as centres of learning and alms distribution, and provided shelter for the oppressed. In other words, their practical and cultural aspects were intertwined and, despite the huge cost of building and maintaining them (and supporting those who worked in them), it is at least a possibility that they represented a net benefit to their societies in purely practical terms. And with the cultural value of unifying society taken into account, the benefits would surely have outweighed the costs.

It is harder to make a positive case for some cultural icons, for instance communist mega-structures and brutalist architecture. Although liberal democracies are perfectly capable of producing brutalist works too, they have not typically done so on the same scale, and not at the careless expense of workers' lives, as was the case for some communist projects. Having said this, the majority of show-piece projects littering Russia and Eastern Europe probably caused little major harm except a waste of resources. However, huge iden-tikit concrete residence blocks, owing as much to ideological motives as to economic ones, probably caused considerable social harm. On a much darker note, there were about 10,000 deaths, along with mass privation and illness, during the first five years alone of construction of Russia's planned city of Magnitogorsk. This was a high price to pay, but at least the city eventually achieved its practical purpose, something that cannot be said of the Soviets' White Sea Canal, which was built too narrow and too shallow because of a determination that it should achieve its cultural purpose as an icon for the regime – it was to be delivered on such a heroic timescale that it would unite the populace behind the communist cause and boost Stalin's personality

* See p. 30.

cult. At least 25,000 people died during its construction.

So, in modern secular cultures, structural icons may progress despite major net downsides, because their practical purpose helps to hide their true purpose. And while a cost–benefit assessment isn't necessarily meaningful when future economic benefit is being measured against lives lost,* a strong cultural bias can make such an assessment seem far more reasonable, also tipping the scales towards building the icon. The overall cost-benefit equation of structural icons is therefore not straightforward, but any comprehensive analysis should consider, as a minimum:

- To what extent do they benefit society, rather than burdening and damaging it?
- To what extent do they satisfy practical rather than cultural purposes, and to what extent is the former merely a cover for the latter?

Only in answering these questions can the cost–benefit equation of cultural structures be assessed.

15.3.2 Hijacking the authority of science

As noted in Chapter 7 regarding the long-lived false orthodoxy about what causes stomach ulcers, even groupthink (culture writ small) is perfectly capable of overcoming evidence. It is therefore well within the power of a major cultural entity to subvert and hijack an entire field of science, and everyone involved in it.

Particularly when a new area of research is perceived to have social importance, public movements can form in support or opposition. The biased beliefs that result can rapidly start to shape attitudes much more strongly than any scientific findings in a field that is, after all, in its infancy. With the support of even a few prominent academics to lend a veneer of scientific respectability, such movements can win widespread public support and morph into full-blown cultures (or be sucked into existing ones), at which point the beliefs will become locked in as an unassailable orthodoxy.

A well-documented historical example is the theory of eugenics.

* Those killed along the way are often dismissed as being unworthy; those that the culture despises anyhow.

This started as a genuine field of scientific inquiry but, at the beginning of the 20th century, a public movement started to form around some of its ideas,[308] and it inevitably became cultural. Some scientists were taken in, and started to communicate those ideas – including some of the more extreme ones - in an increasingly emotive and overly certain manner. However, their support maintained public credibility for the theory. As a result, the theory-cum-cultural-narrative prospered, with public authorities of all kinds eventually signed up to it, believing it had unequivocal scientific backing. In central Europe, it became a component of another cultural entity, the fascist movement. While it was only a part of the full set of fascist beliefs, it was nevertheless important because it was used to justify anti-Semitism and other racial policies. The supposedly certain scientific justifications that eugenics brought to the culture represented a complete hijack of the authority of science.

Because trust in science appears to be a default public position, a culture that can hijack a scientific issue in this way will likely gain widespread support (other things being equal). This does not mean there won't be plenty of (apt) innate scepticism within publics too, but unless the hijack is exposed for what it is – as eventually happened for eugenics when the appalling scenes from Nazi concentration camps were broadcast – support for the culture will grow, and scepticism will be suppressed.

Alliances of science with cultures – even though the cause may seem noble – should be viewed with great suspicion.* The culture, not the science, will typically run the show. How far from reality a culturally hijacked science will stray, and how damaging the acts that it will attempt to justify, depends in part on the strength of the beliefs that are controlling it. In the case of eugenics, the fascist culture of which it became a part was potent, and the hijack took place at a time when society was fragile, having suffered major traumas: initially the First World War, which Germany lost, and then the economic collapse of the early 1930s. As a result, there were few limits on the horrors that could be justified through science.

How far events may go also depends upon the extent to which

* See Section 7.4.

scientists are willing to turn a blind-eye towards spurious cultural claims. Many will perceive the disparity between actual science and what is promoted in its name. However, speaking out risks demonisation, financial disadvantage, cancellation, and perhaps even physical threats. The greater the grip that the culture can exert on society, the greater the risk researchers face, and the greater the likelihood that they will keep quiet.

15.3.3 The use of children

Cultural entities, especially new ones, commonly target children as a source of easily programmable adherents, unless society actively works to stop them.* As a result, once captured, domineering but very inexperienced children can end up dictating the agenda of a culture, shifting it onto a new path and probably more extreme positions.

Worst-case historic examples of this phenomenon make it much easier to see the signs that should serve as a warning in more subtle cases. It is also instructive to see just how excessive the cultural behaviours of children can become, and how terrible is the damage they can cause when societal constraints are eroded or break down completely. However, to repeat, these examples will not be used in a simplistic manner; any comparisons with contemporary scenarios must consider dissimilar motivating narratives and social constraints.

15.3.3.1 The Red Guards

One historic example of domineering children in a culture is the Red Guards,[309] the communist youth movement in China that terrorised the country in the late 1960s. Millions strong, it drew most of its recruits from among older school students (12–18 years), but also had members in elementary schools and universities. It originally arose spontaneously, and later gained the official approval of the Communist Party, which aimed to direct it. However, it eventually span out of all possible control, by the Party or any other body.

Naïvely pursuing absolute cultural purity, the Red Guards aimed to eliminate the 'four olds': old ideas, customs, culture, and habits

* See Section 3.2.4 and Chapter 4.

of mind.* In the process, they took a wrecking ball to society, burning books and trashing museums, along with historical and religious sites. They shamed and abused teachers, intellectuals, and those officials perceived as insufficiently supportive, and eventually moved on to violence and torture, murdering thousands of people they considered 'counter-revolutionaries'. Many more were humiliated in the notorious 'struggle sessions', some going on to commit suicide.

Their activities having been blessed by the Party, the Red Guards met with little resistance, including from the police. This hugely loosened the social constraints on their behaviour, to the extent that even high party officials began to be subjected to abuse and violence. Growing factionalism between different Red Guard groups was soon threatening to spark a civil war. Realising that its position was under threat, the Party hierarchy forcibly disbanded the movement, driving the children into the countryside, supposedly to learn the pure values of rural life, which in most cases simply meant hard agrarian labour. However, by the time order was restored, the Red Guards had already caused profound damage to society and had also changed the direction of Chinese communist culture, and not for the better.

15.3.3.2 The Hitler Youth

Although highly activist, the Hitler Youth did not subject adults to torture and murder. However, they were not beyond betraying adults – and on occasion even their own parents – to the authorities for anti-Nazi behaviour, sometimes with very serious consequences. Being much more a tool of the (Nazi) cultural hierarchy, the Hitler Youth did not degenerate into anarchy in the same way as the Red Guards, and so did not suffer the same fate. They remained a long-term source of recruits for the regime's enforcers: the Stormtroopers and the SS. In this way, their naïve and purist vision, and their willingness to break with the past, helped to steer and sustain the whole of the wider Nazi culture.[310] This is a powerful reminder that children who are captured by a culture can grow up to be very zealous adults.

15.3.3.3 Children are always used by cultures

Cultures all work via the same underlying processes, and the emo-

* This is essentially a millennarian pursuit; millennarian cultures are covered later.

tional power of children in society means that youth movements (formal or informal) are an almost inevitable consequence of their emergence. That said, it is unlikely that youthful adherents generally will become as extreme or militant as the Red Guards or the Hitler Youth, unless the culture from which they arise is itself extreme – perhaps based on narratives of class war or racism – or if societal constraints have already been seriously weakened.

Nevertheless, even without becoming extremists, children captured by a culture can still have a major impact, shaming adults and setting or influencing agendas. And when they grow up, as we have seen, they may well become very zealous adults, committed and convinced far beyond reason. We should therefore be alert to these characteristics and the associated risks. That means, when we see activist children, we should not be fooled into thinking that they haven't fallen under the influence of a culture, even if they are not extremist. In time, they may still become so. Nor should we have our misgivings alleviated by assurances that there is a genuine, non-cultural cause to the behaviours (such as 'saving the planet'). None of this rules out cultural capture, which is a risk, and should be resisted while it is still possible to do so.

When the influence of cultural children becomes sufficiently strong, there are likely to be severe social ramifications. Generally speaking, children think less about consequences than adults, and they can push cultures – already irrational – to extremes.

Even without the influence of the children, cultures can cause great harm to society. The nature and extent of those harms depends on the details of the narrative. Where it targets society or the economy, damage will be inflicted directly, as we saw with the Xhosa and their young prophetess Nongqawuse.* Where this is not the case, damage will be collateral but may still be extensive. In either case, this can include torture and the suppression of liberties, and lead to deaths numbered in millions, often with vulnerable children as victims. Injecting cultures with a high dose of youthful and irrational zeal is therefore highly undesirable. Good outcomes are not likely when leaders bow to the demands of culturalised children.

* See p. 44.

15.4 Comparable features in climate catastrophism

15.4.1 Structural icons

We saw in Chapter 12 that national commitment to renewable energy is primarily a function of cultural factors, not of scientific or engineering ones, or indeed of anything rational. However, they have some genuine utility too; they do actually produce power. So to what extent do renewables, built in support of catastrophist culture, burden and damage society as well as stimulating it? To what extent is power delivery their real purpose, rather than just cover for an underlying cultural imperative?

The practical utility of wind and solar installations is strongly restricted by their persistently high costs,[311] a burden on industry and the public alike. Further problems arise from our inability to store electricity in large quantities and at low cost, and by renewables' inability to produce power at all times. This means that a fully capable network of traditional power sources must operate in parallel. However, that end is hard to deliver, because the presence of subsidised renewables undermines the economics of all other energy generators (and destabilises the grid as well).

Secondary problems include the killing of birds,* the littering of landscapes with steel and concrete, and the environmental and human impact of all the exotic materials extracted to make the turbines. Many of the downsides of renewables are exposed in Michael Moore's film *Planet of the Humans*;[312] which, while coming from a position of cultural belief (climate catastrophe is taken for granted), attacks wind and solar as an 'illusory solution', one which delivers only further damage, not salvation.†

The benefits of reduced emissions are, meanwhile, rather limited, particularly because the big polluters – China and India – show little sign of following the decarbonisation lead of European nations.

Wind and solar installations are therefore undoubtedly structural icons, more akin to cathedrals than to power-stations (the Chapter 12

* Windfarm advocates frequently cite the large numbers of birds killed by cats and collisions with other manmade objects. This is true, but windfarms have a particular impact on rare upland species, and passage migrants.

† Technically, this makes Moore's film a cultural heresy.

measurements tell us this). In most places,* their practical utility is little more than a fig leaf to hide their true cultural purpose. Without the motivating cultural narrative of catastrophe, it is unlikely that they would ever have been deployed on a wide scale, and probable that the world would have seized on adaptation as the only rational response to climate change.

The downsides of renewables will remain obfuscated; it is in the culture's interests for this to be the case. So while there are unlikely to be mass deaths from renewables, as there were in the construction of the communist icons, other social and environmental sacrifices – including loss of life, for example among those who cannot afford to heat their homes, or those who mine for the exotic materials required for the renewables revolution – will remain hidden. An overwhelming cultural bias across society will prevent publics from perceiving the true cost – the true level of reality constraint – in respect of renewables. If it were otherwise, support for these technologies would all but disappear, as the measurements and explanations from Chapter 8–10 reveal.

15.4.2 Hijacking the authority of science

The cultural narrative of certain global climate catastrophe has long-since hijacked the authority of science. The UN elite, presidents and prime ministers, the heads of the major faiths, CEOs, academics and economists, along with the great majority other leaderships and politicians and major influencers around the world, frequently state that certain global climate catastrophe is an unquestionable prediction of 'the science'.† As we have emphasised throughout this book, these statements are false.‡

The hijacking has been even more audacious than was the case with eugenics;§ the gulf between mainstream climate science and what the culture of climate catastrophism claims in its name is much wider. For eugenics there was no consensus at all, while in the climate domain there is one (as expressed in the IPCC technical reports), but

* Off-grid, or in Norway where almost limitless pumped water storage is available to smooth out intermittency, wind and solar may be viable at reasonable cost.

† See Endnote 3 in the CN-Archive for example quotations.

‡ See Section 5.2.2,

§ As described in Section 15.3.2

it is very different to what is claimed in the cultural narrative, saying nothing of global catastrophe.

As a result of the hijack of science, the Catastrophe Narrative is rarely challenged. Those most able to do so, namely mainstream climate scientists, would put themselves at great risk if they spoke out, and the majority seem to turn a blind eye to the excesses of the culture. The small minority that do speak out – academics such as Roger Pielke Jr and environmental commentators such as Bjørn sLomborg – have found to their cost that to quote the IPCC as evidence against claims of certain climate catastrophe merely invites demonisation.

With the hijack in place, cultural irrationality inevitably results. It is difficult to assess how bad things might become, but the historical examples above suggest there are few limits to what can be (falsely) justified by 'the science'. Moreover, climate catastrophism is new, and it is still growing; there is no way to predict its longevity or how strong it might become.

It is some comfort that most of the irrational demands that are currently aired as a result of the Catastrophe Narrative do not actually result in correspondingly extreme policies, but a significant number still do. For instance, while many factors led to the collapse in 2022 of the Sri Lankan economy, and the resulting chaos, violence, and extreme hardship, the climate Catastrophe Narrative played a big part. What turned a crisis into a disaster was the banning of synthetic fertilisers, an extraordinary step in a largely agrarian nation. According to the country's president, the policy was part of a net zero programme, and was principally motivated by a desire to reduce the greenhouse gas emissions associated with fertiliser manufacture and use.[313] Within a season, crop yields had collapsed,[314] and while the ban was later partially reversed, the damage had been done. It was an irrational and cruel act, stemming not from science but from climate catastrophist ideology.

Such policies are not limited to Sri Lanka. At time of writing (2023), extraordinary fertiliser reductions are being demanded, on crash timescales, in Canada, Holland and elsewhere, just when the grain supply has been reduced as a result of war in the Ukraine, and prices driven up further by inflation and high fuel costs. The policies have provoked mass resistance and unrest, just as they did in Sri Lanka.

On the upside, the culture of climate catastrophism does not include explicit racial components, so we shouldn't expect to see anything like gas chambers in action. But giving free licence to cultural narratives will always allow fairy tales and irrationality to prosper. Appalling anti-human and anti-technology memes, of the kind that produced the cruel policy decisions seen in Sri Lanka, are therefore inevitable.

There can also be a variety of consequences that are deleterious, but fall short of outright disaster. Exactly what these are will depend on the particular culture and the biases it produces. So, in the climate domain, we see a bias against the Western nations, which leads to a spurious assumption of their culpability for global catastrophe and yet a 'free pass' given to China, with its massive greenhouse gas emissions. Similarly, some energy policies turn out to have awful consequences, but on the environment rather than directly on human beings. This too suggests that they are the result of cultural bias. Two examples are the clear cutting of forests in the south-western US for manufacture of wood-pellets, which are shipped across the Atlantic, and then burned in the largest power-station in England,[315] and the replacement of natural tropical forest with monocultures of oil palm – another energy crop. It is quite possible that neither process reduces emissions overall, but if rationality ruled, we would not proceed with these policies unless we knew for sure. Indeed, even if we were certain, we might still not proceed if the costs of doing so outweighed the benefits.

Eugenics directly targeted many people. This is not currently happening in the climate domain, but cultural hypocrisy is still off the scale. As a result, people and the environment will still suffer. So while the culture of climate catastrophism remains dominant, and is backed by the hijacked authority of science, we should expect far more damage to occur.

15.4.3 The use of children

In making comparisons with history, we must bear in mind the differences between cultures, as well as their similarities. There are still strong societal constraints on climate catastrophism, which limit the possibility of movements of activist children spiralling out of control. In secular nations, the rational institutions – democracy, the law and science – have been damaged by the new culture, but still provide a check on its progress. In more religious places, the controlling local faith still holds sway.

Additionally, climate catastrophism mostly attacks governments, businesses and other organisations – especially those involved with fossil fuels – as well as systems, such as agriculture, transport, or (via agenda incorporation) capitalism, rather than targeting ethnic or class divisions in society.

As a result of these two factors, a green equivalent of the Red Guards seems a remote possibility.

15.4.3.1 Culturally captured children are undermining society

However, children captured by climate catastrophism are nevertheless a significant problem for society and its future health. They are engaged in an array of activities that are undermining society:

- Giving bold orders to governments and other authorities (think Greta Thunberg), and, despite the demands being irrational and a threat to civilisation, often having them meekly accepted (see Chapter 4).
- Engaging in lawfare – suing governments for alleged climate-related breaches of human rights to life,[316,317] and seeking criminal sanctions and Nuremberg-style trials for those trading in fossil fuels.[318]
- Shaming adults for everyday behaviour, which they believe is causing planetary harm.[319] Even if true, none of this would have been understood by the majority of adults for most of their lives. They are not culpable.
- Putting emotive pressure on parents to acknowledge global catastrophe and act accordingly, perhaps even to the extent of becoming adherents themselves.[320,321]

Far from receiving pushback, the children are generally praised, even by those they are upbraiding. Many adults support the children's aims, and sincerely so: they do not realise the children have been captured by a bizarre culture. Their support legitimises the activists' emotive and irrational pleas, and spreads their influence.

The media is playing a similar supportive role, as a few quotes from national print publications indicate:

> *The school climate change strikes are inspiring – but they should shame us.* 'Children skipping classes to take the moral lead is an indictment of adult complacency'.
>
> Jonathan Freedland.[322]

> *Shame on us for forcing children to wake us up to climate change.* 'We have failed to protect our children – now we are looking to them to protect us. Are we so far sunk into indolence and fatalism that we need our own children to save us?'
>
> Fintan O'Toole.[323]

> *Children are right to youth shame our leaders.*
>
> Tim Soutphommasane.[324]

> *Climate change is not only worrying kids; it's making them angry.*'Students are horrified at what they see as the ghastly ignorance and unforgivable inattention of their elders who wrecked the Earth only to pass it down to their children'.
>
> Alfred Lubrano.[325]

> *A new generation of activists is taking the lead on climate change.* 'Meet the teenagers who are marching in the streets, putting adults to shame and demanding action'
>
> Tessa Stuart.[326]

These supportive stances are typical of governmental and intra-governmental organisations too. For instance, the UN appears to have surrendered its authority to the child activists, even though doing so diverts it from its primary duty to mankind as a whole and leaves it the servant of the new culture. For example, in a statement about youth involvement in climate action, it says:

> Young people's unprecedented mobilization around the world shows the massive power they possess to hold decision-makers account-

> able. Their message is clear: the older generation has failed, and it is the young who will pay in full – with their very futures.[327]

Academia is supportive too. For instance, a developmental psychologist and a young climate activist together wrote that:[328]

> ...the climate crisis is imposing a heavy psychological burden on children and youth, both from experiencing climate-related disasters and from the knowledge that worse is to come. We then describe the global movement of youth demanding urgent climate action. We conclude that health professionals can support young people in many ways, but particularly by supporting their capacity to take action, raising awareness about the impact of the climate crisis on youth mental and physical health, and taking action themselves to work for a secure climate future.

On the upside, they at least acknowledge the psychological burden on children, but their path of cultural affirmation can only make a bad situation worse.

As can be seen in the quotes on the previous page, while some orthodox climate activists consider the tactic of shaming adults to be a good thing, others feel that it is unlikely have much impact on greenhouse gas emissions, and may even backfire. However, they all miss the key point, which is that emissions reduction is not the aim of shaming. The fairy tales of cultural narratives – in this case, the idea that dramatic emissions reduction will deliver salvation from catastrophe – are always false. The real reason for shaming opponents is that it helps to define a cultural boundary, sucking waverers in (through guilt) and emphatically rejecting out-groupers. This is helpful to the cultural entity, but of course also undermines the unity of society, not to mention rational consideration of policy.

15.4.3.2 The climate children knocked off course

Emergent systems are virtually impossible to predict, with or without complex models. This makes it hard to say where the children's movement will end up. However, we do know that the children have been placing increasingly strident demands on adults in general and on authorities in particular. In response, societal leaders have been appeasing them, but this only places them in the precarious position of taking flak for undermining society's infrastructure (which by one route or another will create physical harms), while still being berated by the culturally convinced for doing 'nowhere near enough' to avert certain catastrophe (because no appeasement could ever go far enough to achieve this ideological aim). Children will still pour scorn on them.

We also know, from the historic cases above, that, in the absence of strong constraints, cultures will go to extremes, and they will harness children as a means to take them there. So unless society does something to prevent it happening, there is a real possibility of a clash between youngsters and authority.

For the moment, however, the movement has been knocked off course by the appearance of Covid-19. A *Guardian* article recorded the status of the agenda-setting role of children pre-Covid, and how, on the eve of the pandemic, they had reached a peak of influence:[329]

> Teenagers such as Vanessa Nakate and Greta Thunberg became household names; both of them appeared on the cover of *Time* magazine, the latter as person of the year in 2019. This coincided with a moment in climate politics that was awash with ideas around children, the future, and intergenerational justice. Extinction Rebellion activists used the next generation as a proxy for the future: climate action in the present was a moral necessity for our children and grandchildren. Politicians also adopted this framing. At the same time, young people were taking matters into their own hands. For a time, it seemed that a climate movement was emerging in which children acted simultaneously as the spark, inspiration and energy. This wave seemed unstoppable.

Indeed so; it looked like domineering children, radicalised by the culture of catastrophism, were poised and ready to dictate world policy on climate change. The *Guardian* frames this in a positive way, as though it represented a rational societal response, suggesting that 'climate breakdown is not a future issue, but a devastating present reality for millions in the global south'. However, even if this were true, children are the last people from whom we could expect rational policy ideas on issues that touch every aspect of our civilisation. Moreover, 'climate breakdown', – certain global climate catastrophe – is in any case a cultural narrative (variant). Submitting civilisation to the agenda of radicalised children is irrational on a grand scale, although as we know from historical experience, perfectly normal for cultural entities. The *Guardian* extract is a great description of this chapter of their playbook, right down to the apparent 'moral necessity' and children acting as prime agents.

However, when the virus struck, the wind was taken out of the movement's sails, and its moment seemed to pass. The *Guardian* ascribes this to several factors, of which the main one is the chilling effect of the pandemic on child activism.* While considering this a lost opportunity, it adds that, in a different guise, the children will be back:

> ...instead of using youth as a rallying flag, young people across the country are building movements around radical policies or clear political campaigns. These are both harder to co-opt, and more coherent than the loose association of a movement based on age alone. Given the fragmented politics and the fleeting nature of childhood itself, building youth-led movements around strong campaign foundations – from Stop Jackdaw [oilfield] to Just Stop Oil – is proving more sustainable.

In other words, the radicalised children are already growing up, and will soon become catastrophism's reliable enforcers for the long term, turning it more extreme in the process. The future is less a case of domineering children challenging society head-on, and more one of zealous young adults attempting to undermine it from the inside. The *Guardian* again puts a positive spin on this development:

> It's a good thing that young activists are now viewed less as angelic saviours, and more as political actors in their own right.

Meanwhile, even as Covid issues fade away, another generation of defenceless children is being radicalised into climate catastrophism, bombarded with messages of imminent planetary doom in the media and their school lessons.[330] Once Thunberg is no longer considered young enough to be the true voice of youth, another child prophet

* 'It is worth noting that while the onset of Covid was initially a blow to climate catastrophism, the culture is pivoting to take advantage of societal changes in the pandemic's wake, and the event may end up being a net benefit for it.'

could well emerge, sparking a new global wave of activism. If there is a perception that Thunberg failed, it is likely to be even more strident and uncompromising than the last.

It is unlikely that the children will redirect our societies, in the way that Nonquawuse or the Red Guards did theirs. Societal constraints remain firm for now. Nevertheless, children are believed, sometimes to a simply astonishing degree. As the number of culturally convinced youngsters grows, and as they turn into zealous adults, it would be prudent to wonder whether those constraints will hold in the future. It is high time to divert children away from cultural catastrophism and towards hard common sense, and the findings of mainstream climate science.

15.5 Millennarian cultures

15.5.1 Millennarian cultures in history

Millennarianism is the belief in a coming fundamental transformation of society, after which 'all things will be changed'.[331] Much of society is viewed as 'tainted', and no longer fit for purpose. Millennarians thus aim to sweep aside existing philosophies and beliefs, in order to usher in a new and Utopian way of life. In the most culturally inflamed cases, they may even try to replace the machinery of the economy too.

We have already examined the history of the Xhosa and their prophetess, Nongqawuse, but there are many other millennarian movements. Two interesting cases are the late nineteenth century Mahdi state, and the faith of Babism, a heretical splinter of Shia Islam.[332] There are also strong millennarian roots in the Mormons, the Seventh Day Adventists and Jehovah's Witnesses, and traces can also be found in the mainstream faiths, which sometimes support sects of this more extreme mode of belief. Finally, although in cultural terms they were a flash-fire, which burned fiercely but very briefly, the Chinese Red Guards* also had millennarian goals, which they pursued with almost no restraint. Persuasive millennarian narratives are actually common throughout history.

* See Section 15.3.3.1.

Millennarian cultures often arise in oppressed or stressed societies, notably colonial ones.[333] For example, the Mahdi state was formed in opposition, first to the Ottoman Empire, and then to Anglo-Egyptian rule. Under its leader, Muhammad Ahmad,[334] it achieved the first step of sweeping away the old ways by seizing and then destroying the city of Khartoum; the historic architecture of the hated Ottomans was completely erased.

Millennarian cultures maximise emotive reactions to stress or oppression, or in some cases merely perceived oppression. This enables them to create a common 'identity in adversity' and to become a focus for resistance, in what might otherwise be a population of diverse values. This usually involves demonising the foe (whether human or not), and often results in violence.*

While there are usually enough constraints to prevent millennarian cultures from wholly destroying or dismantling the societies that spawned them, this is not always so; as we saw in Chapter 4, the Xhosa people essentially lost everything, including for many, their lives.

15.5.2 Millennarian elements in climate catastrophism

Few, if any, cultural entities are likely to be millennarian through and through; rather, there will be millennarian elements within a wider spectrum of beliefs (as noted in Section 9.7.1.1, there are normally different levels of belief across a cultural group). The weight of more moderate beliefs will often hold the worst effects of millennarian tendencies in check, although in extreme circumstances that check may fail, as it did with the Xhosa. The 'rustification' of the Red Guards was in essence a last-ditch check, which fortunately worked, and curbed the movement's excesses, albeit after two years of tremendous social damage.

So, while climate catastrophism certainly features millennarian elements, overall it is a broad church. This can be seen in the measurements in this book, which reveal how much of the culture's support is fervent or millennarian, and how much is more moderate. We

* But not always, the Xhosa being a case in point. Calls for war were resisted, perhaps because the colonial power was not the only problem they faced – the lung disease afflicting their cattle, around which their economy was based, was just as important.

know, from the measurements and explanations in Chapters 8 and 9, that the presence of strong reality constraints dramatically reduces belief across all nations. In essence, the complete dismantling of society to make way for a utopian Net Zero replacement is an enormous reality constraint. It is likely that those who would *knowingly* stick to millennarian goals, come what may, are few – fewer even than the numbers revealed in the FC series – the line of core belief.* This is a very small percentage of national populations, especially in religious nations.

Nevertheless, we ignore the millennarian aspect of climate catastrophism at our peril, not least because the primary policy measure being put in place as a result of the movement's efforts is a crash decarbonisation program, which many adherents think will bring about a complete transformation of society,[70] even if others believe it only means switching to electric cars and eating less meat. A key word in the last paragraph, is 'knowingly'; most of the public, whether they lean to climate catastrophism or not, have little idea how hugely impactful a near-term imposition of Net Zero would be; in other words, the true reality constraints are effectively hidden from them. Nor is there any guarantee that the checks holding back millennarian elements in the culture – Extinction Rebellion, Thunberg and the youth organisations, together with extreme elements of Greenpeace, the Green Party and academia – will remain firm.

A good proxy for the combined strength of the millennarian elements in climate catastrophism is the level of pressure to prevent nuclear power becoming a mainstay of Net Zero, or indeed to close it down altogether. Because it can deliver abundant and reliable power, at low cost and extremely low emissions, nuclear power could effectively kill climate catastrophism; as noted in Section 14.2 there is no need for a culture to exist if the supposed issue is solved. The culture of climate catastrophism therefore opposes its use by every means possible.† This is irrational: exactly what we expect where there is a cultural motivation.

Millennarian adherents of climate catastrophism therefore hold

* See Section 9.3.

† Via emotive selection – cultures of course are not sentient.

nuclear power to be as 'tainted' as the rest of the civilisation that they want to transform. For instance, at the time of writing, the war in Ukraine has devastated Germany's energy security* and left its people facing an uncertain winter and the possibility of power cuts. Despite this, the country's leaders are still clinging to their policy of abandoning nuclear power in the near term. Elsewhere, the outlook is somewhat better, but nuclear power's supporters are still demonised, and many of the elite in governments and intergovernmental bodies oppose the technology, having apparently bought wholesale into the millennarian cult of climate catastrophism.

It is worth noting that some aspects of climate catastrophism that we have looked at already may be in conflict with millennarian goals. For instance, building structural icons, especially technological ones, does not necessarily sit well with plans to dismantle society and all of its technical infrastructure too. This contradiction is best seen as more cultural hypocrisy; because cultural entities ultimately rest upon emotions and not rationality, they are perfectly capable of supporting multiple incompatible aims, each pushed by a different sect. So, for instance, a crash Net Zero program might result in the building of many more windfarms, but this could damage society so much as to pivot the culture towards a rejection of all technology. At this point, such projects will become impossible.

15.5.2.1 Self-oppression?

As noted above, millennarian cultures often arise in response to oppression or other social stresses. However, it is clear from the measurements and explanations in Chapters 8 and 9 that core belief in climate catastrophism – and thus in any climate millennarianism – is highest in the irreligious West. This is a puzzle; these are the least oppressed of all nations; their peoples are free and, by and large, justice prevails. Nor are there any obvious social stresses: their populations are the healthiest, wealthiest, best-fed, longest-lived, and suffer the fewest deaths from natural disaster or industrial accident or war. So why have they been gripped so strongly by the culture of climate

* Germany was heavily dependent on natural gas shipped by undersea pipelines from Russia. The Ukraine war led to sanctions on Moscow, which retaliated by first reducing gas supplies and then stopping them altogether.

catastrophism, with its strong millennarian tendencies? An intriguing possibility is that Western societies have convinced themselves that they are oppressed, because they have become overwhelmed by victim cultures.[335]

The core narrative of a victim (or grievance) culture is centred on the idea that a group is oppressed in some way. This leads to highly emotive memes that create a strong shared sense of resistance and an underdog identity, which together tightly bind adherents to the culture. Of course, as is normal with cultures, the oppression doesn't have to be real – historical perspective,* science, and any objective sense of justice will be twisted in order to emotively convince society that the victimhood is real.

Those allocated the role of victims may end up angered, although many will not accept such a role, and might even reject the (largely inappropriate) struggle that is being fought in their name. The supposed oppressors, meanwhile, are not typically defined by their personal actions, but merely by their membership of a defined group. They will suffer guilt and self-doubt as a result. However, the culture will use the former as a 'stick' as well as holding out the prospect of higher social status – in essence, 'ally of the oppressed' – as a 'carrot' – in order to recruit many of them as adherents. Indeed, far more adherents may come from among the supposed oppressors than from the supposed victims, the latter potentially being a very small social minority. Making the social confusion worse, some people will be defined as a victim by one culture, and an oppressor by another.

Victim cultures are now rife in society. Alongside the venerable anti-capitalist grievance culture (everyone except the big bosses is a victim of capitalism), we now have a culture based around Critical Race Theory (CRT; all white people are oppressors and all black people are victims) and Extreme Trans-rights culture (ETR; biological men and women are oppressors of trans people†). There is even a victim theme found as a strand of climate catastrophism, with the developed West cast as the oppressor of the Global South, carefully overlooking the enormous civilisational benefits that industrial

* Genuine historical oppression may be used as justification even where this is no longer a significant issue.

† Unless they explicitly earn allied status, e.g. by stating that trans women are real women.

nations have given the world.[336] This theme sometimes places the West inside the 'Global North', although with no less culpability for all that. An example can been seen in the magazine *Current Affairs*:[337]

> Of course, atrocity isn't a word often used to talk about the climate crisis. But to continue burning fossil fuel is to cause unimaginable suffering and death worldwide. It is an act of depraved violence committed largely by the Global North, the source of most greenhouse gas emissions, against those in the Global South, where the effects of those emissions are, for now, most fully experienced.

I suggest, therefore, that these cultures are fracturing society, undermining prior values and spreading victim mindsets in their place. In this way, the countries of the West may have achieved a kind of self-oppression.

15.5.2.2 How far will climate millennarianism go?

We know from the historic examples above that cultures will go as far as they possibly can, regardless of the consequences. Twin crises – inflation and high energy prices – were both brewing even before war broke out in the Ukraine in February 2021, but proved insufficient to slow the progress of the culture through society or to stop the crash Net Zero program. The war in Ukraine has heightened both issues, and the resultant fossil fuel shortages have threatened basic societal needs – heating homes and maintaining transport and running industry. This seems to have slowed the culture's long march through the institutions, but not to the extent that there is any realistic prospect of abandoning Net Zero.

So while it is hard to say where this all ends, unless new and stronger constraints appear, the culture will continue to blindly follow its irrational course; it will inflict further damage on society and the infrastructure that supports it.

15.6 Combating the culture

15.6.1 We cannot live without cultures

The decline of established religions, particularly Christianity, has left other cultural entities – from age-old nationalism, to elderly communism and fascism, to adolescent climate catastrophism, and the unruly children of CRT and ETR – to fill the gap. It seems that we are unable to live without cultural entities; the group identity they enable is too deeply etched into our brain architecture to be simply set aside, to say nothing of the benefits that group behaviours can bring.*

If we cannot live without cultures then, given the risks, it would seem prudent to encourage them to become more benign; to tame them, so to speak. That is better than destroying them entirely; if we managed to do so, we would have no idea what might spring up instead, and whether it would be better or worse.

But understanding how to tame a culture is not straightforward. We need to work towards an end in which the culture continues to bind society together, with all the benefits that brings, while avoiding most of the potential costs. Examples from history may guide us, but measuring net benefits and even determining the requisite timescales is very difficult. For instance, how do we weigh up the huge death tolls of communism against its lifting of hundreds of millions of people out of poverty and illiteracy?

Climate catastrophism is no easier to assess. It seems harmful at present. The irrational policies and squandering of resources it demands are increasing humanity's vulnerability to real disasters: tsunamis or wars or pandemics, and the damage to vital supply chains that result. Still, if it were tamed, the instinctive sense of stewardship that it fosters could conceivably deliver far better care for nature than rational institutions have done, no small gain in an age of huge technological power and minimal public understanding of complex environmental impacts. First though, the culture and its adherents would have to concede, just as the mainstream faiths once did, that sackcloth and ashes for everyone is not an approach that has much of a future.

15.6.2 Methods of combat: rationality, shaming, and adversity

Whether we attempt to break the culture of climate catastrophism or tame it, how would we go about it anyway? The first step to putting science and public policy back on a rational path is simply to recognise that the culture of climate catastrophism exists. This ought

* See Section 3.3.2.

in principle to be easy, now that its dominant influence on public attitudes and policy can be measured. However, the beliefs and biases that the culture has engendered in people's minds will be hard to overcome. The false idea of global climate catastrophe is now so entrenched in the lives and worldviews of members of the public they will find it hard to give up. They will believe – honestly but unfortunately not rationally – that any attempt to point out that the 'catastrophe' is cultural, not real, must be some kind of denialist ruse. In short, climate catastrophism has a tight grip on society; in the near term, reason is unlikely to prevail.

So, if straightforward rationality is unlikely to make an impact on climate catastrophism, what else might work? The belief in eugenics and the wider culture of which it formed a part were doomed once films of the Nazi concentration camps began to circulate widely. Although it took a generation to fizzle out, the culture was essentially shamed to death.

So, could climate catastrophism be shamed to death too? Or at least shamed into tamer modes of operation? Its negative impacts on humanity and the environment, some of which have been noted above, provide plenty of shameful material, if not on the same emotive scale as the heaps of dead bodies in Auschwitz and Bergen-Belsen.

Currently, much of the support for the culture seems impervious to shame; critics, whether internal or external, are censored or smeared, or simply waved aside. Michael Moore's heretical film is a case in point. Soon after its release, there were widespread calls for it to be banned[338] and these were almost entirely effective; it ended up having little effect on the culture or the narrative of salvation through renewable energy.

But if some adherents of the climate culture are shameless, others may not be. Climate catastrophism globally is hugely dependent on its relationship with the mainstream faiths: the majority of belief across nations comes from this relationship, although it is only shallow allied belief.* However, as the harms associated with the headlong pursuit of Net Zero become more apparent, the relationship with climate catastrophism could become damaging to religious leaderships

* As noted in Section 10.5.

and a source of much discontent in their flocks. Hence, although it seems unlikely at the moment, it is at least possible that the sustained impact of reality constraints, along with resultant embarrassment and shame, could permanently collapse the allied belief of religious publics.

However, the best hope that climate catastrophism will be tamed comes, unhappily, in the shape of adversity. The measurements in this book make it clear that cultural entities yield in proportion to increasingly hard realities. The crash Net Zero program has already created plenty of (self-imposed) constraints, and these will only become more numerous and increasingly onerous. Meanwhile, the war in the Ukraine and the energy crisis have, in the space of just a few short months, exacerbated this situation dramatically. Rationality is not in charge, but the culture is giving a little ground, and will almost certainly give more as these crises continue. We do not want more adversity, self-imposed or otherwise, but it could be the only thing that will relax the grip of this culture.

Appendix A

RELIGIOSITY SCALE DETAILS

A.i Professed religiosity and irreligiosity data

As noted in Section 8.2, I build a straightforward religiosity scale by combining public surveys that probe individuals from opposite angles, namely professed religiosity and professed irreligiosity. For a truly representative picture, nations from different continents and of different faiths are needed, and fortunately the surveys are adequate in this regard.

The religiosity data comes from a Gallup poll covering 150 countries with the question: 'Is religion important in your daily life?'[351] Unfortunately, it dates back to 2009, but since religious commitment tends not to change quickly, this isn't a major concern.

The irreligiosity data is a composite of results from two main sources.[352] The surveys jointly performed by the Worldwide Independent Network and Gallup International Association (WIN/GIA) in 2017, 2015, and 2012 cover about 100 countries, with irreligiosity estimated as the sum of those responding that they were 'not a religious person' and those saying they were 'a convinced atheist'. For a few countries, it is necessary to go back to older surveys, by Dentsu Communication Institute and Zuckerman, from 2006 and 2005.

To form a single scale for religiosity, I took the inverse of the irreligiosity percentages for each nation, then averaged the result with the religiosity percentage. This approach increases robustness and minimises bias effects. Apart from providing figures across the largest possible range of nations, this kind of scale is more generic than typical, more complex approaches, providing *consistent data* across all ethnicities, faiths and nations. See Subsection A.iii below.

A.ii Data gaps

Despite good coverage, the Gallup religiosity data misses both China and Oman. China is a one-party state that suppresses religion, so any data would be biased anyhow, hence it is dropped. Indeed, any one-party state suppressing religion would be expected to be anomalous; Vietnam was shown to be an exception in one of my *Climate Etc.* posts.[353] And as there is representation of various other Islamic/Arab nations, Oman is dropped too.

The irreligiosity data does not cover Egypt. Measurement is problematic anyway; a law was recently proposed (although not passed) to make atheism illegal, despite estimates of 3 million atheists especially among the young. I have filled in the gap using data from a Pew survey from 2012.[354] This is from the early 'Arab Spring', when attitudes were maybe more realistic and religiosity was admitted to be lower (though like the main source, this is a *religiosity* rather than an *irreligiosity* question). Kuwait, Bahrain, Qatar and UAE are also not covered. However, these countries seem to have nothing like the large Egyptian atheist underground, and I assume that the numbers are sufficiently small as to make little difference to the results; I have therefore worked from religiosity data alone.

All sources were sampled (the main sources via the wiki tabulations) in December 2019.

A.iii Choice of religiosity scale

There are many more sophisticated religiosity scales. Apparently, even back in 2007, at least 177 scales had been recognised.[355] There are presumably rather more now. Yet the more sophisticated these become, the more they appear to be tailored towards particular faiths or behaviours and/or cultural regions. For my global analysis, something universal is required. Directly asking participants how important religion is to them is helpfully generic. A host of complex specifics are neatly avoided, such as what activities or theological details, or social interactions are considered important. So, while my method loses these details, we don't need them anyway; we only need data that is consistent across all nations and all faiths. A minor downside is that there's almost certainly more self-assessment bias than in the more sophisticated surveys, although it should be noted that misreporting of behavioural factors, such as frequency of church attendance, is not unknown. The averaging of religiosity and irreligiosity data will minimise the problem, but, as noted below, there does appear to be some systemic self-assessment error in my scale, although it's minor and correctable.

A.iv Error-range in religiosity sources

The Gallup religiosity survey says regarding error range:

> Results are based on telephone and face-to-face interviews conducted in 2009 with approximately 1,000 adults in each country. For results based on the total sample of national adults, one can say with 95% confidence that the maximum margin of sampling error ranges from ±5.3 percentage points in Lithuania to ±2.6 percentage points in India. In addition to sampling error, question wording and practical difficulties in conducting surveys can introduce error or bias into the findings of public opinion polls.

Who knows what that overall error or bias will be? An overall error of about 5 to 10% is not implausible – it could be more. As for the irreligiosity data, the WIN/GIA survey, the source of most of the data, says nothing about its error range.[356] The figure is unlikely to be lower than in the similar Gallup survey. Combining these two surveys should increase confidence and hence reduce error, yet I'd still be inclined not to think in terms of less than 10%, even so. However, that's actually reasonable, although there's more assumption than data in this rough assessment.

Fortunately, across so many different nations, ethnicities and faiths, and with such simple questions too, most errors are unlikely to be systemic, except for self-assessment bias, which is discussed in the next section. Moreover, most series derived using my religiosity scale have robust or very robust results. Had the results been more ambiguous, or poor, examining the error range or perhaps adding further data to reduce it, might have become a project; as things stand this doesn't appear to be necessary.

Finally, errors in the religiosity scale may be a minor issue in the greater scheme of things - some of the surveys of attitudes to climate change might have bigger issues, because of smaller sample sizes, for example.

A.v Removing self-assessment bias from the religiosity scale

As Figure 31 shows, there is a slight elongated 'S' shaped skew in the combined religiosity data described above, when it is plotted against a straight line (i.e. without any reference to climate-change attitudes or any other data).[357] This is almost certainly due to self-assessment bias.

The need to conform to perceived social norms (a subconscious effect) leads individuals to make biased assessments of their own religiosity. So, in highly religious nations, individuals would be likely to overstate their religious belief. Similarly, where low religiosity is the norm, they would probably understate it. This self-assessment bias might be expected to be worse in nations of middling religiosity, where the need to protect a status as the majority in society might

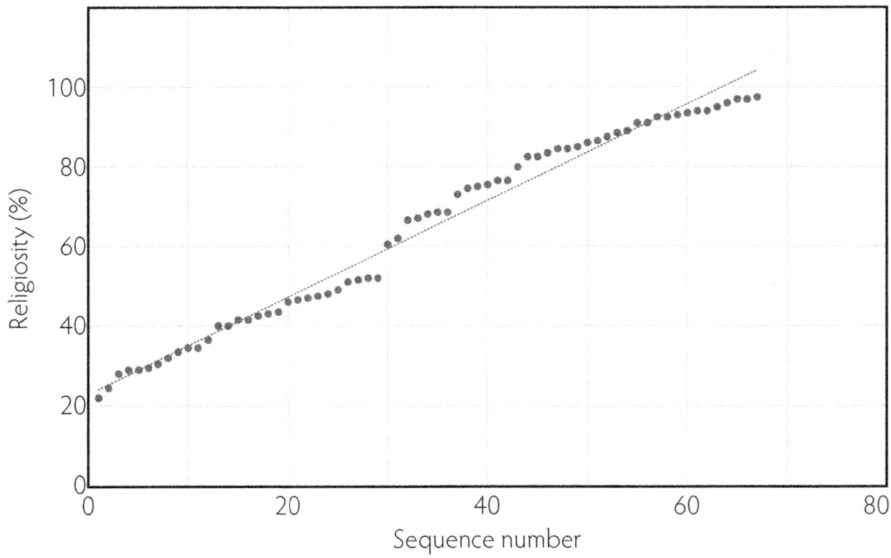

Figure 31. Self-assessment bias in national religiosities.

lead to defensiveness, both among the religious and the irreligious. This explains the observed 'S' shape for raw religiosity, although the curve is symmetrical around a religiosity figure of 60% rather than the 50% that might be expected,

The bias appears to tail off for the most religious nations; as 'actual' religiosity approaches 100%, it is not possible for people to give answers that would bias the figure higher still. In practice, even the most religious nations will have, say, 2 or 3% of disbelievers, unless this is forbidden by law or otherwise discouraged. Religiosity scales based on actual behaviours (e.g. 'going to church') would likely have less self-assessment bias, but, on the other hand, such measures are much less generic across ethnicities and faiths.

Debiasing is simply achieved by finding the residuals for the plot of religiosity against a straight line, and subtracting these from the original religiosity data. Note: this process drives some nations (e.g. Thailand, Nigeria, Morocco) over 100% religiosity, which clearly cannot be real; so these nations are capped at 100%. While some issue is thus created for a very few nations at the very top of the scale, this is smaller than the issue being solved by the procedure.

In practice, because the 'S' shape straddles the trend fairly symmetrically, this exercise has almost no impact on the R values in the main analyses of climate attitudes and national religiosity set out in Chapter 8. For example, it produces an increase of 0.92 to 0.94 for the SA series from Figure 3, and an even smaller decrease in the WA series. Hence the debiased religiosity scale makes the data-spread rather more even, but doesn't materially impact the trends and so the assumptions based upon them. For this reason, I was confident about implementing the debias procedure, although further work to prove out the 'S' shape causation would be valuable.

See also Online-Appendix Avii.

Appendix B

EXAMPLES OF FULL-DATA PLOTS

B.i For some of the Basic series

Figure 32 shows the full data for the WA (black, 24 nations) and SC (grey, 16 nations) series, as examples of typically results at a detailed level. The trend lines for these series feature in Figures 5 and 8. With an R value of only 0.57, the SC series is the least robust of all the primary linear series reported here (a summary of all the primary series is given in Table 23, Appendix G); in part this is due to the low number of data points, but also the fact that they are all relatively low on the *y*-axis, which increases the effect of noise. The series is

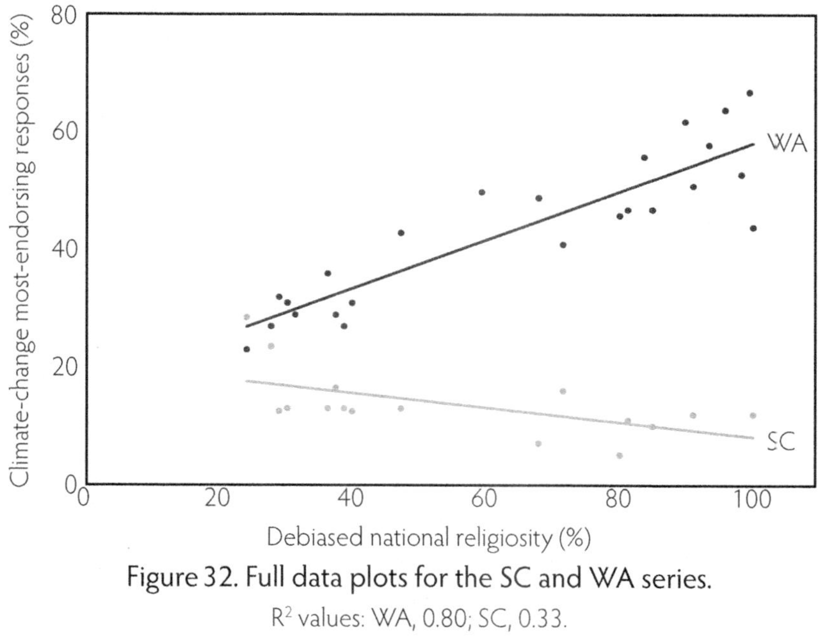

Figure 32. Full data plots for the SC and WA series.

R^2 values: WA, 0.80; SC, 0.33.

actually a composite of two series, which fortunately have the same question wording; strongly constrained series over many nations are hard to find. See the Excel-Ref for details, sheet 'Main Trends'. Note: The full data for the SA and WC series are shown in the main text, as Figures 3 and 4 respectively.

See Appendix C for the question text, trend parameters and data sources associated with these series.

B.ii For some 'Resistive' series

Figure 33 shows the SA series from Figure 3 and its resistive opposite Res 1. Figure 34 shows the WA series, and its resistive opposite, Res2. See Appendix C for the question text, trend parameters and data sources associated with these series.

Note: the equivocal response 'a fair amount', shown as the dashed grey series in Figure 34 (trend only), has moderate anti-correlation

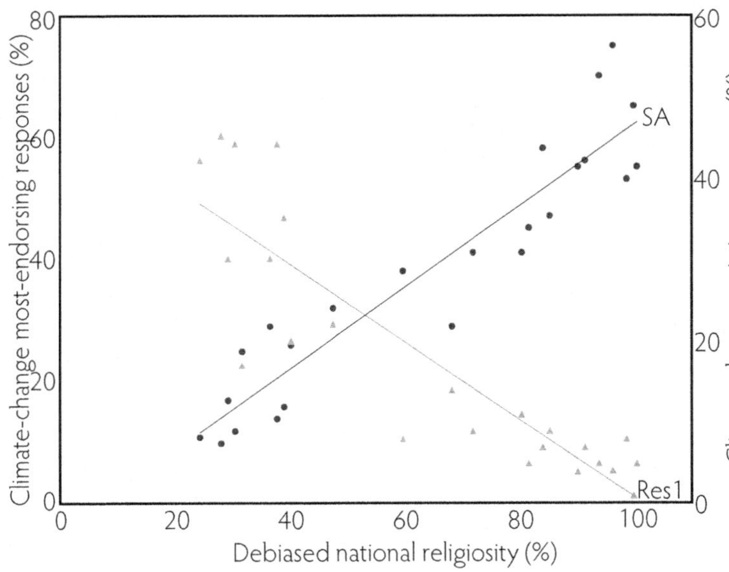

Figure 33. The SA series and its resistive opposite, Res1.

The question asked about the personal impact of climate change. The SA series shows those who said 'a great deal', while Res1 represents those who said 'not much' or 'no impact'. R^2 values are: SA, 0.87; Res1, 0.79.

with national religiosities. While still (modestly) endorsing of climate change, it nevertheless contradicts the Catastrophe Narrative; see Section 10.1.5.

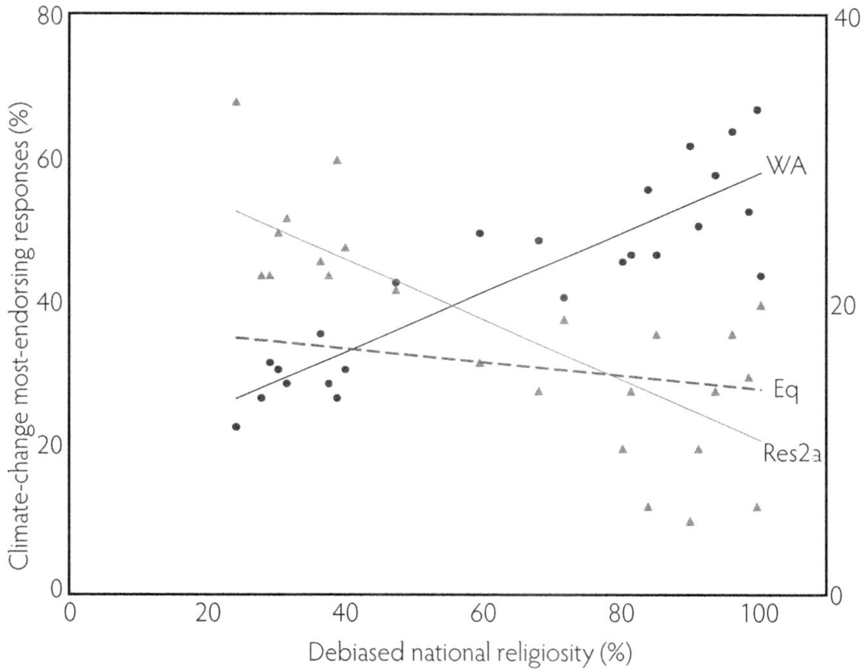

Figure 34. The WA series and its resistive opposite, Res2a.

The dotted line (Eq) is the trend for responses 'a fair amount', with the results plotted on the left-hand axis. Refer to main text for details. R^2 values: WA, 0.80; Res2a, 0.61, Eq, 0.18.

Appendix C

INFORMATION LINKS

Table 20–22 provide links to data, full data plots, survey question text, and the relevant figure(s) in the main text.

Table 20. Information links for unconstrained questions.

Series name	Data source (listed below)	Excel-Ref chart (listed below)	Questions and results tables	Figures
Half-SA	1	a	See Aux1 below	Not shown
SA	1	b	7, 8 For resistive data, see §	3, 5, 8
MSA	7	k	7, 8	5, 8
MWA	8	l	7, 8	5, 8
WA	1	c	7, 8 For resistive data, see‡	5, 8, 31
WA1	6	m	7, 8	35
WA1+O2	6	d	12	8, 13, 35

§ Resistive equivalent is Res1 (in Figure 33 and also the Excel-Ref, sheet 'main trends').
‡ Resistive equivalent is Res2a (in Figure 34 and also the Excel-Ref, sheet 'main trends').

Table 21. Information links for reality-constrained questions.

Series name	Data source (listed below)	Excel-Ref chart (listed below)	Questions and results tables	Figures
FC	2,3	e	7, 8	Not shown
SC	4	f	7, 8	3, 5, 8
MC	Not measured	n/a	n/a	5, 8 (intuited)
WC	5	g	7, 8	4, 5, 8
WC1+O1	1	h	12	8, 36

Table 22. Information links for extremely weakly aligned questions.

Series name	Data source (listed below)	Excel-Ref chart (listed below)	Questions and results tables	Figures
Extremely Weakly Aligned1	1	j	Appendix F	8, 37
Extremely Weakly Aligned2	1	i	See Aux2 below	8

C.i Data sources and original Excel charts

C.i.a Survey data-sources

1. 2019 YouGov international survey: https://yougov.co.uk/topics/science/articles-reports/2019/09/15/international-poll-most-expect-feel-impact-climate.

2. European Perceptions of Climate Change (EPCC) 2016 survey: https://orca.cf.ac.uk/98660/7/EPCC.pdf.

3. UK government 2015 public attitudes tracker: https://www.gov.uk/government/uploads/system/uploads/attachment_data/file/424515/DECC_Public_Attitudes_Tracker_wave_13_excel_tables.xlsx.

4. YouGov 'What the world thinks' (2016): https://d25d2506sfb94s.cloudfront.net/cumulus_uploads/document/a7dgdmj6is/Results_160208.pdf. Composite with 'Special Eurobarometer 459' (2017): https://ec.europa.eu/clima/sites/clima/files/support/docs/report_2017_en.pdf.

5. The huge 2015 UN 'My World' poll: http://web.archive.org/web/20190802231507/http://data.myworld2015.org/. Note: This archive web-copy is very slow. It does yield the data though. Incredibly, the UN deleted their interactive site to explore results from ~10 million participants across many nations.

6. Climate questions in the Reuters/University of Oxford 'Digital News Report 2020' survey.

7. http://www.digitalnewsreport.org/survey/2020/how-people-access-news-about-climate-change/. 2015 Pew survey on concerns about Climate Change: https://www.pewresearch.org/global/2015/11/05/1-concern-about-climate-change-and-its-consequences/.

8. IPSOS 2021 concern about 'tipping points' in climate: https://www.ipsos.com/sites/default/files/ct/news/documents/2021-08/global-commons-tables-final.pdf.

C.i.b Original charts and data

(a) See Chart '3xy' here: https://curryja.files.wordpress.com/2020/04/one-datafile.xlsx. 24 nations, x/y reversed, raw not debiased religiosity on the *x*-axis. Delete the inapplicable US and Vietnam rows. This series only pulls in a subset of Catastrophe Narrative supporters; the 'doomsters' only.

(b) See Excel-Ref, sheet 'main trends', Chart 'SA'.

(c) See Excel-Ref, sheet 'main trends', Chart 'WA'.

(d) See Excel-Ref, sheet 'WA1+O2 and WC37', Chart 'WA1+O2'.

(e) See Excel-Ref, sheet 'main trends', Columns E, G and H, Rows 112 to 145 and Chart 'Trends'.

(f) See Excel-Ref, sheet 'main trends', Chart 'SC'.

(g) See Excel-Ref, sheet 'main trends', Chart 'WC' (rf, rr, rr1).

(h) See Excel-Ref, sheet 'main trends', Chart 'WC1+O1'.

(i) See Chart '3yx' here: https://curryja.files.wordpress.com/2020/05/datafile.xlsx. Non-linear response. Pink cones correspond to grey on Figure 8 summary here. See Section 10.1.3 for explanations of non-linear responses to extremely weakly-framed questions.

(j) See Excel-Ref, sheet 'main trends', Chart 'VWtrends'. Non-linear response. Green cones correspond to grey on Figure 8 summary and Figure 37. See Section 10.1.3 for explanations of non-linear responses to extremely weakly-framed questions.

(k) See Excel-Ref, sheet 'Main Trends', Chart 'MSA(blue)', Pew 2015: Concern about Personal Harm from climate-change.

(l) See Excel-Ref, sheet 'Main Trends', Chart 'MWA(blue)', IPSOS 2021: Attitudes about human caused 'tipping points' in 'climate or nature'.

(m) See Excel-Ref, sheet 'WA1+O2 and WC37', Chart 'WA1 & O2'.

C.i.c Auxiliary questions and responses

Aux1 (Strong Alignment but partial scope): 'How likely do you think it is that climate change will cause the extinction of the human race?' and response 'Very likely'.

Aux2 (Extremely Weakly Aligned2): 'The climate is changing and human activity is mainly responsible' and response 'Affirmative'.

C.i.d Weblinks for MM series data

The surveys that generated the MM1, MM2 and MM3 series can be found at the websites of the respective pollster companies as follows:

- MM1, from Pew Research: https://www.pewresearch.org/global/wp-content/uploads/sites/2/2019/02/Pew-Research-Center_Global-Threats-2018-Report_2019-02-10.pdf.
- MM2, from Pew Research: https://www.pewresearch.org/global/2015/11/05/1-concern-about-climate-change-and-its-consequences/.
- MM3, from The Lloyd's Register Foundation World Risk Poll: https://wrp.lrfoundation.org.uk/2019-world-risk-poll/data-resources/.

Appendix D

FAITH SIGNS UP FOR CLIMATE CATASTROPHISM

In Section 9.5, we looked at how religions had started to endorse the culture of climate catastrophism, looking at attitudes in recent years and also in 2005-2009, representing a putative 'pre-signup' period. However, the 'sign-up' process for religious leaderships would in practice have extended over many years; it was emergent among the religious hierarchy, rather than 'ordered', top-down, by the supreme leader of each faith. An earlier dataset would be preferable, but would have a different problem in that there would be less climate catastrophism for religious people to react to, and less effect to measure. There may not be such a dataset anyhow; attitudes to climate change weren't widely surveyed outside the West back then.

By the end of the historic period – say 2008–2009 – the survey results hint that attitudes were already transitioning in the direction of those seen today, but results averaged across the whole 2005–2009 window are as expected: there is a very stark contrast with attitudes today. Additionally, and most conveniently, 2009 was a big turning point, because of the impact of the crucial Copenhagen climate conference, and of Climategate, both events which intensified public polarisation on the climate issue.

Although biased towards English language sources and possibly Western Christianity as well, the Yale School of the Environment has maintained a list of faith statements supportive of climate change.[358] Most include either an explicit or implicit allusion to climate catastrophe/crisis/emergency/chaos. The contents can be summarised as follows (but note that such official 'declarations' are the end point of the process of religious leaderships allying to the culture of climate catastrophism, not the beginning.

1. *Catholicism*: 26 statements between 2001 and 2021, the 16th being *Laudato Si* in May 2015. *Laudato Si* is long and formal, and contains much that is rational and reasonable, such as a desire to reduce 'traditional', so to speak, pollutants. Nor is it as overtly catastrophic, as some other papal statements or commentary have been.* However, the statement has many nods to catastrophism, such as 'The destruction of the human environment', a quote of Patriarch Bartholomew's 'disfigurement and destruction of creation', 'The work of the Church seeks not only to remind everyone of the duty to care for nature, but at the same time "she must above all protect mankind from self-destruction"', 'the spiral of self-destruction which currently engulfs us', and on policy, 'Halfway measures simply delay the inevitable disaster'. Given prior Catholic positioning, this statement effectively puts the Catholic church behind climate catastrophism from this date, if not before.

2. *Protestantism*: 45 statements from various sects of Protestantism, dating from between 1990 and 2021. Precisely how much each statement aligns to the culture of catastrophism varies from sect to sect. However, all of them call on the faithful to join the climate fight, and many frame climate change as a moral issue, thus opening the emotive gate for allied belief. Some statements also cite near Net Zero policies (even from before 'Net Zero' was coined), which are essentially cultural. A few nod to *Laudato Si* or other catastrophe-aligned statements; indeed some formal joint statements with the Catholic church are very clearly aligned to the catastrophic.[359] The word 'crisis' is frequent, and some statements are strongly aligned to the catastrophic, such as those from the current[360] and former archbishops of Canterbury.[361] Some Protestant church leaders or organisations have taken part in, or sanctioned, activism against climate change (largely anti-fossil-fuel campaigning), which blurs their position with extreme activism. Given this flood of support and its general nature, beyond about 2012–2015 I think it would be impossible for most in the Protestant flocks to distinguish their

* See CN-Archive Group 1v)i), also from 2015.

leaderships' position from the general Catastrophe Narrative as propagated by authority sources worldwide.

3. *Orthodox Christianity*: 8 statements between 2002 and 2018. These are all from Patriarch Bartholomew, who is the 'first in honour' or 'first among equals' of all the Eastern Orthodox bishops.* At the Synaxis of the Heads of Orthodox Churches in 2008, Bartholomew sought to unite the faithful in confronting the 'ecological crisis that now threatens our planet', in which endeavour he does not appear to have been opposed. His message on World Environment Day 2009 included: 'We experience the results of this neglectful and avaricious behavior today, as weak-willed spectators of the repercussions of catastrophic climactic changes'. This effectively puts the Orthodox church behind climate catastrophism from that date.

4. *Islam*: 6 statements between 2003 and 2021. The Islamic Declaration on Global Climate Change in August 2015 commits the faith's leadership to climate catastrophism. The statement includes: '...the destruction of the very conditions that have made our life on Earth possible', 'Our species...has been the cause of such corruption and devastation on it that we are in danger ending life as we know it on our planet'. It emphasises a 'tipping point' into 'catastrophic climate change' that is close at hand, and which is very unlikely to be avoided if we continue with 'business as usual'.

5. *Hinduism*: 3 statements between 2009 and 2021. The Hindu Declaration on Climate Change, of November 2015, commits the faith's leadership to climate catastrophism: 'It is a matter of survival.' The statement says it superseded a prior declaration in 2012, but that one appears to have first been presented at the Convocation of Hindu Spiritual Leaders (perhaps as a draft) in 2009. This version included statements that 'Hindus recognize that it may be too late to avert drastic climate change', and that 'powerful forces' challenge 'the very concept that unnatural climate change is occurring'. These elements don't appear in the later statement, which includes

* The Orthodox church is decentralised, so Bartholomew does not have jurisdiction over other patriarchs, but he has the right to convene extraordinary synods, and is often referred to as the spiritual leader of the Orthodox church.

a roll-call of symptoms, and states a 'desperate need' for 100% clean energy, and stresses the dharmic duty of all believers to reduce the 'pain, suffering, and violence' of climate change.

6. *Buddhism*: 8 statements between 2009 and 2016. The Buddhist Declaration on Climate Change, of November 2015, committed the faith's leadership to climate catastrophism.

7. *Jainism*: 2 statements between 2003 and 2019. The Jain Declaration on 'The Climate Crisis', of October 2019, fully commits the faith's leadership to the fight against climate change. While more balanced, and not explicitly framing climate change as 'catastrophic', (although a sequence of 'natural catastrophes' is mentioned), the overall statement is sufficiently cultural that I think those of the faithful following its guidance wouldn't be able to distinguish between the position it advocates and the Catastrophe Narrative as propagated by authorities worldwide.

8. *Judaism*: 10 statements between 1997 and 2020. The Rabbinic Letter to All Communities on the Climate Crisis, of May 2015, was inspired by the Pope's statement, and referenced 'climate chaos', among other terms, effectively committing the faith's leadership to climate catastrophism.

9. *Baha'i*: 3 statements between 2008 and 2015. The last, a statement of the Baha'i International Community to the United Nations Climate Change Conference in Paris, appeared in November 2015, and fully committed Baha'i leadership to the fight against climate change. This statement was, however, more positive than those of the other religions. However, it remained cultural, emphasising global salvation through, to paraphrase, 'the cooperation of one human world', above the explicit aspect of threat to the biosphere. That said, the 2008 statement does mention the 'climate crisis', suggesting it was the biggest challenge of the 21st century; overall the position is somewhat more ambiguous.

10. *Indigenous* (mostly American Indian): 9 statements between 1998 and 2019. The Indigenous Elders and Medicine Peoples Council Statement to the United Nations COP21 Convention on Climate

Change, Paris, of November 2015, effectively commits the leaderships of American indigenous spiritual tradition to climate catastrophism. The statement includes:

> There is no more time for discussion on preventing 'Climate Change'. That opportunity has passed. 'Climate Change' is here. The Air is not the same anymore. The Water is not the same anymore. The Earth is not the same anymore. The Clouds are not the same anymore. The Rain is not the same anymore. The Trees, the Plants, the Animals, Birds, Fish, Insects and all the others are not the same anymore. All that is Sacred in Life is vanishing because of our actions. The truth is we have moved beyond climate change to survival.'

11. *Inter-religious*: 35 joint statements made between 2002 and 2021. One example was triggered by a multi-faith open letter to both sides in an Australian election, and confirms that all faith leaderships within the country are committed at that date to climate catastrophism.[362]

Appendix E

Visualisation/Detail for Lifted Series

Lifted series are discussed in Section 10.1.2. This appendix gives full-data representations, and further details.

E.i WA1+O2

The WAO1+O2 series in Figure 8 comes from a Reuters/Oxford University survey question: 'How serious a problem, if at all, do you think climate change is?' Originally, I only had data that aggregated two different endorsing responses: 'Very serious' and 'Extremely serious', with a trend line that was very high on the y-axis, but with a cultural gradient too. I therefore speculated that one of the responses was giving a weakly-aligned trend and the other, a neutral offset. Fortunately, Reuters kindly provided the disaggregated data, enabling me to prepare the individual plots (Figure 35). My assumption was correct. The neutral response 'very serious' provides the lift. It threads a line between the culturally supportive and culturally resistive attitudes to climate catastrophism – between belief and innate scepticism. The 'Extremely serious', on the other hand, provides all the cultural gradient.

Note: where an 'Extremely serious' option isn't available for other survey questions, some cultural support might appear in the 'Very serious' responses, but would probably be far weaker. The very presence of an 'Extremely serious' option causes both higher cultural support and opposition too (in response to the appropriate options, such as 'Not serious'). These options may also invoke slightly different responses when used in reference to the older term 'Global Warming' rather than the newer 'Climate Change'.

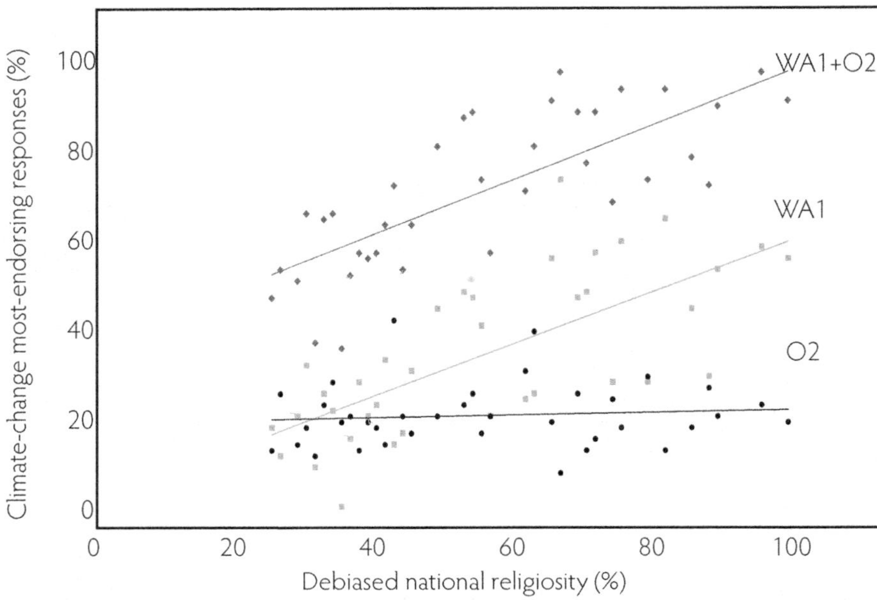

Figure 35. Lifted series WA1+O2: full data.

Refer to main text for details of question. Statistical details per Table 28.

E.ii WC1+O1

Figure 36 shows the full-data values for the WC1+O1 series in Figure 8. WC1+O1 is almost certainly lifted upwards on the *y*-axis for similar reasons to WA1+O2. However, I have been unable to isolate the specific elements causing the neutral offset and cultural trend.

The WC1+O1 series relates to the question: 'Which countries, if any, do you think have had the most negative impact on global warming and climate change?', and reflects the proportions of people in each country who responded 'China'. Respondents could choose from among five nations, so there are five series in total.

The primary subject matter is the global impact of climate change, so cultural responses should be invoked. However, the act of comparing the total emissions of countries is a reality constraint. We should therefore expect a similarity to the reality-constrained trends in Figure 8, which anti-correlate with national religiosities. This is indeed what we see. The gradient is about the same as that of the WC series, because the consideration of relative emissions is a weak constraint.

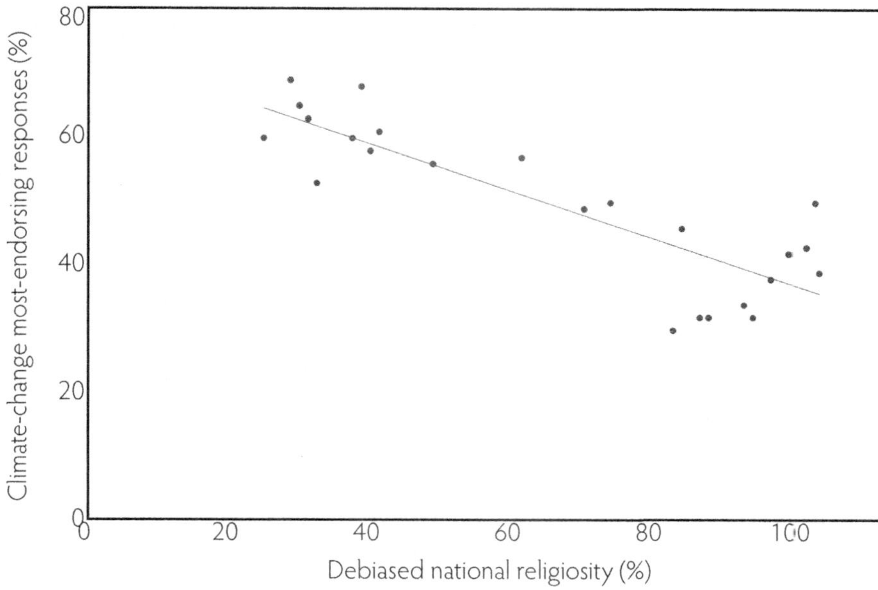

Figure 36. Lifted series WC1+O1: full data.

Refer to main text for details of question. Statistical details per Table 28.

What's different to the normal constraint situation is that providing the (correct) answer doesn't *explicitly* imply support or rejection of climate catastrophism. This is apparently enough to lure more respondents across all nations to give 'China' as their answer. Some actual knowledge is required to do this, but knowledge of China's huge population was well-established in world citizenry long before climate change was an issue, and the fact of its huge and growing industries have likewise been 'common knowledge' for decades. So the information itself is very likely not contested – that is, *culturally neutral* – across nations. The confidence in this 'common knowledge' answer, and the assumption of no cultural identity commitment in providing it, is probably what gives the culturally neutral 'lift' on the *y*-axis. One can't make a direct comparison with the WA1+O2 example, because reality-constrained responses not only slope in the opposite direction, but also vary in different ways with strength. It's probable the question and its one-word response are again triggering two psychological effects, but I don't know how to prove this. And why is the 'lift' on the *y*-axis about 20, and not say 10 or 35 or whatever? I don't know.

Sometimes, the components of a series are available, but don't obviously match the clear separation into trend and lift that we see for WA1+O2. The components may need to be subdivided (if possible) to properly resolve the psychological effects, because the questions/response-options offered will not normally do this. Series 'MSAe + Offset4' ($r = 0.77$) in the 'Extra' sheet of the Excel-Ref is a Medium-Strong lifted series with two identifiable sub-components, but which *both* have a cultural trend (although all of the 'lift' comes from one sub-component). See Appendix E.iv.

E.iii Lifted series: 'Priority of climate change in economic recovery after Covid'

Another interesting lifted series represents respondents' expressed priority for climate change in the economic recovery in the aftermath of Covid. The results come from an IPSOS survey,[363] taken in 12 nations from April 2020 (the US and China are excluded for reasons given in the main text*), which was only about three months into the pandemic.

The results are reasonably strong ($R = 0.74$ and $p = 5.8 \times 10^{-3}$), but because 12 nations is a small sample, the result may not be reliable over a larger dataset. However, it does appear to fit into the same framework that explains all the other independent data series trends. Assuming the result is reliable...

The question wording assumes there *will* be an economic recovery, effectively implying climate change doesn't threaten growth. It's unlikely many will doubt this assumption, contributing I think to the large neutral offset. While the question also provokes a cultural trend, the gradient is shallow; it's a very-weakly-aligned trend, because the wording is not at all emotive. Even for the most irreligious nations where innate scepticism is high, majorities appear to have no issue with 'a priority' that doesn't in isolation clash with other policy priorities, and so doesn't trigger their innate scepticism. See the Excel-

* See footnote p. 207.

Ref, sheet 'Extra', series 'VWA1 + Offset3: Economic Priority of CC post Covid', for the original data/source.

E.iv Lifted series: EIB/BVA survey on impact of climate change

The 'Medium-Strong 'Lifted Series' series in the 'Extra' sheet of the Excel-Ref is from a 2020 European Investment Bank/BVA survey (23 nations, but Europe only). R = 0.77. The climate-change resistive trend is shown too; as there's no 'don't know' or other opt-out option, this mirrors the above supportive trend (R = −0.77). This survey is post Covid-19, suggesting (as Appendix E.iii also does) that the basic relationship between national religiosities and climate catastrophism has survived the pandemic up to that date, at least for the unconstrained/aligned trends; this may not be so robustly the case for reality-constrained trends, where Covid is a new entry to threat lists. However, I don't have the pair of surveys – one from before Covid and one from afterwards – with identical wording (or almost so) that would allow comparisons to be made.

E.v Lifted series: EDF Obscop 2020 survey on the effects of climate change

The question was: 'On the whole, would you say that, in your region, you have already noticed the effects of climate change?' The results were ranked according to the responses for the total of both 'Yes' options ('Yes absolutely', and 'Yes somewhat'). Plotting these against national religiosities reveals a VWA-type 'lifted series', with R = 0.73.

Intuitively, one might think components of this series would be a base VWA cultural trend, and a culturally neutral 'lift', the latter provided by the 'Yes somewhat' component. However, the 'Yes absolutely' series is a WA-type series, i.e. with higher gradient, and robustly so at R = 0.75. As well as 'lifting', the 'Yes somewhat' series dilutes this cultural gradient, because it anti-correlates somewhat with national religiosities (and with R = 0.4 it is a real, albeit weak trend). The fact that pretty much all adverse weather events are now conflated with climate change in the public mind (this passes for 'common knowledge') almost certainly produces the very large lift, while the negative gradient is probably due to 'Yes, somewhat' representing an expression of mild resistance towards the Catastrophe Narrative.*

See the chart 'VWA2 + Offset5: EDF 2020 Global Survey' in the 'Extra' sheet of the Excel-Ref.

E.vi Lifted series: UN 'My World 2030' survey

The WC series comes from responses to the UN 'My World 2015' survey. Its successor, 'My World 2030', has very different questions. Early responses suggest these are going to lead to a lifted series (though still weakly-constrained). See Online-appendix F for a chart and explanation.

* These elements are emotively processed; notwithstanding the immediate time frame, a true catastrophe wouldn't have only 'somewhat' of an effect.

Appendix F

Extremely Weakly Framed Non-Linear Responses

The question 'Do you think you could be doing more to tackle climate change?' is extremely weakly-framed, therefore failing to invoke an emphatic response, either reality-constrained or unconstrained. However, with the public having no significant knowledge of climate change, supportive responses (in this case 'can personally do more') can only be cultural, so simply drift between the normal two response modes (dark grey and light grey lines in Figure 37), in the area indicated by the grey cones.*

'Doing more' can be interpreted in many different ways, from adopting a life off-grid for the zealous, to a switch to public transport for the less ardent, and perhaps only installing low-energy lightbulbs for the mostly disengaged. In other words, questions about 'doing more' are simply not sufficiently emotive to generate clear cultural responses.

Intriguingly, it has been reported that while those highly concerned about climate change are most supportive of government climate policies, they are least likely to take action themselves. Sceptics of policy measures, on the other hand, are more likely to engage in 'pro-environmental behaviours'.[364] In the absence of strongly emotive triggers in survey questions, such secondary or tertiary effects may be important in determining attitudes, and thus will tend to lead to non-linear patterns of responses.

The exceptions of Spain and Italy, at the top of Figure 37, might be caused by unusually high youth influence. Both nations have a

* These cones are wider than the equivalents in Figure 8, by an amount roughly representing noise. The two response modes are centred on the (estimated) dashed trendlines for a very weakly-aligned series (dark grey) and a very weakly-constrained one (light grey).

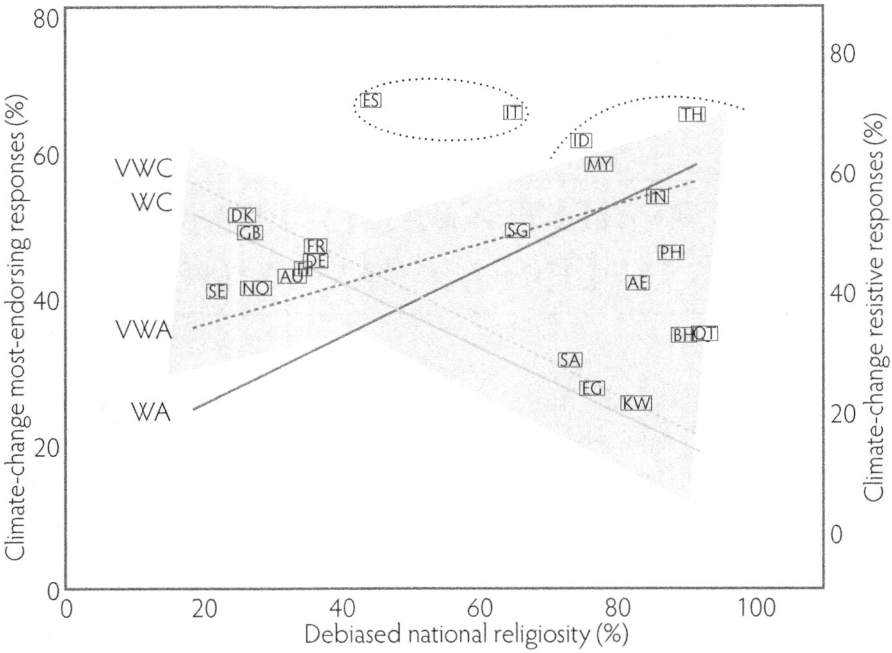

Figure 37. Responses to a weakly-framed question: full data.

Question: 'Do you think you could be doing more to tackle climate change?', response 'can personally do more'. The dotted trendlines are estimates. Note: as the pale-grey cones are notional, Thailand and Indonesia could actually be inside them.

higher 'Children's Strike Weekly' ranking than initially seemed likely for their religiosity level. For Spain at least, this seems to be due to an unusually high religiosity gap between children and adults (see Section 13.2.1). Both nations also have very high youth unemployment, making them fertile ground for cultural entities. So, when weakly-framed questions don't invoke more potent effects, might irreligious and disaffected youth wield sufficient influence to shift national attitudes? I don't know, but it is possible.

Something similar is seen with another extremely weakly-framed question, 'The climate is changing and human activity is mainly responsible'.* Although not shown here, the responses for several countries – Italy, Spain, Indonesia and India – again fall outside the

* See Table 22 on page 324 for further details.

cones. I wonder if the latter two countries may have a religiosity gap between young and old similar to that in Italy and Spain. Certainly, in Indonesia, where conservatism and Islamist elites are gaining power, there has been a wave of youth protests against the perceived threat to democracy and liberal values.[365] Links to climate concern have also been noted.

India's demographics are weighted towards youth, with 50% of its population below the age of 25. It too has seen mass youth protests in recent years.[366] These started back in 2016,[367] and although they are now focused on citzenship laws, they are ultimately an expression of opposition to conservative (Hindu) religious power,[368] once again perceived as a threat to liberal values and democracy.

So are the attitudes of the young a systemic explanation for the four nations that have 'escaped' the pale-grey cones? They may have an influence, but the 'escapee' nations are not exceptional when questions are *not* weakly-framed; that is, they conform to the pattern of all the other countries. It is hard to think what culturally uncontested 'common knowledge' or other 'lift' effects (see Section 10.1.2) could possibly occur *for these particular nations alone*, and only when the questions asked of them are weakly-framed? There are more mysteries than answers for these exceptions.

Note: WA and WC are in slightly different positions for this 22-nation plot than for the Main Trends plot in Figure 8. Hence so are the estimates VWA and VWC.

Appendix G

SUMMARY AND REFERENCES FOR ALL DATA-SERIES

This book presents 20 primary linear series, all of which are plots against national religiosities. Seventeen represent international (non-US) attitudes to climate change. Two represent international activism on climate change. One represents international policy on climate change (the deployment of wind and solar energy installations across nations). There is also a Post-Covid policy series.

Table 23 lists the main correlation parameters for each of these 20 series, plus references to where they are charted in the book and the accompanying Excel-Ref archive, alongside the same details, where applicable, for three groups of auxiliary series.

In all of the tables, the subscript in the leftmost column is the series abbreviated name. ER means 'Excel-Ref'; the corresponding series are mentioned but not actually charted in the book, although the charts can be seen in the Excel-Ref. The letters in the rightmost column refer to Excel-Ref worksheets. A key is found at the foot of the final page of the table. The second column from right, lists either the chapter/section (e.g. '10.1'), or appendix/sub-appendix (e.g. 'Dv') or online-appendix/sub-appendix (e.g. 'OAiii'), where the corresponding series is first discussed.

- Basic series: These are the linear series featured in Figures 5 and 8, and represent the basic model of cultural causation. They are generated from climate-change most-endorsing responses.
- Lifted series: These are the additional linear and 'lifted' series featured in Figure 8, which, along with the two representative non-linear series in the same figure, represent the full model of cultural causation. They are generated from climate-change

most-endorsing responses.

- Extra series: These are five further climate-change most-endorsing linear series, of which four are also lifted:
- Resistive series: These are three linear series generated from climate-change resistive responses. They are each related to a most-endorsing series, as described.
- Activism and policy series: Unlike all the other series, these do not reflect general public attitudes.

Five of these are also presented in an 'alternative series' version, each of which features a different number of data points and/or different encoding of secondary information than in the original.

There are also four primary non-linear series. One represents an extremely weakly-framed attitude to climate change, three others represent mixed-mode attitudes. These series are generated from climate-change most-endorsing responses, and fit the same single model of cultural causation as the linear series.

The 'Post-Covid' series cover 12 survey questions taken more than a year after the onset of Covid. I haven't fully evaluated how the pandemic affects attitudes to climate change, but these series appear consistent with pre-Covid patterns, despite some minor unresolved features. One policy series, the commitment to electric vehicles, was also measured long after the onset of Covid.

All these series have robust correlations or anti-correlations with national religiosity or, in the case of the non-linear series, 'envelope' boundaries. All fit a single model, with a straightforward explanation based upon cultural causation. I'm not aware of any literature tying such a wide range of climate-change attitude/activism/policy expressions into any kind of unified model, let alone one so straightforward. Online-Appendix C provides a description of the main climate-change attitudes, as revealed by the primary series, in the shape of text-based 'rules', with brief explanations for each based upon the cultural causation model.

The 'historic' series referenced in Section 9.5, the eight 'religion only' series from Section 9.6, and the ten 'US-only' series from Chapter 11 are not included here.

Table 23. Statistical summary and references for main series.

Figure	Series type	R^2	p	Ref	Sheet
Basic series					
$3/5_{SA}$	Strongly aligned	0.87	2×10^{-11}	8.3	M
5_{MSA}	Medium-strong aligned	0.45	1.7×10^{-4}	8.6	M
5_{MWA}	Medium-weak aligned	0.66	7.4×10^{-5}	8.6	M
5_{WA}	Weakly aligned	0.80	4.6×10^{-9}	8.6	M
5_{WA1}	Weakly aligned	0.49	1.5×10^{-6}	8.6	M
5_{SC}	Strongly constrained	0.33	2.1×10^{-2}	8.6	M
$4/5_{WC}$	Weakly constrained	0.58	4.1×10^{-10}	8.3	M
Lifted series					
8_{WA1+O2}	Weakly aligned & lifted*	0.59	3.8×10^{-8}	10.1.2	M
8_{WC1+O1}	Weakly constrained & lifted	0.72	1.6×10^{-7}	10.1.2	M
Extra series					
$ER_{VWA1+O3}$	Very weakly aligned & lifted	0.55	5.8×10^{-3}	E.iii	E
ER_{SCe}	Strongly constrained (Europe)	0.35	2.6×10^{-3}	10.1.8	E
$ER_{MSAe+O4}$	Med-strong aligned & lifted	0.59	1.7×10^{-5}	E.iv	E
$ER_{Levi2021}$	Weakly aligned & lifted	0.41	9.4×10^{-4}	OAiii	E
$ER_{VWA2+O5}$	Very weakly aligned & lifted	0.53	1.8×10^{-5}	E.v	E

Figure	Series type	R^2	p	Ref	Sheet
Resistive series					
32_{res1}	Resistive opposite to SA*	0.79	8.5×10^{-9}	ER	M
33_{res2}	Resistive opposite to WA*	0.61	7.2×10^{-6}	ER	M
ER_{RMSAe}	Resistive opposite to above*	0.59	1.9×10^{-5}	E.iv	E
Activism and policy series					
26_{XR}	XR group density per nation	0.75	1.3×10^{-9}	13.1	X
27_{CSW}	CSW group density per nation	0.64	2.1×10^{-7}	13.2	X
23_{RCR}	Renewables commitment rank	0.42	2.5×10^{-5}	12.5	R
Alternative series					
35_{WA1+O2}	With/without offset, full data	Same as matching series above		10.1.2	W
13_{WA1+O2}	Full data + GDPpc coding			10.1.5	W
36_{WC1+O1}	24 nation version, full data			10.1.2	M
$31_{WA/SC}$	Weakly-A/Strongly-C, Full data	Similar to above		8.6	M
14_{WC37}	37 nation WC + GDPpc coding			10.1.5	W
Non-linear series					
$37_{VWtrends}$	Extremely weakly framed	n/a		E	M
9_{MM1}	NL mixed-mode (WA/WC)	n/a		10.1.4.1	E
10_{MM2}	NL mixed-mode (VWA/VWC)	n/a		10.1.4.1	E
11_{MM3}	NL mixed-mode (MSA/WC)	n/a		10.1.4.1	E

Figure	Series type	R^2	p	Ref	Sheet
Post-Covid series					
ER_{PCWA}	WA–MSA; needs more data	0.34	2.2×10^{-2}	10.1.4.2	P
ER_{LMM}	Linear mixed-mode (VWA/ VWC)	0.36	3.1×10^{-5}	10.1.4.2	P
ER_{LMM1}	Linear mixed-mode (SA/FC)	0.46	2.8×10^{-4}	10.1.4.2	P
28_{Y1}	Linear mixed-mode?	0.64	1.1×10^{-6}	13.3	P
29_{Y2}	Weakly constrained	0.48	9.9×10^{-5}	13.3	P
ER_{Y4}	Weakly aligned & lifted	0.74	9.9×10^{-5}	13.3	P
ER_{Y5}	NL mixed-mode (WA/SC)	n/a		ER	P
ER_{Y6}	NL mixed-mode (MSA/FC)	n/a		ER	P
ER_{sce1}	Strongly constrained (Europe)	0.32	4.6×10^{-3}	10.1.8	P
ER_{oecd1}	Weakly constrained & lifted	0.68	2.5×10^{-5}	ER	P
ER_{oecd2}	Very weakly aligned & lifted	0.46	2×10^{-3}	ER	P
ER_{oecd3}	Resistive	0.54	4.9×10^{-4}	ER	P
Post-Covid policy series					
ER_{ev1}	EV commitment rank	0.46	3.9×10^{-4}	12.7	P

Key to rightmost column: M, 'Main Trends'; E, 'Extra'; X, 'XR & CSW'; R, 'Relig DB and Renewables'; W, 'WA1+O2 and WC37', P = 'PostCovid'.

Appendix H

GDP PER CAPITA AND RENEWABLES COMMITMENT

As noted in Section 12.6, directly plotting the renewables commitment of nations against national religiosities (instead of directly against the WC series responses that represent a cultural attitude) is likely to produce a less robust correlation. Indeed R reduces from 0.73 to 0.65, even when using the *rank* of renewables commitment (which pulls in outliers somewhat, e.g. Germany), and to 0.6 using GDP-normalised MW-per-capita values. This is in part due to more noise when using religiosity, which is just a proxy for the cultural attitude that motivates renewables commitment, and in part due to the systemic variance about trend from religio-regional GDP-per-capita (rrGDPpc), as explained in Section 10.1.5. Figure 38 depicts this variance.

Nations with low rrGDPpc (represented by hollow shapes) are expected to be above the main trendline (or near it – inside the shaded area). Nations with high rrGDPpc (represented by filled shapes) are expected to be below or near the trendline. Twenty-three nations conform to this expectation, but three don't (Spain, Bulgaria, and Lithuania, labelled). Nine nations don't have suitable religio-regional peers in this chart, in order to assess rrGDPpc variance (data-point for these is a light-grey cross). However, four of these conform (any in the grey box are bound to conform). So, 89% of nations conform without including the un-peered cases, and worst-case conformance including un-peered nations is 77%. The more likely conformance with un-peered is ~83% (say 2 of 5 unknowns conforming).

See Section 10.1.5 for more detail on rrGDPpc variance and a representation of the effect on the WA+O2 and WC series. As noted in that section, all data points on this chart could be expressed as a function of *religiosity only* (i.e. which religio-regional group, and national distance from that group's average religiosity).

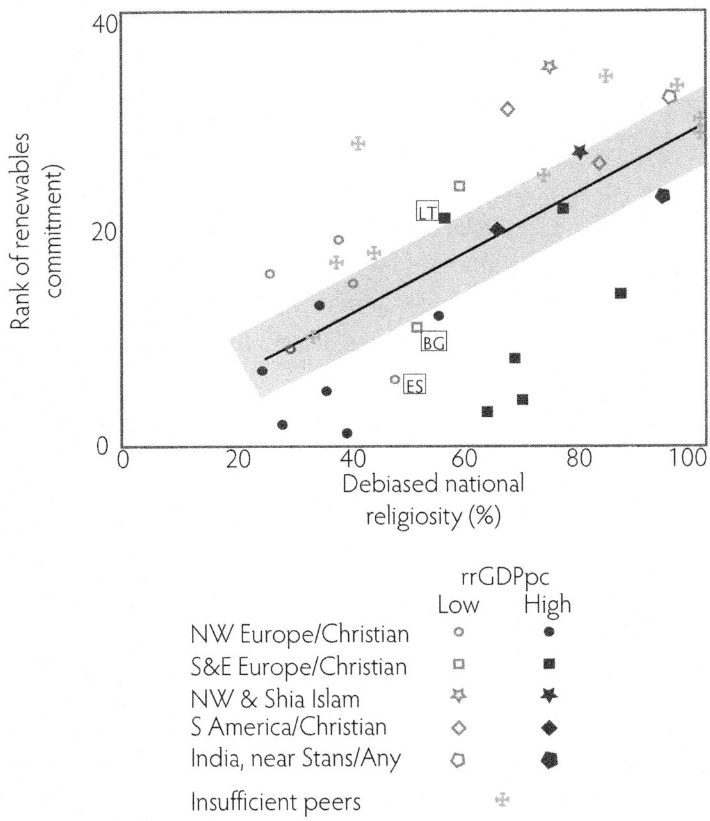

Figure 38. Rank renewables commitment, religiosity, and rrGDPpc.

Rank renewables commitment per nation, against national religiosities, showing rrGDPpc variance.

Appendix I

DATA SAMPLING NOTES FOR CHAPTER 13

I.i In relation to Extinction Rebellion groups

I originally obtained the figures for XR group presence across nations from their website[369] on 10 January 2020. This was prior to the spread of Covid-19, and pre-Covid data later remained my preference because the pandemic might impact group activity. However, at that time I only scraped data for 22 nations common to the two main surveys I was using.[370] This unfortunately left me very short of data for countries of between 50% and 80% religiosity. Going back in 2021 to fill-in this gap, I found that the website appeared to have undergone a major re-organisation and update shortly after my first visit, with national boundaries disappearing from their group maps for a very long time. Hence, even via archive files (which unfortunately didn't capture the site in January 2020 or long before), it proved impossible to retrieve reliable data for the same date, or close to it. Because the reorganised website[371] was not both stable and archived again until 2 March 2021, well over a year after my original sampling, rather than fill-in with data from this date, I chose instead to go for consistency and use all XR group data from 2 March 2021. The point of all this detail is that my data comes from a year after the onset of Covid. However, at this point, the pandemic doesn't appear to have impacted the expected trend across nations that would indicate the cultural nature of XR support.

This also explains my choice of nations. For simplicity, I stayed with my original 22, which are marked in normal text on Figure 26. The original reason was only their inclusion in the two above-mentioned surveys of attitudes to climate change. Those marked in bold are the first seven nations that have a national religiosity of between 50% and 80%, from the list of those that deploy wind turbines which is used in Section 12.3; in other words, a relatively arbitrary pick to fill the sparse patch.[372]

I.ii In relation to Children's Strike Weekly groups

The CSW group presence for my original 22 nations, used when measuring Extinction Rebellion above (and labelled in normal text on Figure 27), was taken from the School Strike for Climate website[373] on 10 January 2020. The CSW group presence for 7 further nations (again the same as used for XR above, labelled in bold text) was from 10 May 2020. This mixed data is because the same issue noted above, of a 'sparse patch' for nations between 50 and 80% religiosity, also occurred for my data sampling here.

However, in this case, data from a fairly close date (10 May 2020, from a 14 May WayBack machine archive) was available, and I used this to fill-in. While this date is during the early pandemic, there's probably a very modest impact (it's only a few weeks into lockdowns, and strikes running at the time were probably maintained on the website anyhow, even if impacted in practice). In addition, the original 10 January data is still pre-Covid-19, yet also very close in date to 14 May. Given that school strikes are even more vulnerable to actions countering the pandemic, my preference is to stick with this data, although it comes from different dates.

To ensure consistency, only 'Weekly' strikes, the most common by far within the measured nations (as the group name implies), are used throughout. As is the case for the UK, the founding nation of Extinction Rebellion, Sweden, as the founding nation of CSW, has a much larger group presence than other countries.

Appendix J

SOCIAL PSYCHOLOGY OF CONFLICTED DOMAINS

Although the social psychology of the climate-change domain has been my overwhelming focus, some of the findings revealed in this book apply outside of this domain too. With that in mind, the summary of culturally conflicted domains in Section J.i is generic: the points made are in principle true for a conflict about science-related issues precipitated by any culture. However, where examples are cited, they still come mainly from climate catastrophism. The rest of this appendix returns to the climate-change domain.

J.i Cultural conflicts over science issues

J.i.a Cultural analysis of conflicted science

Where there is a public conflict over a scientific issue, cultural analysis cannot say what is true or who is right. But it *can* tell us whose support stems mainly from cultural belief. Cultural beliefs are based upon fairy tales, and so are necessarily wrong. However, those whose motivation is cultural might – for reasons of cultural alliance or opposition – be aligned with those who are on the right side of the conflict. Such people are on the right side, but for the wrong reasons.

J.i.b Mature or not mature, that is the question

If the contested science is mature, it should be reproducible. However, cultural inertia that means disputes can continue for generations after maturity is achieved. In these circumstances, a cultural analysis should explain what is happening, validating the approach, and giving us confidence in applying it in newer fields in which there are disputed claims about the maturity of the science.

J.i.c Apt and inapt innate scepticism are the same behaviour

In a culturally conflicted domain, innate scepticism (or 'cultural disbelief') can either be apt or inapt. Apt innate scepticism resists a cultural consensus; it may be resisting a scientific untruth, or a scientific truth that is inappropriately entangled with a culture. Inapt innate scepticism resists a scientific truth, having been triggered to defend existing values that this challenges. In essence, it involves mistaking the scientific truth for the kind of unquestionable consensuses that cultures generate. Society tends to laud the apt variety of innate scepticism and stigmatise the inapt, yet they come from the same behaviour; you cannot have one without the other. See Chapter 7.

J.i.d Cultural belief bypasses innate scepticism

My working proposition about innate scepticism is that the mechanisms producing it are enabled 'by default', although they still have to be triggered by the factors noted in Chapter 7. However, belief unsurprisingly appears to disable or supress innate scepticism of the believed culture. Triggers are ignored (or their impact is at least diminished) by cultural believers.

It is well established that cultural belief results in strong biases; sometimes these are described as bypassing adherents' rationality.* So we could say that cultural belief bypasses not only rationality but innate scepticism too.

J.i.e Scepticism and 'denialism'; conflation and confusion

The term 'denialism' is so abused there may not be much utility left in it. However, in principle, inapt innate scepticism may be denialism, albeit it has a very different character to the popular image. To put this a different way, only a tiny corner of inapt innately sceptical behaviour will be close to the popular image of denialism.

There cannot be bulk rational scepticism in publics, because they are not domain-literate; rationality needs knowledge. There can be bulk *innate* scepticism, because this is instinctive and cultural-value dependent and thus requires no domain knowledge.

However, at best, the innate and rational forms of scepticism are conflated, obscuring clear analysis. At worst, the former is tarred as

* Which in brain architectural terms, might be literally true.

'denialism', even when it is apt, and sometimes also when it is both apt and factually correct.

J.i.f Most claims of 'denialism' are out-group demonisation
In order to reinforce a culture and police its narrative, ardent adherents demonise those outside the cultural consensus, including those who merely ask legitimate questions. In recent years, academic backing for the popular but deeply flawed framing of 'denialism', has allowed such questioners to be stigmatised, but in a manner that both disguises and legitimises the age-old practice of out-group demonisation, which might otherwise have been more widely recognised for what it really is. See Chapter 6.

J.i.g Conspiracy and dishonesty are not the problem
If behaviours within a conflicted domain can be shown to be largely cultural, a hoax or conspiracy can be decisively ruled out as the cause of the conflict. There will be some bad actors and opportunism of course – these are found everywhere – but cultures are emergent, and work primarily via subconscious mechanisms, whereas conspiracies involve wholly conscious motivation and planning.

Cultures engage the strongest coordination mechanisms humans possess, which can make them look as though they are part of some nefarious and highly engineered plan, encompassing a vast number of conspirators. Nothing could be further from the truth; it is simply that adherents have all caught the same emotive cultural virus, a 'cultural cold' so to speak; their actions – narrative policing for example – are the resulting behavioural symptoms. In fact, at all levels of society, from the elite to grass roots, the overriding motivation is an honest belief in the emotive central narrative (although an extreme fringe may descend into noble-cause corruption, leading to dishonesty, but not about the belief).

J.i.h All cultural entities act similarly
The surface impressions of cultures can be very different, depending on what narratives emerged: the 'story' of the culture. However, beneath the stories of 'what is important', such as God, or no ownership of private property, or racial purity, or certain global catastrophe, all cultures work via the same mechanisms, and so in principle

their basic features and adherent behaviours will be similar. I do not say 'the same', because deriding a culturally disfavoured group on social media is not the same, say, as subjecting them to genocide; but the point is that the cultural disfavour stems from mechanisms common to both cases. How far this goes depends on how much of a grip the culture gets, and what competing cultures or non-cultural processes limit its actions.

Section 9.6 demonstrates that probing attitudes across religious cultures reveals a consistent pattern, which is the same pattern seen when probing attitudes to climate change. This speaks to the common mechanisms beneath all these cultural entities.

J.i.i Cultural entities interact with each other

Successful cultural entities occupying the same social space for significant lengths of time will interact. They all have the same underlying mechanisms and are competing for adherents from the same pool of humans, which makes interaction inevitable. However, straight competition isn't the only outcome. For instance, two cultures may co-operate and compete simultaneously, or co-operate far more than they compete if this works better for their joint memberships. None of this is planned; relationships are emergent.

As illustrated in this book, it is such interactions that allow us to see so clearly the presence of a culture dominating public attitudes to climate change. Imagine trying to understand what's happening in Figure 8, for example, without a religiosity axis.* Across many nations and all the main faiths, attitudes to climate change have a simple formulaic relationship with national religiosity. As explained in Chapter 9, this relationship is consistent with the interaction between two strong global cultures, namely religion and climate catastrophism.

J.i.j Cultural entities can exert tremendous influence

The behaviours described in this appendix, and the measurements of the cultural impacts on policy outcomes and levels of activism,† tell us a great deal about the power of cultural entities in general, and cli-

* While there's some conception of this axis inside the US due to political tribalism, which also has a religious weighting (see social psychologist Dan Kahan's work for example), regarding the rest of the world this is where analysis efforts are at now.

† See Chapter 12.

mate catastrophism in particular. Successful cultures can dominate the attitudes of authorities, policy elites and the general public. While this can lead to emotive rejection as well as acceptance, there can still be major real-world impacts, including behavioural changes across the whole of society, and enormous spending programmes (think cathedrals). Both serve the cultural narrative.

J.i.k Rising cultural entities use and abuse children
Children are primed to receive cultural templates, which they embed into their cognitive processes and their maturing worldviews. Children therefore play important roles in a new rising culture, as a source of new adherents, as 'innocent' proselytisers, and, potentially, for certain individual children, as high-profile cultural prophets. See Chapter 4 for a detailed view of these roles in the climate domain.

Children may suffer damage from a rising culture, most notably when they mistakenly interpret inculcated fear as being real and not just a cultural fairy tale. This is as true in cultures that are potentially net positive in their overall impacts as in ones that are unquestionably harmful.

J.i.l Cultural narrative is a population of emotive variants
Cultural entities are sustained by a wealth of narrative variants of a central emotive theme. The theme typically has existential implications for the cultural group or its way of life, or for the whole of society, and invokes a wide range of emotions, including fear, hope, anxiety and joy. A cultural consensus forms around the narrative, and is policed by adherents. Nevertheless, the narrative evolves, new variants appearing and fading away. The variant population as a whole represents the optimal mix for delivering responsiveness to new challenges, effective propagation, emotive persuasiveness, and minimisation of emotive rejection. Chapter 5 provides an overview of the population variants for climate catastrophism.

J.ii The unique opportunity of climate catastrophism

Climate catastrophism represents a unique opportunity. For the first time in history, the spread and development of a well-bounded cultural entity with global reach has been almost entirely recorded on the Internet. This provides unparalleled source material on how cultures work; it is a golden opportunity for researchers. All the cultural factors listed above can be seen in operation, and their effects measured.

The observations in this book reveal much about religion too. Measuring cultural reactions in the main faiths is commonplace, and some studies find reactions that are common to all of them. But I know of no other case where the results are so straightforward, so robust and unambiguous, as they are for the common reaction of all faiths to climate catastrophism, as measured here. That could be my lack of knowledge. However, I believe that climate catastrophism could be the only global (and globally consistent) cultural entity with a core narrative sufficiently distanced from religion to not be a 'relative', so to speak. It may therefore be the only candidate culture having a reaction with religion that, even in theory, could be invariant across all faiths and all regions. Communism and Fascism have historically never been globally consistent, and Communism's outright rejection of religion is not conducive to probing the more subtle rules of cultural interaction either.

J.iii The social psychology of climate catastrophism

We can now paint a complete picture of the social psychology of the climate-change domain. This revolves around the cultural concept of 'certain catastrophic global warming' (to use the original phrase rather than the emergent term 'climate change').

J.iii.a Catastrophism dominates attitudes to climate change

Attitudes to climate change across nations are predominantly cultural (Chapters 8 and 9), and shaped by the Catastrophe Narrative (Chapter 5). Like a slow-motion virus, a 'cultural cold', this fairy-tale about imminent catastrophe has infected all nations. And like all strong cultural narratives, it is false, as confirmed by mainstream science, which certainly does not support certain global climate catastrophe (Section 5.2.2). Nevertheless, over decades, this emotive narrative has seeped deeper and deeper into our institutions: governments, the law, business, charities, academia, schools and more, bending or changing legislation and even our moral landscape along the way (Chapter 14). To the detriment of rational policymaking, it has become a lodestone for many policy elites, especially in the secular West (as evidenced by renewables policy, Chapter 12).

J.iii.b   Neither climate change support or resistance is rational
Cultural adherents who believe the Catastrophe Narrative and propagate it further are mainly honest, often passionate, and wrong. Bulk resistance to Catastrophe Narrative within national publics does not come from rational scepticism, but from innate scepticism, an instinctive disbelief in a culture.

J.iii.c   Neither climate change support nor resistance is constant
Given climate change is interpreted culturally, many people, in some nations even a majority of people, are not exclusively supportive of nor resistive to climate change, but can lean one way or the other depending on the scenario they face (reality-constrained or unconstrained). The precise nature and size of the gulf between these two main scenarios, for any nation, depends on the membership of other cultures, and for most nations the critical cultural membership in this regard is religion.

J.iii.d   Climate catastrophism interacts consistently with religion
Due to faith leaderships adopting a catastrophist stance, religion has a shallow cultural alliance with climate catastrophism, but beneath the surface, still competes (Chapter 9). This relationship is constant across all the main faiths. At the national level and outside the US, this interaction determines overall cultural attitudes to climate change. In other words, religion and climate catastrophism are the only cultural players that matter; they determine the trends measured in Chapter 8–10.

Primarily, religious belief disables innate scepticism of the Catastrophe Narrative, causing climate-change most-endorsing responses to unconstrained survey questions to *rise* with national religiosity. But whenever any reality constraint appears, competing values rooted in religious faith re-enable innate scepticism of the Catastrophe Narrative, such that climate-change endorsing responses then *fall* with national religiosity.

J.iii.e In the US, political tribes matter too

The same cultural 'rules' are applicable inside the US, but the uniquely strong political polarisation of the public there adds two more cultural players, the Rep/Con and Dem/Lib 'tribes'. The Dem/Lib tribe has formed a strong cultural alliance with climate catastrophism, so the Rep/Con tribe naturally opposes it. Yet while this also means the Rep/Cons oppose a fundamental untruth (certain imminent catastrophe), which falsely claims the backing of science, this opposition is not rational but cultural. Further, the climate-change hoax or conspiracy narrative promoted by the more ardent Rep/Cons is also wrong, being the product of a cultural reaction.

Chapter 11 confirms two expectations. Firstly, the attitudes to climate change expressed by Dem/Lib supporters in response to various strengths of reality-constrained and unconstrained questions proportionally match the attitudes in other countries (albeit amplified, as *two* cultures are contributing). Secondly, the allied belief of Rep/Con supporters is much lower than in other countries, because innate scepticism of climate catastrophism is enabled even for religious Rep/Cons. This must be the case given their opposition to the Dem/Libs.

J.iii.f What matters everywhere in the world, is cultural identity

This US scenario matches social psychologist Dan Kahan's rule of thumb: US attitudes to climate change (and other polarised topics) are not about *what people know*, but *who they are*, in terms of their cultural identity. It's generally assumed that alignment to Rep/Con or Dem/Lib tribes constitutes most of this identity. But it appears that beneath the obvious political polarisation, climate catastrophism and religion contribute to identity *in their own right*. Not only that, but for all other nations, they are the only players that matter. This means that Kahan's rule of thumb actually applies *everywhere*, but it so happens that the formula for cultural contributions is simpler outside the US. See Section 9.2.

J.iii.g Conscious efforts serve cultural biases

The climate-change domain is driven primarily by subconscious mechanisms: support or resistance within publics is therefore predominantly shaped by the cultural narrative, and not by (mainstream) science or objective policy considerations. However, this doesn't mean a lack of conscious efforts to judge or act, but these are subject to strong cultural biases of one sort or another.

This is not an illness or a mental flaw or a delusion or mendacity; it is a feature of all human cognition. It is therefore true of individuals on both sides of the climate conflict. Some have been caught in the grip of the culture, others have reacted against it. Who adopts which attitude simply depends on their individual prior cultural values, which confer susceptibility, relative immunity, or perhaps provoke outright resistance to the narrative of catastrophism.

A few people may learn insulation from cultural bias, for instance via the scientific method. However, it is possible for cultures to undermine this tendency, for example by hijacking the cloak of science, as climate catastrophism does (see Chapter 15).

J.iii.h Policy serves the cultural agenda

Like activism (Chapter 13), policy (Chapter 12) also serves the cultural agenda, even though it claims to follow 'the science'. Because public authorities and elites are not immune to cultural capture, to an extent proportional to the (weakly reality-constrained) cultural support in each country, the agenda of climate catastrophism simply bleeds through to policy implementation. Yet the reality constraint is essentially everywhere weaker than it ought to be; objective public knowledge about the downsides and ineffectiveness of most policies (including the measured case of renewables) would likely destroy all support for them. But what the public and indeed public authorities 'know' is largely cultural.

J.iii.i 'Climate psychology' is blinded by the culture agenda

People are smeared as 'climate-change deniers' on a scale that could be termed 'industrial', supported not only by the misframing of 'denialism' (Chapter 6), but by much of 'climate-psychology'. This discipline, if such it may be called, seems to spend most of its time assembling long lists of tenuous reasons why large swathes of publics reject a proposition that the discipline itself has never evaluated: the proposition of certain global climate catastrophe. Whether or not the

IPCC is subject to groupthink, as scientific climate-sceptics claim, it *isn't* IPCC/mainstream science that's being rejected. It's the *cultural* narrative of certain catastrophe (and coupled salvation), as can be measured by attitudes to climate change across national publics (Chapters 8–10).

While correctly determining that public scepticism cannot be rational, climate psychology tends to denigrate it as 'denialism' or at least inspired by 'denialists', when in fact it is *apt* innate scepticism *correctly* resisting a cultural narrative. The findings in this book not only render obsolete the literature for social predictors of international attitudes to climate change, they fatally undermine most of the outputs of climate psychology (with the exception of Kahan's and similar theories, as noted above).

J.iii.j Climate catastrophism uses and abuses children

Considering the rising phenomenon of children who are genuinely fearful of an impending catastrophe (Chapter 4), the mass recruitment of children by climate catastrophism is a major concern, although not unexpected for a strong culture. Proselytising adults amplify this issue by constant use of the Catastrophe Narrative variant 'Engaging anxiety for children' (Section 5.3.3). In addition, the culture has surfaced Greta Thunberg as the ultimate voice of innocence (Section 5.3.8). She is in essence a modern Nongqawuse, a mouthpiece for a millennarian culture, likewise calling for termination of the main means by which our civilisation is sustained (Chapter 14). Many children now don't have to 'catch' the cultural cold in order to proselytise. Unlike the biological equivalent, they can be raised with it; climate catastrophism has been installed as a major component of their *original* cultural value-set.

J.iii.k Culture rules; the climate and climate science, don't matter

None of this has anything to do with physical climate change and whether, in coming decades, that will be benign or bad. It has little to do with climate science and the wrangles between minority sceptics (and lukewarmers) and the mainstream; all of these positions contradict the Catastrophe Narrative. And no one, from the highly climate concerned to the very sceptical, should want a culture effectively in charge of the domain. Cultural entities cause all resources and thought and even morality to be bent towards their sustenance, while explicitly *not* solving the ostensible issue. Allowing this to happen would kill the culture.

Appendix K

SOCIAL PREDICTORS FROM THE LITERATURE

I set out here the details of the social predictors of attitudes to climate change as reported in the literature. These are discussed and charted in Section 10.3.5.

1. KVn = Kvaløy et al. (2012)[374]

The paper gives positive predictors for 'seriousness of global warming', across 47 nations.

- KV1 = Education, 21%
- KV2 = Left political stance, 15%.
- KV3 = God important, 10%.

2. Tn = Tranter et al. (2015)[375]

Data from 2010. The paper outlines predictors for climate change scepticism across 13 countries (excluding the US results). It is a multi-level predictor model, including 12 demographic and attitudinal variables, from age, gender, degree, to left-right alignment and distrust in government, plus belief that science can solve environmental problems. Model predictive values from reported pseudo R^2 (Nagelkerke).

- T1 = 6% (Germany) to 29% (Sweden).
- Others: AUS = 26, AUT = 11, CAN = 17, DEN = 18, FIN = 24, FRA = 9, NZ = 21, NOR = 24, SPA = 9, SWI = 9, UK = 26. (For completeness, the US is 36.)

3. Mn = McCright et al. (2015)[376]

Predictors for 'perceived seriousness of climate change', across 14 western European nations.

- M1 = Political ideology, 9%
- M2 = Perceived understanding, 10%.

4. LCn = Lo and Chow (2015)[377]

Across 33 nations.

- LC1= GDP per capita, 37%, for 'perceived importance' of climate change.
- LC2 = GDP per capita, 38%, negative, for 'perceived danger' of climate change.

5. Hn = Hornsey et al. (2016)[378]

A meta study for belief in climate change across 56 countries (including the US) and 171 studies. The strongest 6 of 27 predictors are given (assuming the Table 1 'correlation' column lists standard R coefficients from their multivariate analysis, and for the purposes of rough comparison treating these as though they are single variable values to produce R^2; depending on correlations between input variables, this will likely be favourable to individual predictors, not unfavourable).

- H1 = 'New ecological paradigm', 24%
- H2 = Willingness to prioritize environment over economy, 15%
- H3 = Trust in scientists, 15%
- H4 = Perceived scientific consensus, 12%
- H5 = Experience of local weather change, 11%
- H6 = Policy support, 10%.

Note, some of these are in essence complex intermediate attitudes in their own right.

6. Ln = Levi (2021)[379]

Data from 2016–19. Country-level predictors for 'belief that climate change is human-caused', across 60 nations.

- L1 = Environmental protection, 7–11%
- L2 = Civil liberties, 7%
- L3 = Exposure to climate impacts, 4–7%.

7. Kn = Kácha et al. (2022)[380]

Pre-print, data from 2016–17. Spans 22 European countries and Israel. A multi-level predictor model, including age, gender, household income, level of education, left-right ideology, and 'class' (engaged, pessimistic, indifferent, doubtful). Model predictive values from reported pseudo R^2 (Nakagawa's conditional R-squared).

- K1 = Activist behaviour, 22%
- K2 = support increased taxes on fossil fuels, 14%
- K3 = support subsidised renewable energy, 11%.

Appendix L

LIST OF COUNTRY CODES

Table 24. List of country codes.

Code	Country	Code	Country	Code	Country
AE	United Arab Emirates	GR	Greece	NZ	New Zealand
AF	Afghanistan	HK	Hong Kong	PH	Philippires
AT	Austria	HU	Hungary	PK	Pakistan
AU	Australia	ID	Indonesia	PL	Poland
BE	Belgium	IE	Ireland	PT	Portugal
BG	Bulgaria	IL	Israel	QA	Qatar
BH	Bahrain	IN	India	RO	Romania
BR	Brazil	IQ	Iraq	RS	Serbia
CA	Canada	IR	Iran	RU	Russia
CG	The Congo	IT	Italy	SA	Saudi Arabia
CH	Switzerland	JP	Japan	SE	Sweden
CL	Chile	KR	Korea (South)	SG	Singapore
CZ	Czech Rep.	KW	Kuwait	SI	Slovenia
DE	Germany	KZ	Kazakhstan	SK	Slovakia
DK	Denmark	LT	Lithuania	TH	Thailand
EE	Estonia	LV	Latvia	TN	Tunisia
EG	Egypt	MA	Morocco	TR	Turkey
ES	Spain	MX	Mexico	TW	Taiwan*
FI	Finland	MY	Malaysia	UA	Ukraine
FR	France	NG	Nigeria	VE	Venezuela
GB	Great Britain	NL	Netherlands	ZA	South Africa
GH	Ghana	NO	Norway		

*The status of Taiwan as a country is disputed. The codes used for it is not therefore official ISO-2 status.

Endnotes

1. Michael Shermer has suggested there have been 100,000 past religions (see citation in Dean Boening's *The Extinct Cognitive Christian*, iUniverse, 2000). But at later dates he quoted lower numbers, which seem more consistent with anthropological data.

2. *Climate Etc* can be found at https://judithcurry.com/. My original posts were as follows:

- Chapter 2: 'Climate culture' https://judithcurry.com/2015/11/20/climate-culture.
- Chapter 4: 'Child prophets and proselytizers of climate catastrophe', https://judithcurry.com/2019/07/29/child-prophets-and-proselytizers-of-climate-catastrophe/.
- Chapter 5: 'The catastrophe narrative', https://judithcurry.com/2018/11/14/the-catastrophe-narrative/.
- Chapter 6: 'The denialism frame', https://judithcurry.com/2016/04/21/the-denialism-frame/.
- Chapter 7: 'Innate skepticism', https://judithcurry.com/2017/02/20/innate-skepticism.
- Chapter 11: 'Cultural motivations for wind and solar renewables deployment', https://judithcurry.com/2020/11/19/cultural-motivations-for-wind-and-solar-renewables-deployment/.

3. A list of comparisons of the climate-change domain to religions/the religious, can be found at A West, 'Climate culture'. *Climate Etc* blog, 2015. https://judithcurry.com/2015/11/20/climate-culture/. There have been many more comparisons since that date.

4. H von Storch and W Krauss, *Die Klimafalle* [The Climate Trap]. Hanser, Carl GmbH, 2013. Translated quote from *NoTricksZone* blog, http://notrickszone.com/2013/01/28/new-book-by-hans-von-storch-climate-scientists-took-on-role-of-prophets-completely-in-over-their-heads/.

5. Michael Shellenberger. *Apocalypse Never*. Harper, 2020.

6. This is a fictional publication, but I think it really *ought* to exist. See Dan Kahan's blog of the same name: http://www.culturalcognition.squarespace.com/blog.

7. In this context, reductionism is the notion that complex social phenomena (such as cultural attitudes and behaviours) are due in large part to simpler processes, which themselves have causes that are simpler still. While some people object to

the idea of complexity arising out of simple processes, it is an expectation for an emergent system, for instance the immensely complex output of the simple process of natural selection.

8. Blackmore, one of the 'three memeteers': https://www.susanblackmore.uk/memes-and-tremes/.

9. British biologist and author Richard Dawkins: https://www.richarddawkins.com/.

10. C Booker, 'The BBC's "dirty little secret" lands it in a new scandal', *The Telegraph*, 2012. https://www.telegraph.co.uk/comment/9684775/The-BBCs-dirty-little-secret-lands-it-in-a-new-scandal.html.

11. Climategate references:

- J Curry, 'The legacy of Climategate', *Climate Etc* blog, 2012. https://judithcurry.com/2012/05/07/the-legacy-of-climategate/.
- J Curry, 'The legacy of Climategate: Part II', *Climate Etc* blog, 2012. https://judithcurry.com/2012/05/09/the-legacy-of-climategate-part-ii/.
- J Curry, 'The legacy of Climategate: Five years later', *Climate Etc* blog, 2014. https://judithcurry.com/2014/12/01/the-legacy-of-climategate-5-years-later/.
- J Curry, 'The legacy of Climategate: Ten years later', https://judithcurry.com/2019/11/12/legacy-of-climategate-10-years-later/.

12. The Hockey Stick graph depicts long-term proxies for global temperature. See A Montford, *The Hockey Stick Illusion: Climategate and the Corruption of Science* (Anglosphere Books, 2015) for a critical take on this icon, and M Mann, *The Hockey Stick and the Climate Wars* (Columbia University Press 2013) for the take by the author of the Hockey Stick graph itself.

13. It was challenged via various technical criticisms. Also, in the eyes of non-technical folks 'the pause' was undermining its depiction of a monotonic steep rise in the 'blade' of the Hockey Stick. For 'the pause', see J Curry, 'Causes and implications of the pause', *Climate Etc*, 2014. https://judithcurry.com/2014/03/04/causes-and-implications-of-the-pause/.

14. See MS Gazzaniga, *Who's in Charge*. Harper Collins, 2012.

15. There is great insight on cultural group selection and gene-culture coevolution in: J Henrich 'Cultural group selection, coevolutionary processes and large-scale cooperation', *Journal of Economic Behavior & Organization* 2003; 53(1): 3–35.

16 P Richerson et al. 'Cultural group selection plays an essential role in explaining human cooperation: A sketch of the evidence'. *Behavioral and Brain Sciences* 2014, 39, E30.

17 TM Waring and ZT Wood. 'Long-term gene–culture coevolution and the human evolutionary transition'. *Proceedings of the Royal Society B: Biological Sciences* 2021; 288(1952): 20210538.

18. Richerson et al. say:

> The idea that cultural variation fell under group selection at the scale of tribes is a modernization of a hypothesis first proposed by Darwin in *The Descent of Man*… humans developed a social psychology organized around culturally acquired social rules ('norms' to psychologists, 'institutions' to sociologists). People came to take on social identities that tied them emotionally to their social groups. We became exquisitely sensitive to social boundaries symbolically marked by language, dress, ritual, and other stylistic differences between 'us' and 'them'.

- PJ Richerson et al., 'Gene-culture coevolution in the age of genomics', *PNAS* 2010; 107(Suppl 2): 8985–8992.

19. David Sloan Wilson says that:

> …culture appears to represent a group-level mechanism for human groups to adapt to environmental challenges…and, particularly, changes over time…

- R O'Gorman et al. 'For the good of the group? Exploring group-level evolutionary adaptations using Multilevel Selection Theory' *Group Dynamics: Theory, Research, and Practice* 2008; 12.1: 17.

20. The consistent selection of these genes across humanity is contingent upon them mingling between groups (in the longer term), which naturally occurs through events such as wars or natural disasters or the collapse of cultures as others arise; these tend to stir some individuals into crossing cultural boundaries.

A relatively high level of immigration does not normally dissolve the integrity of a cultural group; if this was not the case, most cultural groups, and so the major effect of competition between them, would collapse. Cultures reject the values of outsiders, so immigrants must learn the new culture, or perhaps live at the fringes, without proper acceptance, or else try their luck elsewhere, all options tending to preserve the integrity of the host culture. So, provided they don't practice their old culture, immigrant genes can mingle; those individuals who take on board a new culture may breed freely within it, and this doesn't disrupt its integrity.

21. Multiple sources for gene-culture coevolution:

- PJ Richerson et al., 'Gene-culture coevolution in the age of genomics', *PNAS* 2010; 107(Suppl 2): 8985–8992.
- T Waring and Z Wood, 'Long-term gene–culture coevolution and the human evolutionary transition', *Proceedings of the Royal Society B* 2021; 288(1952): 20210538.
- J Henrich 'Cultural group selection, coevolutionary processes and large-scale cooperation', *J,ournal of Economic Behavior & Organization* 2003; 53(1): 3–35, (modelling gene-culture interaction and group selection; cultural evolution will favour pro-sociality, whether starting from a kin-selection or group-selection perspective).

For the classic example of gene-culture co-evolution, see Endnote 212.

22 According to DH Stein et al.:

> Rituals increase internal commitment…a primary function of collective rituals is to signal that the ritual-performers are committed to the group. This function of signaling commitment is based on the idea that it is costly to perform rituals (due to the time, effort, and/or resources they require), making their performance a stronger signal of commitment toward a group than verbal expressions, which are susceptible to deception. The costliness of rituals ensures that only those individuals who are devoted to the group will participate in the ritual, distinguishing cooperators from defectors and ultimately fostering trust among members.
>
> …in addition to this external signaling function, group rituals also serve the function of increasing the ritual-performer's internal commitment (i.e. the strength of an individual's attachment to their group). Thus, even performing a ritual alone with no one watching, when it has no signaling power, still enhances the performer's internal commitment to their group.

- DH Stein et al., 'A sacred commitment: How rituals promote group survival' *Current Opinion in Psychology* 2021, 40: 114–120.

23 Watson-Jones and Legare suggest that:

> …rituals serve four core functions that address the adaptive problems of group living: They (a) provide reliable markers of group membership, (b) demonstrate commitment to the group, (c) facilitate cooperation with social coalitions, and (d) increase social group cohesion. We propose that the capacity to engage in ritual is a psychologically prepared, culturally inherited behavior geared toward facilitating social group dynamics.

- RE Watson-Jones and CH Legare, 'The social functions of group rituals', *Current Directions in Psychological Science* 2016; 25(1): 42–46..

24 Gregory Bonn says:

> …it is clear that even a few words on sensitive culturally related topics can produce emotive responses in people, which may range from say pride in the national flag to nationalist hostility, to outrage and insults from merely mentioning a simple biological fact (courtesy of extreme trans-rights culture), to religious joy in salvation, to racial hate, to revulsion at out-group references, to extreme loyalty, to hope, to despair (courtesy of climate catastrophism), to guilt (about sin, or a carbon footprint) and many more. In other words, culturally-framed phrases easily push our emotive hot-buttons. While the academic literature explores many specific cases of emotive cultural expression, for instance in a particular political culture or as prompted by a specific religious conflict, generally in great detail, there appears to be far less material on the generic invocation of emotion by any culture. At the most basic level, it is acknowledged that the root of all cultural identity lies in a set of primary emotions (see F-link), but I've yet to find a more sophisticated 'map' of emotions that could in principle result from equivalent prompts within different cultures (i.e. a neat descriptive list of the hot-buttons). This may be due to my lack of familiarity with the literature. Some

hot-buttons are made clear simply from the elementary nature of the narrative variants that target them, for example see 'hope and fear' and 'anxiety for children' in Chapter 5. However, for more complex variants it is not always so clear what cocktail of emotions they are appealing to.

- G Bonn. 'Primary process emotion, identity, and culture: cultural identification's roots in basic motivation'. *Frontiers in Psychology; Social and Evolutionary Neuroscience* 2015; 6. https://doi.org/10.3389/fpsyg.2015.00218.

25 Social anthropologist Charles Whitehead says:

> ...you could say that it is the 'job' of human culture to falsify our perceptions of ourselves and the world we live in. 'Collective deceptions' were at one time necessary to coerce our social but selfish ancestors into collaborating in a non-selfish system, and western science has not yet freed itself from them.

- The original link is http://www.consciousness.arizona.edu/abstracts.htm. However, this has now gone dark. The same quote can be seen at: http://www.psybertron.org/archives/1260.

26 Anthropologist Chris Knight, whose work I appreciate, yet whose activism I do not, notes the collective deception of culture as part of his studies on language. For example:

> ...far from embodying self-evident truth, symbolic culture may be better understood as a world of patent fictions held collectively to be true on some deeper level. Myths, dramatic performances, art and indeed all expressions of human symbolic culture may in this light be understood as 'collusion in deception'.

- From: C Knight, 'Ritual/speech coevolution: a solution to the problem of deception', in: JR Hurford et al. (eds), *Approaches to the Evolution of Language*, Cambridge University Press, 1998. http://www.chrisknight.co.uk/ritual-speech_coevolution/.

Richard Dawkins' better known portrayal of religion as a delusion makes the same point, but seems to me to skirt rather too close to interpreting it as pathological, especially in the context of his emotive brand of atheism; culture is not delusional, in the sense of an illness or mental debilitation.

27 For an overview of memes as a population, see Balkin's *Cultural Software*.

Memetics tends to underemphasise the behavioural impacts of memes that are also cultural narratives, focusing on those that directly affect meme propagation. Other approaches recognise that such memes (although they would not use the term) have a variety of effects on people, for example triggering altruism and cooperation, but don't tend to see them as a family of variants that obey rather simple mathematical rules.

I believe it is possible to combine the two approaches simply by considering the impact each has on the other.

- JM Balkin. *Cultural Software: A theory of ideology*. Yale University Press, 1998.

28 See, for example, Maunder's *Our Lady of the Nations*, particularly Table 5.2. While limited to Catholic Marian visions, the total number of seers across Europe throughout the 19th and 20th centuries is heavily weighted towards under-18s, and also to females rather than males. Maunder says that children and adolescents were afforded a special status as visionaries because of their alleged innocence.

- C Maunder, *Our Lady of the Nations: Apparitions of Mary in 20th-Century Catholic Europe*, Oxford University Press, 2016.

29 This is the field of Cultural Consensus Theory (CCT). See http://www.public.asu.edu/~huanliu/sbp09/Presentations/paperpresentations/William Batchelder-CCTSBP092009%281%29.pdf, and a do-it-yourself CCT pack at https://mran.microsoft.com/package/CCTpack.

30. For instance, in an exchange between Enlil, the head of the Mesopotamian pantheon, and the other great gods, there is an emphasis on the religiously imposed duty of maintaining irrigation works and ditches, to water the earth and make plant life flourish, bringing piles of harvested grain.

- For the full verse and context see: D Bokovoy, 'KAR 4: An example of Mesopotamian creation motifs in Genesis', Patheos, 2014. https://www.patheos.com/blogs/davidbokovoy/2014/12/kar-4-an-example-of-mesopotamian-creation-motifs-in-genesis/.

31 Jack Balkin says:

> Because cultural software [i.e. the narrative] is transmitted from person to person, there is a natural analogy between cultural software and viruses. The human mind is susceptible to memes just as the human body is susceptible to infection from particular viruses. The study of cultural evolution is a study of comparative epidemiology. Some memes are more contagious, or 'catching', than others in a population and thus spread more widely and successfully. The metaphor of susceptibility to viruses helps us understand the deep connections between the power of human intelligence and its vulnerabilities.

- JM Balkin. *Cultural Software: A theory of ideology*. Yale University Press, 1998.

32. David Sloan Wilson website: https://thisviewoflife.com/profile/david-sloan-wilson/.

33. See this useful teaching video by Alex Barrientos, which outlines the spectrum of opinion on the evolutionary origins of religion: https://www.academia.edu/video/kzw0gj?email_video_card=description-read-more&pls=RVP.

34. Some people speculate that omnipresent 'Big Gods' provide an authoritative source that helps to hold the community together. However, it is the narrative triggering of behaviours arising from simply being 'in the cultural club' that really holds the group together. Big Gods are only a surface characteristic of some specific religious types; some religions don't feature them, and neither do secular cultures.

35. The main means to reinforce gas-mask discipline was repetition, through a requirement for frequent drills at (UK) schools. However this was reinforced, and sometimes excessively so, through live tear-gas tests: 'The aim was twofold: to test the fit of gas masks, as any leak or gap would be immediately obvious; and, through forcing the children to breath a little of the gas before leaving, to train them rather brutally in the importance of gas protection.'

For further detail, see:

- G Moshenka, 'Gas masks: material culture, memory, and the senses', *Journal of the Royal Anthropological Institute* 2010; 16(3): 609–628.
- G Moshenka, 'Government gas vans and school gas chambers: preparedness and paranoia in Britain, 1936–1941', *Medicine, Conflict and Survival* 2010; 26(3): 223–234.

36. For discussion of the instilling of fearful Easter messages into Christian children, see: S Evans, 'Let's stop confusing education with religious inculcation', *National Secular Society* blog, 2015. https://www.secularism.org.uk/opinion/2015/04/lets-stop-confusing-education-with-religious-inculcation.

For similar Islamic messages see: A Kamguian, 'The impact of religion on children's development', *National Secular Society* blog, 2004. https://www.secularism.org.uk/33004.html.

37. See:

- D Carrington, 'School climate strikes: 1.4 million people took part, say campaigners', *The Guardian*, 2019. https://www.theguardian.com/environment/2019/mar/19/school-climate-strikes-more-than-1-million-took-part-say-campaigners-greta-thunberg.
- *School Strike for Climate* website: https://www.schoolstrike4climate.com/.

38. Greta Thunberg bio: https://www.britannica.com/biography/Greta-Thunberg.

39. Greta Thunberg's UN speech: https://www.youtube.com/watch?v=VFkQSGyeCWg&feature=youtu.be.

40. Malala Yousafzai bio: https://www.nobelprize.org/prizes/peace/2014/yousafzai/biographical/.

41. Malala Yousafzai's speech to the United Nations: https://www.youtube.com/watch?v=5SClmL43dTo.

42. https://www.biography.com/news/black-history-birmingham-childrens-crusade-1963-video.

43. N Rothman. 'Child soldiers in the culture wars', *Commentary* magazine, 2019. https://www.commentarymagazine.com/politics-ideas/child-soldiers-in-the-culture-wars/.

44. In relation to Greta Thunberg's Nobel prize nomination, Madeline Grant in the UK newspaper *The Telegraph* (see link below) reports a belief in this false meme:

The Nobel Committee is unwittingly adopting a common trope in contemporary debate, attributing, in biblical fashion, special insights to children despite their limited knowledge of complex subjects – 'out of the mouth of babes and sucklings', as the saying goes.

- M Grant, 'Can we please stop garlanding children for being wrong?', *The Telegraph*, 2019. https://www.telegraph.co.uk/news/2019/03/16/can-please-stop-garlanding-children-wrong/.

Grant is right that climate is a 'complex subject', and that we would not expect children to understand it better than adults. However, the meme implies that children – and in this case Greta Thunberg – can 'see through' artificial conventions or groupthink to the truth beneath, along the lines of The Emperor's New Clothes. Current research (see below) suggests this is extremely unlikely: in general, children may be less aware of the consequences of publicly contradicting a cultural norm, but on average they're no more likely to see through it than adults, and probably less. See:

- VC Vaden and JD Woolley, 'Does God make it real? Children's belief in religious stories from the Judeo-Christian tradition'. *Child Development* 2011; 82(4): 1120–35.
- KH Corriveau et al., 'Judgments about fact and fiction by children from religious and nonreligious backgrounds'. *Cognitive Science* 2015; 39(2): 353–82.
- JD Woolley and M Ghossainy, 'Revisiting the fantasy-reality distinction: children as naïve skeptics'. *Child Development* 2013; 84(5): 1496–510.

45. This doesn't imply *personal* irrationality or mental illness, or a lack of intelligence or integrity. Cultural adherents are committed to grand-scale groupthink, emotive belief, which can result in the bypassing of reason. The mechanism exists in all humans and is perfectly normal. The irrationality is essentially carried in the cultural narrative.

46. Both reality-based and cultural pitches (which don't necessarily have to come from children) can benefit if they create emotional bias. However, *cultural* fears are based on emotive conviction rather than reason, so they easily resonate with a suitably aligned pitch, increasing the likelihood of irrational (but culturally consistent) responses. Note: emotional bias in the climate domain is systemic:

- A West, 'Contradiction on emotional bias in the climate domain', *Climate Etc* blog, 2015. https://judithcurry.com/2015/04/24/contradiction-on-emotional-bias-in-the-climate-domain/.

47. According to *Our World In Data,* extreme poverty has fallen dramatically in the last 30 years (see OWID1-link below), and is projected to fall still further. However, it is a stubborn problem in some regions, notably Sub-Saharan Africa. Moreover, 'virtually half of the people living in extreme poverty are under 18 years of age' (see OWID2-link), of which the great majority are under 15. Malala is correct that eliminating poverty would also remove what is currently a major barrier to provision of a proper education for, currently, hundreds of millions of children.

- OWID1-link: https://ourworldindata.org/extreme-poverty.

• OWID2-link: https://ourworldindata.org/children-and-poverty-results-from-new-data.

Literacy by country (OWID3-link) shows that poverty and Sub-Saharan Africa are not the only problem areas; countries elsewhere, particularly those with religious (Islamic) regimes, also have a current problem of poor education.

• OWID3-link: https://ourworldindata.org/grapher/literacy-rate-by-country.

Thus Malala is referring to *current and real* problems, which are clear in the data.

• See: M Kugelman, 'Why Pakistan hates Malala', *Foreign Policy*, 2017. https://foreignpolicy.com/2017/08/15/why-pakistan-hates-malala/.

48. Malala's ask is big, although she hasn't dictated specific policies or set a timescale. However, the spread of prosperity in recent decades means that the key indicators are already heading in the right direction. As noted in Endnote 47, extreme poverty has fallen dramatically, and is projected to continue falling for some years. And although there is extremism/cultural resistance, the numbers of those without education is falling almost everywhere (see link below), and this trend is projected to continue. This doesn't mean improving the situation is easy, and indeed in 2050 some places are still expected to have significant minorities of children receiving no education. But it is not unreasonable to ask for a major acceleration of programs to eliminate poverty and expand education, and to ask the UN to pressure regimes that resist them.

• Link: https://ourworldindata.org/grapher/projections-of-the-rate-of-no-education-based-on-current-global-education-trends-1970-2050?country=BGD+BRA+CAF+TCD+CHN+EGY+IND+IDN+NGA+PAK+RUS+GBR+USA.

49. 'Being here with such honourable people is a great moment in my life…Honourable Secretary General… Dear fellows…Dear sisters and brothers…'

• Full text at: https://malala.org/newsroom/archive/malala-un-speech.https://www.un.org/News/dh/infocus/malala_speach.pdf.

50. From Greta's short UN speech; see the full text at the link below. See also the extract from her pitch to the UK Parliament in Endnote 53. http://kismetgirls.com/conservationists/Greta_Thunberg's_Speech_COP24_Cimate_Change_Conference_United_Nations.htm.

51. See 'already suffering the consequences' in the extract in Endnote 53.

52. Her solution is: 'We need to keep the fossil fuels in the ground'. Her UN speech implies that about 60 years away (her 75th birthday) would be far too late for this, from which we deduce that most fossil fuel usage must cease decades before then. In her speech to the UK parliament, she states that (presumably worldwide) emissions must already be down by 50% by 'around 2030' (see the extract in Endnote 53), i.e. only 11 years after the speech. While these two narratives are broadly consistent, the latter emphasises the idea of major progress on emissions reductions in the very near term, as does her comment to Sky TV about the UK goal of Net Zero by 2050

(see Endnote 55). She also appears to make common cause with Extinction Rebellion (see XR-link), whose goal is Net Zero by 2025 (see NZ-link, although it is not always clear whether this applies to only to the UK, or globally).

- XR-link: J Conley, '"We are the ones making a difference": Greta Thunberg addresses Extinction Rebellion in London', *Common Dreams* blog, 2019. https://www.commondreams.org/news/2019/04/21/we-are-ones-making-difference-greta-thunberg-addresses-extinction-rebellion-London.
- NZ-link: C Farand, 'Extinction Rebellion goes global with call for net zero emission by 2025', *The Energy Mix* blog, 2018. https://theenergymix.com/2018/11/26/extinction-rebellion-goes-global-with-call-for-net-zero-emission-by-2025/.

53. Greta's speech to the UK Parliament found in full at *The Guardian*: https://www.theguardian.com/environment/2019/apr/23/greta-thunberg-full-speech-to-mps-you-did-not-act-in-time. Some important extracts are:

> I know many of you don't want to listen to us – you say we are just children. But we're only repeating the message of the united climate science…
>
> …You lied to us. You gave us false hope. You told us that the future was something to look forward to. And the saddest thing is that most children are not even aware of the fate that awaits us. We will not understand it until it's too late. And yet we are the lucky ones. Those who will be affected the hardest are already suffering the consequences…
>
> Around the year 2030, 10 years 252 days and 10 hours away from now, we will be in a position where we set off an irreversible chain reaction beyond human control, that will most likely lead to the end of our civilisation as we know it. That is unless in that time, permanent and unprecedented changes in all aspects of society have taken place, including a reduction of CO_2 emissions by at least 50%.
>
> And please note that these calculations are depending on inventions that have not yet been invented at scale, inventions that are supposed to clear the atmosphere of astronomical amounts of carbon dioxide…
>
> These projections are backed up by scientific facts, concluded by all nations through the IPCC. Nearly every single major national scientific body around the world unreservedly supports the work and findings of the IPCC.

54. In her speech (see https://www.youtube.com/watch?v=ESDpzwWrmGg) to the French National Assembly, Greta also cites the IPCC's claims (link below) about the carbon budget that it says remains if temperature increases are to be kept to under 1.5°C.

She goes on, as in her previous speeches, to castigate adults for not being mature enough to grasp these issues, leaving the issue instead to children. She also claims that business as usual will 'likely' lead to 'tipping points' and 'irreversible climate breakdown', and she raises the strawman that those who challenge (her, the children, and) the IPCC, are challenging the conclusion of a climate emergency and all this entails. Yet her judgement of climate breakdown/emergency/end of civilisation represents a very different position to that expounded by the IPCC in the technical chapters of its scientific reports. Her proposed solution: 'has to include everything and everyone'.

- IPCC, Special Report on 1.5 Degrees, Chapter 2, p. 108 Table 2.2. https://www.ipcc.ch/site/assets/uploads/sites/2/2019/05/SR15_Chapter2_Low_Res.pdf

55. When Sky News put it to Greta (see link below) that the official UK target of Net Zero by 2050 is 'immensely ambitious', she says:

> ...You could argue that is better than nothing but I think it's actually doing more harm than good...[and therefore regarding UK politicians]...If they don't act now, then in the future they will be seen as some of the greatest villains in human history...

She goes on to suggest that the UK policy 'sends a signal' so that people will think they can 'continue like now' for maybe about '20 years'. Presumably she'd want some very significant part of the Net Zero reduction to be delivered within the next twenty years or sooner. This is confirmed by her speech to the UK Parliament, in which she implies a target of at least 50% reduction by 2030.

- H Thomas-Peter, 'UK carbon emissions target "doing more harm than good", teen climate activist Greta Thunberg tells Sky News', Sky News, 2019. https://news.sky.com/story/uk-carbon-emissions-target-doing-more-harm-than-good-teen-climate-activist-greta-thunberg-tells-sky-news-11764916.

56. At COP24, she told the audience, 'You are not mature enough to tell it like it is. Even that burden you leave to your children.' She also told the UK Parliament (Endnote 53) that they may be 'seen as some of the greatest villains in human history' for adopting a target of 'Net Zero by 2050'.

- G Thunberg, Speech to COP24: https://unfccc.int/sites/default/files/resource/COP24_HLS_ENGO.pdf.

57. See 'you lied to us', from the extract in Endnote 53.

58. Nongqawuse bio: https://www.sahistory.org.za/people/nongqawuse.

59. The main Xhosa homeland at the time was British Kaffraria. Numbers relating to the population drop appear to vary somewhat with source. Price says 40,000 were starving.

Price also says:

> 20,000 moved into the Cape colony to avoid that fate by becoming agricultural and domestic laborers. The result was that the population of British Kaffraria declined... ...from 105,000 to 37,000.

- See: R Price, *Making Empire: Colonial Encounters and the Creation of Imperial Rule in Nineteenth-Century Africa*, Cambridge University Press, 2008.

This disagrees slightly with Pieres, who says 27,000 remained in Kaffraria (from which my drop of ~78,000). Due to a lack of census data, all figures are estimates in any case.

- JB Pieres, 'The central beliefs of the Xhosa cattle-killing', *Journal of South African History*, 1987; 28: 43–63. http://smu-facweb.smu.ca/~wmills/course322/Central_Beliefs_Cattle_Killing.pdf.

60. There are summaries of the story at Wiki (see Wiki-link below) and, much longer, at South African History Online (see SAHO-link below). Some complexities and ongoing debate include: the extent of Christian influence inspiring the movement, the precise role and input of Nongqawuse's uncle (he'd spent some time among the Europeans apparently), the differing reports of colonial and indigenous witnesses, and the extent to which Europeans helped the Xhosa or exploited their plight. Some colonists certainly tried to prevent disaster and to help afterwards, but others took advantage of the situation to seize land and obtain cheap labour. 'However, as noted in Endnote 61, historians do view this movement as a *cultural* one, a millennialist response to social stress.

- Wiki-link: https://en.wikipedia.org/wiki/History_of_the_Cape_Colony_from_1806_to_1870#Xhosa_cattle-killing_movement_and_famine_(1854-1858).
- SAHO-link: https://www.sahistory.org.za/article/cattle-killing-movement.

61. Historians now view this movement as a millennialist response (millennialism and millennarianism are essentially equivalent terms), both directly to the lung disease spreading among Xhosa cattle, and to the stress to Xhosa society caused by the continuing loss of their territory and autonomy. Peires also says 'This fusion of Xhosa and Christian prophecies created an apocalyptic tradition which outlasted the Cattle-Killing and remained potent well into the twentieth century.'

- JB Peires, *The Dead Will Arise*, Indiana University Press, 1989.

62. The single largest value judgement regarding these checks is whether science supports the certainty of imminent (decades) global climate catastrophe (absent dramatic emissions reduction), because conflict on this issue is not only within the public domain but within the enterprise of science too. There are at least four viewpoints which can all boast at least some support from scientists with relevant experience and qualifications: the sceptics, the Lukewarmers, the mainstream, and the catastrophists, the latter objecting to the IPCC for being too conservative, while sceptics and Lukewarmers think it is too alarmist. The mainstream point of view has by far the most formal support, probably by much more than an order of magnitude. Taking this as the gold-standard, as noted in Section 5.2.2, this certainly does not support a certainty of imminent global climate catastrophe.

63. In the historic case of the Xhosa cattle killings, the people and the relevant leadership (the council of chiefs) were also strongly primed. From *South African History Online* (https://www.sahistory.org.za/article/cattle-killing-movement):

> …Prior to Nongqawuse's prophecy, came forth a number of similar prophecies. By 1855, more than five prophets had emerged in British Kaffraria saying that they were in contact with the black nation across the sea that would come to the

> aid of the Xhosa. The black nation was reference to the Russians who had killed George Cathcart in the Crimean War. The prophets preached that the Xhosa people should stop cultivation and kill their cattle.'

George Cathcart had been a governor of the Cape Colony during the last stage of the 8th Xhosa war, about three years before the start of the cattle killing movement. Similarly to Nongqawuse, one of the prophets he mentioned had even ordered the building of new cattle enclosures, which would be filled when the resurrected and 'unsullied' cattle appeared at the fulfilment of the prophecy. These earlier prophets took a more anti-white stance than Nongqawuse, but their activities fizzled out in early 1856, after the Crimean war ended and the British made peace with the supposed Russian allies.

However, there was a lasting impact:

> When Nongqawuse made her prophecy against the backdrop of impending destitution, it fell on fertile ground to be received by the Xhosa nation, weakened in their resistance against colonialism.

This prior cultural priming featured both apocalypse and salvation.

64. In a Sky News interview, she said:

> ...why should we care about our future, educating ourselves, if you don't care about our future. I think that is a very powerful message, and many people feel guilty when children say that.

• Full interview at: H Thomas-Peter and D Mercer, 'UK carbon emissions target "doing more harm than good", teen climate activist Greta Thunberg tells Sky News', Sky News, 2019. https://news.sky.com/story/uk-carbon-emissions-target-doing-more-harm-than-good-teen-climate-activist-greta-thunberg-tells-sky-news-11764916.

65. These characteristics can unite opposite parts of a spectrum of belief. Those who are inclined towards belief yet are suspicious of cynical agendas when political heavyweights push it, are emotively convinced when the prophet is innocent and apolitical and 'above questioning'. Those at the opposite end of the spectrum who already have strong belief, may nevertheless not get behind political heavyweights either, because they worry that these may betray or delay or dilute the 'high purpose' for reasons of political gain or maintaining appeal to the doubters. Such ardent believers can similarly back a young girl without having any qualms.

66. Encyclopedia.com says:

> Millenarianism, known also as millennialism, is the belief that the end of this world is at hand and that in its wake will appear a New World, inexhaustibly fertile, harmonious, sanctified, and just. The more exclusive the concern with the End itself, the more such belief shades off toward the catastrophic; the more exclusive the concern with the New World, the nearer it approaches the utopian.

This source goes on to categorise types of millennarianism, from which I believe

Greta and Nongqawuse's expressions would be 'transformative', but Greta's also tending more to the catastrophist end of the scale. The prominence of women leading millennarian movements is also noted, along with the issue that why this is so has not been given appropriate scholarly attention.

67. In Greta's case this is very explicit:

> We have not come here to beg world leaders to care. You have ignored us in the past and you will ignore us again. We have run out of excuses and we are running out of time. We have come here to let you know that change is coming, whether you like it or not.

- Full text: http://kismetgirls.com/conservationists/Greta_Thunberg's_Speech_COP24_Cimate_Change_Conference_United_Nations.htm

68. As explained in detail in Endnote 63 above, in both the Greta/modern case and the Nongqawuse/historic case, society and leadership were primed for decades by the relevant cultural narrative. Such narratives are not 'imposed' from the top down, although when they take hold within influential sections/organs of society, and especially within leadership, their ability to propagate more effectively to mass audiences is much increased.

69. Environmentalist and nuclear advocate Michael Shellenberger points out the 'renewable energy advocates like Greenpeace, AOC and Thunberg' oppose nuclear energy (see Forbes-link below). Anti-nuclear sentiment stretches back long before current climate catastrophism and, according to Shellenberger, grew at least in part from (fossil fuel) vested interests, although instinctive fears connected to nuclear weapons and fall-out are likely a big factor with or without such exploitation. Cultures often sweep older fears into their narratives, despite this potentially creating logical disconnects or outright contradictions (nuclear has no emissions!).

- Forbes-link: M Shellenberger, 'Why renewables advocates protect fossil fuel interests, not the climate', *Forbes*, 2019. https://www.forbes.com/sites/michaelshellenberger/2019/03/28/the-dirty-secret-of-renewables-advocates-is-that-they-protect-fossil-fuel-interests-not-the-climate/.

70. Heyward and Rayner say:

> Millenarian rhetoric ultimately aims to promote behavioural change. Humanity is admonished for its current failings and encouraged to pursue a different path. Most millenarian accounts prescribe that material goods must be redistributed, in some cases rejected outright. Conventional activities aimed at securing those goods must cease. In religious apocalypticism, this change is regarded as essential preparation for the new era of very different forms of social and spiritual relations. In narratives of cataclysmic forewarning, such as green millenarianism, the rejection and redistribution of key material goods is necessary in order to avert the impending catastrophe. In either case, the momentous nature of the changes required means that resistance is to be expected, but it is permissible to overcome it in order to achieve the desired ends' [emphasis mine].

- C Heyward and S Rayner, *Apocalypse Nicked!,* Climate Geoengineering Governance Working Paper Series: 006 (2013). http://web.archive.org/web/20200302091251/http://geoengineering-governance-research.org/perch/resources/workingpaper6heywardraynerapocalypsenicked.pdf.

71 HM Hallam, 'Greta: Germany making "mistake" by ditching nuclear for coal'. *Deutsche Welle*, 2022. https://www.dw.com/en/greta-thunberg-germany-making-mistake-by-ditching-nuclear-power-for-coal/a-63406732.

72 I McTaggart, 'Greta Thunberg: It's time to transform the West's oppressive and racist capitalist system'. *The Telegraph*, 2022. https://www.telegraph.co.uk/news/2022/11/02/greta-thunberg-time-overthrow-wests-oppressive-racist-capitalist/.

73. There is already a great deal of tension among adherents of climate catastrophism regarding support for nuclear power. This is epitomised by Naomi Oreskes' article in *The Guardian* (see G-link below), calling out James Hansen, Kerry Emanuel, Ken Caldeira and Tom Wigley among others as 'denialists', because of their strong support for nuclear as the backbone of emissions reduction. Hansen is not only a prolific propagator of the Catastrophe Narrative, he is perhaps the best-known climate scientist in the public domain.

- G-link: N Oreskes, 'There is a new form of climate denialism to look out for – so don't celebrate yet', *The Guardian*, 2015. https://www.theguardian.com/commentisfree/2015/dec/16/new-form-climate-denialism-dont-celebrate-yet-cop-21.

74. Though as per Endnote 66, female representation at the top of millennarian cultures is more likely.

75. B O'Neill, 'The cult of Greta Thunberg', *Spiked Online*, 2019. https://www.spiked-online.com/2019/04/22/the-cult-of-greta-thunberg/.

76. P Neuding, 'Self-harm versus the greater good: Greta Thunberg and child activism', *Quillette*, 2019. https://quillette.com/2019/04/23/self-harm-versus-the-greater-good-greta-thunberg-and-child-activism/.

77. R Clarke, 'The trouble with Greta Thunberg', *The Spectator*, 2019. https://blogs.spectator.co.uk/2019/04/the-trouble-with-greta-thurnberg/.

78. See M Newman, 'Climate change signals part of socialist plot ', *The Australian*, 2019. https://www.theaustralian.com.au/commentary/climate-change-signals-part-of-socialist-plot/news-story/a29d692e6efed92606cdf4e56315c0be (paywall), and A Seaman, 'We need a grown-up debate about climate change ', *Spiked Online*, 2019. https://www.spiked-online.com/2019/04/29/we-need-a-grown-up-debate-about-climate-change/.

79. https://www.biography.com/news/black-history-birmingham-childrens-crusade-1963.

80. H Thomas-Peter and D Mercer, 'UK carbon emissions target "doing more harm than good", teen climate activist Greta Thunberg tells Sky News', Sky News, 2019. https://news.sky.com/story/uk-carbon-emissions-target-doing-more-harm-than-good-teen-climate-activist-greta-thunberg-tells-sky-news-11764916.

81. As part of which, it's likely that the children's climate strikes including anti-authority chants and dressing up, are actually a fulfillment of 'rites of passage', in the context of a cultural identity within climate catastrophism. That is, becoming a recognised member of this cultural group who advocate for 'emergency' action, who fight to gain authority and voice against older cultural norms. See CS Alcorta, 'Adolescence and religion: An evolutionary perspective', in: JR Liddle and TK Shackleford (eds), *The Oxford Handbook of Evolutionary Psychology and Religion*, Oxford University Press, 2016. Abstract follows:

> Throughout the world adolescence is deemed the appropriate life stage to 'learn religion'. Nearly three-quarters of societies conduct adolescent rites of passage transmitting sacred rituals and beliefs. Neurophysiological changes that occur during adolescence render this an 'experience-expectant' period for the transmission of religious schema and values. Brain regions critical to emotional, social, and symbolic processing mature, creating a plastic neural substrate for imbuing social and symbolic schema with emotional meaning and reward value. Religion in general, and adolescent rites of passage in particular, are optimally adapted for this task. Music-based ritual and emotionally evocative elements of religion optimize reinforcement learning. The costly and autonomically arousing ordeals of many rites ensure fear conditioning. Such learning shapes maturing neural networks, impacting choices and behaviours. Evolutionary anthropologists view religion as a costly signal of group commitment. Adolescent rites of passage are a powerful proximate mechanism for creating and maintaining co-operative, cohesive groups.

82. D Runciman, 'Democracy is the planet's biggest enemy', *Foreign Policy*, 2019. https://foreignpolicy.com/2019/07/20/democracy-is-the-planets-biggest-enemy-climate-change/.

83. The most obvious example of group cultural behaviour is provided by religions/religious activity. According to Britannica (see Brit-link below), the oldest burials that attest to belief in life after death are around 30 to 50 thousand years old. Some scholars push the date of nascent religious behaviour back to 100,000 years ago (the earliest undisputed burial), or even beyond 300,000 years ago.

- Brit-link: https://www.britannica.com/topic/prehistoric-religion#ref52336.

84. JL Barrett, *Born Believers,* Atria Books, 2012.

85. JL Barrett, *Born Believers*, Atria Books, 2012, p. 9:

> Perhaps you remember a preschool shape-sorter toy that is a nearly round, hollow, red and blue plastic object with lots of different yellow shapes that fit into matching holes. Ordinary child development provides children with a number of

> conceptual holes that have particular shapes. One of these holes is a god-shaped hole. Children are naturally ready to receive the shape – the cultural idea – that fits well into the hole: gods of various sorts. Some gods fit the hole better than others, but many fit just fine.

Barrett names his underlying concept as 'natural religion', supporting god concepts that he claims are detectable in children very early indeed.

According to a *New York Times* article (NY-link), Barrett believes in God. This seems a curious position for someone arguing as to why we are so prone to such beliefs (albeit especially as children). This may explain why his advice later in the book, for promoting atheism against the instincts of religious belief, comes across very weak if not dangerous; see Endnote 86.

- NYT-link: R Henig, 'Darwin's God', *New York Times*, 2007. https://www.nytimes.com/2007/03/04/magazine/04evolution.t.html.

86. Barrett says (p. 218):

> ...this counterreligious indoctrination could include offering children alternative ways to interpret their perception of design and purpose in the natural world and their detections of non-human agency. Chance, government, natural selection (as an intentional, directional agent of change), and other pseudo agents could do nicely here.

But there is great danger in this approach. If the desired concepts are reframed too much to fit, they will end up triggering the same cultural convictions that we are trying to avoid with religion! It is not the details of religious set-up that cause the convictions (there is a vast range of these, and more lost to history than currently exist). It is their underlying existential and emotive payloads, which can potentially subvert *any* topic. For instance, Barrett's quote above already lets slip this danger, via his word 'intentional'. Natural Selection is not in any way intentional, and to imbue it with that characteristic would soon evoke the whole emotive paraphernalia that accompanies strong culture. Already in history, via Eugenics, the theory of evolution came to inappropriately legitimise a very negative cultural movement. Adapting science and reason to 'fit' cultural profiles that we (and especially children) are sensitised to easily adopt, is a bad way to go.

87. From climate scientist Mike Hulme:

> In this new mood of climate-driven destiny the human hand, as the cause of climate change, has replaced the divine hand of God as being responsible for the collapse of civilizations, for visitations of extreme weather, and for determining the new twenty-first-century wealth of nations. And to emphasize the message and the mood, the New Economics Foundation and its partners have wound up a climate clock that is now ticking, second by second, until December 1, 2016, when human fate will be handed over to the winds, ocean currents, and drifting ice floes of a destabilized global climate: 'We have 100 months to save the planet; when the clock stops ticking we could be beyond the climate's tipping point, the

point of no return.' Such eschatological rhetoric offers a post-2016 world where human freedom and agency are extinguished by the iron grip of the forces of climate. Such a narrative offers scant chance for humans to escape a climate-shaped destiny.

- Link: M Hulme, 'Reducing the future to climate: a story of climate determinism and reductionism', *Klima* 2011; 26(1): 245–266.

88. One doesn't need technical knowledge of renewables to note that out of a large range of assessments, only the most wildly optimistic (see for example, J-link below) claim that renewables *on their own* could replace fossil fuels to power civilisation in anything like its current form. And if one truly believes in imminent global climate catastrophe, then refusing the huge benefit of zero emissions nuclear power and committing to renewables alone (even allowing for technical advance) is illogical at best and more likely a blind article of faith. In practice, we know for sure that the motivation behind renewables is faith – a strong cultural belief (see Section 12.7).

- J-link: M Jacobson et al., 'Low-cost solution to the grid reliability problem with 100% penetration of intermittent wind, water, and solar for all purposes', *PNAS* 2015; 112(49): 15060–15065.

89. See for example:

- Emotive dying polar bear video, originally and wrongly attributed to catastrophic climate-change: https://www.youtube.com/watch?v=_JhaVNJb3ag&feature=youtu.be.
- Emotive climate catastrophist video, *Will the World End Because of Climate Change?*: https://youtu.be/w-inEu9T1m4.

90. A teacher toolkit approved by the Senior Policy Advisor at the (UK) National Association of Headteachers, starts off with the heading, 'Headline facts about climate breakdown' [emphasis mine], and lists: 'We have 12 years to cut greenhouse gas emissions by 45% to avoid catastrophe', and 'More than one million species face extinction'.

- L-link: Iwill, *Global Action Plan: 'Climate Chaos Response Toolkit' for schools.* https://www.iwill.org.uk/global-action-plan-climate-chaos-response-toolkit-for-schools.

91. Australian Parents for Climate Action: https://www.ap4ca.org/. Also E Snaith, '"More than 2,000" mothers and families march through London demanding urgent climate action', *The Independent*, 2019. https://www.independent.co.uk/news/uk/home-news/mothers-climate-march-london-parents-rise-up-konnie-huq-rosamund-kissi-debrah-a8910811.html.

92. M van Brunnersum, 'Child bullied by peers and teacher for not participating in strikes', *Daily Mail*, 2019. https://www.dailymail.co.uk/news/article-7062517/Swedish-girl-bullied-not-participating-Greta-Thunberg-climate-strike.html.

93. Bill Nye 'the Science Guy', *The Planet is on Fucking Fire!!!* https://www.youtube.com/watch?v=IFgBFYkBZ6E. Highly inappropriate and false emotive 'lesson' on climate-change (Animaplates): https://youtu.be/pIXYdDrfa90.

94. D Lawson, 'Politicians and the police must stop indulging Extinction Rebellion', *The Times*, 2019. https://www.thetimes.co.uk/edition/comment/we-do-not-need-an-ecological-pol-pot-vr5tgbh06 or free extract here (Dominic Lawson): https://www.thegwpf.com/dominic-lawson-we-dont-need-an-ecological-pol-pot/.

95. For example:

- A di Santolo, 'Disgusting moment school kids chant "Theresa May f*****g wh*re" at climate change march, *The Express*, 2019: https://www.express.co.uk/news/uk/1100647/Theresa-May-chant-Climate-change-march-London-school-strike
- C Davis, 'Student climate change protest turns SOUR as pupils chant "f*** May' outside Parliament', *The Express*, 2019. https://www.express.co.uk/news/uk/1087970/westminster-climate-change-protest-strike-students-London-march-Parliament-Theresa-May.

96 Watson-Jones and Legare say that:

> Recent work has found that young children are sensitive to cues to social conventions such as rituals and imitate ritual actions with higher fidelity than instrumental behavior...Other research has found that engaging in collective rituals increases preferences for in-group members...and that the motivation to affiliate with social groups may underlie children's imitation of ritual actions...

- RE Watson-Jones and CH Legare, 'The social functions of group rituals', *Current Directions in Psychological Science* 2016; 25(1): 42–46.

97. Writing at *CapX*, Marian L. Tupy airs this view of climate catastrophism as a standard religion, referencing generic research about religions from psychologists at North Dakota State University, and also citing the book *The End of the World As We Know It: Faith, Fatalism, and Apocalypse in America*, by Daniel Wojcik of the University of Oregon:

- Link: ML Tupy, 'The planet's future is too important to leave to the catastrophists', *CapX*, 2019. https://capx.co/the-planets-future-is-too-important-to-leave-to-the-catastrophists/.

98. Fears inculcated by cultural entities are *not real*; likewise for hopes (e.g. of salvation from that which causes the fear) joys, anxieties etc. Cultural narratives are essentially emergent group deceptions fulfilling the connective role of holding the group together. They trigger a range of behaviours to do this, including narrative policing and out-group demonisation. The system arose via gene-culture co-evolution: '...genes and culture depend on each other for the evolution and expression of deceptive and self-deceptive adaptations' [GS-link below]. Yet exactly how group self-deceptions actually deceive, isn't easy to determine.

- GS-link: G Gorelik and TK Shackelford, 'Culture of deception', *Behavioral and Brain Sciences* 2011; 34(1): 24–25.

While cultural adherents generally don't behave as though their core narratives are literally true (e.g. see Endnote 99), *except for* minority cases where there *is* literal belief (per Section 4.4.2 and Endnote 101), and aren't conscious of resultant contradictions, how the brain accomplishes this appears not to be understood.

99. There seems to have been considerable effort over the years in exploring how literal or otherwise are religious beliefs. Most people appear to believe metaphorically or symbolically rather than literally, and don't interpret religious fears as real. For instance, Cranney et al. (C-link below) note:

> ...if 13% [those who actually thought they would be going to hell] of a population thought it was more likely than not that they would spend the last 10 years of their life being brutally tortured [i.e. an equivalent, or actually much less as it's only 10 years, reality-based fear], one might expect a crippling amount of anxiety from this group. Furthermore, if they knew that they could evade this fate by following religious dictates, this group would undoubtedly score very high in terms of religious practice and observance.'

In other words, commensurate fear/anxiety and a corresponding strict conformance are simply not seen.

- C-link: S Cranney et al, 'Hell anxiety as non-pathological fear', *Mental Health, Religion & Culture* 2018; 21(9–10): 867-883.

100. For example, ex-adherent Fran Ugalde eventually realised that her climate fears were 'a religious belief': https://judithcurry.com/2019/06/13/extremes/#comment-893861.

101. Some while ago, theory held that those with Autism Spectrum Disorder (ASD; sometimes 'neurodiverse people') were less likely to hold religious or spiritual beliefs than neurotypical (NT) people. For instance, see SRT-link below. Later work (EO-link) indicates that:

> ...factors more related to religious or spiritual perception did not reveal an inferior sensitivity in the neurodiverse or autistic sample compared to the control group. Furthermore, a spiritual factor with items like belief in ghosts, the supernatural and the paranormal had higher prevalence in autism and neurodiversity.'

And from the ASD community itself, there's plenty of anecdotal expression of belief (ASD1-link), questioning (ASD2-link), and more formal reservation (ASD3-link), which all contradict the earlier theory.

- SRT-link: C Caldwell, 'Why are high-functioning autistics more likely to be atheists or agnostics?', *Science and Religion Today*, 2011. http://web.archive.org/web/20200202180147/http://www.scienceandreligiontoday.com/2011/09/26/why-are-high-functioning-autistics-more-likely-to-be-atheists-or-agnostics/.

- EO-link: L Ekblad and L Oviedo, 'Religious cognition among subjects with autism spectrum disorder (ASD): Defective or different?' *Clinical Neuropsychiatry* 2017; 14(4). https://www.researchgate.net/publication/311900234_Religious_cognition_among_subjects_with_Autism_Spectrum_Disorder_ASD_Defective_or_different.
- ASD1-link: C Bonello, 'Asperger Syndrome and religion: reconciling logic with faith', *autisticnotweird.com*, 2017. http://web.archive.org/web/20211018012024/https://autisticnotweird.com/religion/ (Chris Bonello, autistic teacher, author and speaker).
- ASD2-link: J Elder Robinson, 'Religion and autism: are they together or apart?', *Psychology Today*, 2014. https://www.psychologytoday.com/us/blog/my-life-aspergers/201409/religion-and-autism-are-they-together-or-apart.
- ASD3-link: O Bustion, 'Autism and Christianity: An ethnographic intervention', *Journal of the American Academy of Religion* 2017; 85(3): 653–681.

Moreover, we know both from research and the ASD community itself that confusion about literal meanings – e.g. of slang and/or common metaphors – is a primary characteristic of ASD (see PT-link and AC-link below). And yet 'normal' group adherence to a strong culture, of which religions are the most familiar example, *depends on* a (subconscious) non-literal interpretation (see also the Chris Knight quote at the bottom of this endnote). To make things worse, when the culture is secular and has hijacked the authority of science as a cloak to hide its nature, as happened in the early twentieth century with eugenics and is now happening in the climate domain, even many NT people can be lured into emotive belief through confidence in science. It seems plausible at the very least that people with ASD, and especially children, will be more susceptible to adopt an emotive and *literal* belief in global climate catastrophism (even if, later, the cultural nature of the narrative is eventually discerned). As a result, they'll suffer the corresponding and *genuine* fears that literal belief in imminent global catastrophe must bring.

- PT-link: I Stuart-Hamilton, 'People with autism spectrum disorder take things literally, *Psychology Today* 2013. https://www.psychologytoday.com/us/blog/the-gift-aging/201304/people-autism-spectrum-disorder-take-things-literally.
- AC-link: N Frances, 'Literal language in Asperger's Syndrome', *Asperger Child*' website, 2003. http://web.archive.org/web/20161001112629/https://www.aspergerchild.com/aspergers-blog/literal-language-in-aspergers-syndrome.

Professor of anthropology Chris Knight says:

> ...all expressions of human symbolic culture may in this light be understood as 'collusion in deception' – collaboration in the maintenance of fictions which have social support. Trust in the founding fictions is not given lightly. Durkheim (1965) indeed showed long ago that a community will place ultimate confidence only in those fictions which are emblematic of itself. If all collude, then on another level the deceptive signal may constitute a performative, constructing its own truth. Ritual specialists may assume the burden of sustaining such circular

'truths' on which group identity depends. Note, however, that ingroup/outgroup polarity is central here: one group's most sacred truths may be another's transparent deceits….An ability to handle fictional representations, then, is the essence of human symbolic competence. Distinguishing between surface and deeper meanings poses a major cognitive challenge; involvement in 'pretend play' during childhood is crucial to the development of the necessary cognitive skills.'

- Link: C Knight, 'Ritual/speech coevolution: a solution to the problem of deception', in: James R Hurford et al. (eds), *Approaches to the Evolution of Language*, Cambridge University Press, 1998.

102. E Willis, 'Extinction Rebellion protests: youth activists break down in tears at Heathrow as demonstrations over climate change continue in London', *Evening Standard*, 2019. https://www.standard.co.uk/news/uk/extinction-rebellion-protests-police-surround-climate-change-activists-at-heathrow-airport-as-a4122006.html.

103. C Harris, 'Watch: Girl, 11, makes tearful climate plea: "I love our planet and I don't want it to ever stop"', *Euronews*, 2019. https://www.euronews.com/2019/02/17/watch-girl-11-makes-tearful-climate-plea-i-love-our-planet-and-i-don-t-want-it-to-ever-sto.

104. P Cockburn, 'Tears outside PM's office as students skip school to demand climate action again', ABC News, 2019. https://www.abc.net.au/news/2019-05-03/students-strike-around-the-country-for-climate-action/11077022.

105. Original link now dead (https://www.treehugger.com/culture/no-kidding-one-in-three-children-fear-earth-apocalypse.html), see Wayback Machine archive: https://web.archive.org/web/20191216033208/https://www.treehugger.com/culture/no-kidding-one-in-three-children-fear-earth-apocalypse.html.

106. T Bridge, 'Climate strikes and the youth mental health crisis', *The National Observer*, 2019. https://www.nationalobserver.com/2019/05/02/opinion/climate-strikes-youth-mental-health-crisis.

107. R Clark, 'Child climate change protestors aren't truants, they're traumatised', *The Spectator*, 2019. https://blogs.spectator.co.uk/2019/02/child-climate-change-protestors-arent-truants-theyre-traumatised/.

108. A Scher, '"Climate grief": The growing emotional toll of climate change', NBC News, 2018. https://www.nbcnews.com/health/mental-health/climate-grief-growing-emotional-toll-climate-change-n946751.

109. D Johnson, 'Climate change scenarios scare, and motivate, kids', *Washington Post*, 2007 (Paywall): http://www.washingtonpost.com/wp-dyn/content/article/2007/04/15/AR2007041501164.html.

'For many children and young adults, global warming is the atomic bomb of today.'

110. 'Anxiety about my future because of climate change'. Original post deleted, but comments still there: https://www.reddit.com/r/climatechange/comments/bqt2rk/

anxiety_about_my_future_because_of_climate_change/. Here is another example: https://judithcurry.com/2019/12/14/the-toxic-rhetoric-of-climate-change/.

111. See:

- L Holson, 'Climate change is scaring kids. Here's how to talk to them', *The New York Times*, 2019. https://www.nytimes.com/2019/06/27/science/climate-change-children-education.html.
- L Blair, 'Does your child have eco-anxiety? Here's how to tackle it', *The Daily Telegraph*, 2019 (paywall): https://www.telegraph.co.uk/health-fitness/mind/does-child-have-eco-anxiety-tackle/.
- *Parent Map*: https://www.parentmap.com/article/dont-allow-climate-change-anxiety-ruin-your-life.

112. M Pearl, '"Climate despair" is making people give up on life', *Vice*, 2019: https://www.vice.com/en_ca/article/j5w374/climate-despair-is-making-people-give-up-on-life.

113. There are many articles in which therapists or psychologists unwittingly conspire with the fear of their patients by failing to point out that this fear, essentially of climate doon, is unfounded. Instead, they encourage ways in which to essentially accommodate the fear, to 'live with it' in a less debilitating manner. Doing this, will actually tend to confirm the fears, whether or not sharing them or joining Extinction Rebellion helps in a modest way. See, for example, V Knight, '"Climate Grief": Fears about the planet's future weigh on Americans' mental health', KHN, 2019: https://khn.org/news/climate-grief-fears-about-the-planets-future-weigh-on-americans-mental-health/.

114. R Randall, 'Should we be working with children about climate change?', Rosemary Randall blog, 2011. https://rorandall.org/2011/03/23/should-we-be-working-with-children-about-climate-change/.

Note: In case of doubt, Randall's theory of projection is from someone who fully believes that climate-change is a very serious and urgent global problem that…

> …will make some parts of the world uninhabitable and others inhospitable', meaning 'we can't continue to live like we do – our economic systems, social practices and personal lifestyles are unsustainable. Everything we are used to, much of what we hold dear and many of our dreams and aspirations have to go.'

- From: 'Interview with Rosemary Randall on psycho-analysis and climate change', Manchester Climate Monthly, 2013. https://manchesterclimatemonthly.net/interviews/interview-with-rosemary-randall-on-psycho-analysis-and-climate-change/.

115. See S Asayama et al., 'Why setting a climate deadline is dangerous' *Nature Climate Change* 2019; 9: 570–574 (emphasis mine):

> Pushing hard to meet a deadline may also cause (unintentionally) dangerous political side effects. For example, deadline-ism incubates the political opportunism

> of declaring a climate emergency. It is no surprise that new political movements calling for the declaration of a climate emergency in parliaments, cities, schools and universities have arisen in the months after the release of the IPCC SR15... The rhetoric of emergency emerges from *the worldview of millenarianism* and its conception of 'compressed time' that calls for immediate actions before it is too late. However, regardless of the original intentions, an empty call for emergency actions can be interpreted in myriad ways. In the worst case, the emergency rhetoric could become 'stolen rhetoric', used as justification for solar geoengineering and potentially for more authoritarian forms of governance and regulation.

- Paywall-link: https://www.nature.com/articles/s41558-019-0543-4.epdf.
- Free-link: https://www.academia.edu/41380192/Why_setting_a_climate_deadline_is_dangerous.

116. In mid-2019, some French right-wing politicians called for a boycott of Thunberg's speech to the National Assembly (see BBC-link). But otherwise there is minimal pushback.

- BBC-link: https://www.bbc.co.uk/news/world-europe-49092653.

117. As Lewandowsky acknowledges when talking about the spread of emotive misinformation (see 'Misinformation and its correction: Continued influence and successful debiasing', *Psychological Science in the Public Interest* 2012; 13.3: 106–131) emotional response is rewarded with more retransmission than is veracity:

> But we have also noted that the likelihood that people will pass on information is based strongly on the likelihood of its eliciting an emotional response in the recipient, rather than its truth value (e.g., K. Peters et al., 2009).

While Lewandowsky's paper is about *misinformation,* as the quote implies, emotion wins out over veracity for information more generally in certain contexts (e.g. high uncertainty), and where both occur within the same narrative block. Moreover, the narrative of high certainty of imminent global climate catastrophe *is* misinformation, if we adopt *mainstream* climate science per the AR5 Working Group Chapters as the gold standard for reality.

118. JL Drake et al., *Communicating Climate-Change and Natural Hazard Risk and Cultivating Resilience: Case studies for a multi-disciplinary approach.* Springer, 2016.

119. The mainstream science position is most faithfully represented by the IPCC Working Group Chapters of AR5, rather than, say, the condensed IPCC Summaries for Policymakers. This is because, while neither express a certainty of imminent global catastrophe, there appears to be significant tension within the IPCC between raw scientific input at one end, which is presumably the least biased input, and more emotively biased output plus press/quotes from same, at the other end. Hence there's a consequent gradation of messaging for layers in-between. Caleb Rossiter describes this scenario (see RCE-link):

- RCE-link: C Rossiter, 'The UN's terrifying, but ever-receding, human-caused climate catastrophe', *RealClear Energy* blog: https://www.realclearenergy.org/articles/2018/10/31/the_uns_terrifying_but_ever-receding_human-caused_climate_catastrophe_110360.html.

120. Steve Koonin bio: https://www.energy.gov/seab/contributors/steven-e-koonin.

121. Steve Koonin's GWPF 2021 lecture: https://www.youtube.com/watch?v=6Tz1MiX1p5I.

122 Out of 39 Catastrophe Narrative examples in Group 1 of the CN-Archive, a little over half employ the terms 'catastrophe' or 'catastrophic'. However, many that don't use the exact terms invoke even worse consequences. While 'catastrophe' means great damage or calamity or cataclysm or disaster, even this kind of event would leave remnants of civilisation intact. However, there is rather less expectation of survival in alternative phrasings, such as:

- Example 1c)i]: 'five minutes after midnight', referring to the doomsday clock and presumably also the already stored consequences of current emissions.
- Example 1h): 'future generations will be roasted, toasted, fried and grilled', which dire and lurid prospect does not lend itself to considerations of survival.
- Example 1k): 'what is at stake is the future of the planet, the future of life', a simple statement that's about as existential as one can get regarding not just humanity, but its home.
- Example 1s): 'two decades to save the world', simpler still yet no less existential, and with urgency too.
- Example 1v): 'we are at the limits of suicide', a term meaning self-termination for an individual still implies terminal when extrapolated to the human race.

Other alternatives employ phrases that are merely broad equivalents to 'catastrophic'. For example: 1z) 'calamitous', 1c) 'dramatic damage' and 'devastating consequences', where the context is global for people or the planet and is sometimes quite explicitly framed, such as 1y) 'so far-reaching in its impact and irreversible in its destructive power, that it alters radically human existence'. Other phrases such as 1x) 'committing the world to a drastically different place' or 1r)ii] 'we have 500 days to avoid climate chaos' or 1j)ii] 'killing our planet', are also hard to interpret as meaning anything other than the end of world.

123 Group 3, example c). E Chasan and M Chediak, 'Jerry Brown is the face of America's climate-change resistance', *Bloomberg*, 2018. https://www.bloomberg.com/news/articles/2018-09-11/jerry-brown-challenges-trump-with-demand-for-100-clean-energy.

124 For instance, a study from 2014 including this result was: N Smith and A Leiserowitz, 'The role of emotion in global warming policy support and opposition', *Risk Analysis* 2014; 34(5).

Also see 'The psychology of climate change', (The Breakthrough Institute, 2014.

https://thebreakthrough.org/issues/energy/the-psychology-of-climate-change), which says:

> A growing body of scholarly and scientific studies finds that fear-based appeals around climate change backfire, resulting in increased climate skepticism and fatalism among much of the public.

More recently, Stephan Lewandowsky has noted (see LW-link below):

> Some past attempts to use visual imagery to communicate climate change have evoked negative emotions, such as fear, through conveying apocalyptic visions of the future. Unfortunately, these may actually demotivate audiences, triggering denial or apathy instead of engagement.

- LW-link: S Lewandowsky and L Whitmarsh, 'Climate communication for biologists: When a picture can tell a thousand words', PLoS Biology 2018; 16(10): e2006004.

125 Group 4, example j). P Krugman, 'Wind, sun and fire', *The New York Times*, 2016. http://www.nytimes.com/2016/02/01/opinion/wind-sun-and-fire.html?_r=0%20.

126 Group 1, example n). H Clinton, 'America must lead at Paris climate talks', *Time* magazine, 2015. http://time.com/4128624/hillary-clinton-climate-change/.

127 Group 5, example ac). From a letter in response to science communicator Joe Duggan's question, 'how do you feel about climate change?' (2016): http://www.isthishowyoufeel.com/this-is-how-scientists-feel.html#stefan.

128 Group 5, example bb). Via Envisionation: http://www.envisionation.co.uk/index.php/blogs/nick-breeze-blogs/153-rowan-williams-interview.

129 Group 2, example m). I Dunlop, 'If business leaders want to regain our trust, they must act on climate risk ', *The Guardian*, 2018. https://www.theguardian.com/environment/commentisfree/2018/mar/15/if-business-leaders-want-to-regain-our-trust-they-must-act-upon-climate-risk.

130 Group 5, example db). E Gosden and H Samuel, 'Paris climate summit: David Cameron warns 'earth is in peril' – as it happened', *The Telegraph*, 2015. https://www.telegraph.co.uk/news/earth/paris-climate-change-conference/12024206/Paris-climate-change-conference-LIVE-world-leaders-meet-for-UN-talks.html.

131 Group 5, example dd). Frack Off, a UK anti-fracking organization. From their website (sampled July 2018, the webpage says 2015): http://frack-off.org.uk/fracking-hell/climate-chaos/.

132 Group 2, example u). P Whish-Wilson, 'Cape Grim will send a message to the world on climate change', at the Australian Green Party website (May 2016): https://peter-whish-wilson.greensmps.org.au/articles/cape-grim-will-send-message-world-climate-change

133 For example, Editorial, 'Exxon's big court win exposes major malpractice in the New York Attorney General's Office', *New York Post*, 2019. https://nypost.

com/2019/12/14/exxons-big-court-win-exposes-major-malpractice-in-the-new-york-attorney-generals-office/.

134 Group 1 s) i], Group 7 j) i] and j) ii], Group 8 d).

135 Group 7, example da). Corey Bradshaw has a PhD in Zoology and was 2008-2014 Director of Ecological Modelling, 2014-2017 Sir Hubert Wilkins Chair of Climate Change, both at the University of Adelaide. Via Joe Duggan's Is this how you feel site: (2014): https://www.isthishowyoufeel.com/this-is-how-scientists-feel.html#corey.

136 Group 5, example ea). 64 children from 20 countries, attending Children's Climate Conference in Sweden. In addition to presenting a communiqué to the Swedish Environment Minister (Ms. Romson) to take to COP21, three children from the conference travelled to Paris themselves to present the document to world leaders:

> The communiqué, or 'Children's Demands,' was written and signed through thumbprints by the 64 children who attended the conference. The document called on the adults of the world to 'act like a kid'.

- Link: R Burridge, 'Wiky youths chosen to present childrens' climate central report to Sweden's minister of the environment', *Manitoulin Expositor*, 2015. https://www.manitoulin.com/wiky-youths-chosen-to-present-childrens-climate-central-report-to-swedens-minister-of-the-environment/.
- Radio Sweden article: https://sverigesradio.se/sida/artikel.aspx?programid=2054&artikel=6311895.

137 Group 5, example h). R Read, 'Climate breakdown, civilisation breakdown?', *GreenWorld* blog, 2018. https://greenworld.org.uk/article/climate-breakdown-civilisation-breakdown.

138 Group 6, example z). T Goreau, '350 PPM is a death sentence', Briefing to AOSIS at the Copenhagen Climate Change Conference (Dec 2009): https://www.scribd.com/doc/126755874/350-PPM-is-a-Death-Sentence-AOSIS-Briefing.

139 Group 7, example ha). Via Joe Duggan's Is this how you feel site: https://www.isthishowyoufeel.com/this-is-how-scientists-feel.html#Katrin/.

140 Group 6, example h) ii]. In a National Public Radio interview with Guy Raz (April 2017): https://www.npr.org/transcripts/522856713.

141 Group 6, example v). G Polya, submission to Australian Senate select committee inquiry on climate policy. https://www.aph.gov.au/~/media/wopapub/senate/committee/climate_ctte/submissions/sub273_pdf.ashx.

142 Group 6, example w). G McPherson, 'Three paths to near-term human extinction', *Nature Bats Last* blog, 2011. https://guymcpherson.com/2011/08/three-paths-to-near-term-human-extinction/.

143 Group 7, example bc). E Holthaus, 'Should climate change stop us from having babies?', *Vice*, 2015. https://www.vice.com/en_us/article/kwpmdy/should-climate-change-stop-us-from-having-babies-305.

144 See J Goodell, 'What's another way to say "We're f-cked"?', *Rolling Stone* magazine, 2018. https://www.rollingstone.com/politics/politics-news/climate-change-sea-level-rise-737012/.

145 D Spratt and I Dunlop, *What Lies Beneath: The understatement of existential climate risk*. Breakthrough, National Centre for Climate Restoration, Australia, 2018.

146 J Curry, 'CAGW: a "snarl" word?', *Climate Etc* blog, 2018. https://judithcurry.com/2018/11/26/cagw-a-snarl-word/.

147 Stephan Lewandowsky says (SL-link below): 'Nonetheless, being human, scientists operate with the same cognitive apparatus and limitations as every other person'. His paper attempts to make the case that memes from 'denial' seeping into the climate-science community significantly bias this community away from orthodox/catastrophic conclusions. I have explained why the case does not stand up (see AW-link). But Lewandowsky's point about scientists is well made. And indeed, it works both ways. Given that there's a measurable and dominant culture of climate-change catastrophe within the public domain, this could hardly fail to inject some bias into the climate-science community.

- SL-link: S Lewandowsky, 'Seepage: The effect of climate denial on the scientific community', *Shaping Tomorrow's World* blog, 2014. http://www.shapingtomorrowsworld.org/lewandowskyseepage.html.
- AW-link: A West, 'A key admission regarding climate memes', *Climate Etc* blog, 2015. https://judithcurry.com/2015/07/03/a-key-admission-regarding-climate-memes/.

148 AJ Mohammed, Speech to UNFCCC Technical Expert Meeting, 16 May 2017. https://youtu.be/54WEz2Q-sL8. Group1, example b).

149 F Macron, Speech to Paris Climate Conference. See: L Friedman, 'World leaders open Paris climate talks', E&E News, 2015. https://subscriber.politicopro.com/article/eenews/1060028636. Group 1, example k).

150 B Obama, 'A world that stands as one' (Speech in Berlin 2008). See https://www.telegraph.co.uk/news/worldnews/barackobama/2455237/Senator-Barack-Obama-A-world-that-stands-as-one.html. Group 1, example u) ii].

151. UN. 'Secretary-General's remarks to the opening of the high-level session of the COP21', United Nations 2015. https://www.un.org/sg/en/content/sg/statement/2015-12-07/secretary-generals-remarks-opening-high-level-session-cop21.

152. M Schultz, '"Make our planet great again": Macron appeals to Trump on Paris accord', *New York Post*, 2018. https://nypost.com/2018/04/25/make-our-planet-great-again-macron-appeals-to-trump-on-paris-accord/.

153. 'Blair urges climate change action': BBC News website, 2006: http://news.bbc.co.uk/1/hi/uk_politics/6068226.stm.

154. P Ghosh, 'Hawking says Trump's climate stance could damage Earth', BBC News, 2 July 2017. http://www.bbc.co.uk/news/science-environment-40461726.

155 P Diethelm and M McKee, 'Denialism: what is it and how should scientists respond?', *European Journal of Public Health* 2009; 19(1): 2–4.

156 The paper has 434 citing articles on Google Scholar.

157 The Wiki article as originally referenced within my 2016 Climate Etc. post, is at https://en.wikipedia.org/w/index.php?title=Denialism&oldid=711203148. The latest version, which may have changed again since writing this chapter (2021), is here: https://en.wikipedia.org/wiki/Denialism.

158 According to Google Ngram Viewer, overall usage of the term 'denialism' has risen steadily throughout this century (up to 2019 where the Viewer ends when last sampled for this note). According to Google Trends, there was a big spike in usage during 2009, with a smaller one in 2010. Because both D&M2009 and Michael Specter's book (see MS-link) came out in 2009, one can't distinguish between the contributions they each made. But the former may have contributed more within academia and the latter more to the popular conception. Diethelm and McKee's follow-up paper came out in 2010. Generic Google searches (conducted at the time of my *Climate Etc* post) reveal a large proportion of hits that either directly or indirectly reference Diethelm and McKee, Wiki (or variants/copies thereof), or Specter's book, or occasionally Hoofnagle.

- MS-link: M Specter, *Denialism: How Irrational Thinking Hinders Scientific Progress, Harms the Planet, and Threatens Our Lives*, Penguin, 2009.

159 Captured 5th March 2016, from the first link at Endnote 157. Mark Hoofnagle (https://rationalwiki.org/wiki/Mark_Hoofnagle) is also attributed.

160 The so-called 'Slayers' are a case in point. This group of climate sceptics (the name comes from the title of their book *Slaying the Sky Dragon*, i.e. slaying the theory of greenhouse gases; see SSD-link below) were briefly prominent after 2010. They rightly oppose the (cultural) certainty of imminent global climate catastrophe, but – according, not only to the climate orthodox, but also the great majority of climate sceptics – they do this for the wrong reasons, which stem from a theoretical misunderstanding. Their argument is purely technical and their impact on the domain is very modest.

- SSD-link: J O'Sullivan et al. *Slaying the Sky Dragon*, ebookpartnership.com, 2010.

161 Noble cause corruption occurs when cultural adherents are such ardent believers in what they feel is a righteous cause, they bend or break rules, possibly even the law, in order to support their (cultural) cause. For an example, see IPA-link below. The secondhand smoking domain 'inherited' righteousness from the clear evidence about first-hand smoking, and this may have exceeded reasonable bounds (see U-link). I haven't investigated this domain and have no particular reason to question the consensus of danger. Yet there are articles (see TA-link) that raise the possibility that strong noble cause corruption may have gripped the consensus side,

with ill effects. The linked article references both Diethelm and McKee as authorities within the domain who are being challenged by their peers.

- IPA-link: J Marohasy, 'Misbehaving models and missing mammals' (Review of the book *The Virtuous Corruption of Virtual Environmental Science*', by Aynsley Kellow), *IPA Review*, 2008. https://ipa.org.au/wp-content/uploads/archive/1210898777_document_marohasy.pdf.
- U-link: S Ungar et al., 'Silencing science: partisanship and the career of a publication disputing the dangers of secondhand smoke'; *Public Understanding of Science* 2005, 14(1), 5–23.
- TA-link: M Siegel, 'Response to article comparing challengers of secondhand smoke conclusions to holocaust deniers published in *European Journal of Public Health*'. *The Rest of the Story* blog, 2022. http://tobaccoanalysis.blogspot.co.uk/2009/02/response-to-article-comparing.html.

162 Social consensuses frequently bias or wholly derail science; this problem is acknowledged, yet I think is greatly underappreciated. For context on this issue, see PA-link and CE-link. For examples of such consensuses in medicine, see Med-links below.

- PA-link: K Cherry, 'What is groupthink?', *VeryWellMind* blog, 2020. https://www.verywellmind.com/what-is-groupthink-2795213.
- CE-link: J Curry, 'Distinguishing the academic from the interface consensus', *Climate Etc* blog, 2014. https://judithcurry.com/2014/09/18/distinguishing-the-academic-from-the-interface-consensus/.

Med-links on consensuses prior to:

- Discovery of circulation of the blood: see MW Rampling, 'The history of the theory of the circulation of the blood', *Clinical Hemorheology and Microcirculation* 2016; 64(4): 541–549.
- Implementation of antiseptic practice: https://en.wikipedia.org/wiki/Ignaz_Semmelweis.
- Consensus prior to the discovery that *Helicobacter pylori* causes ulcers: see J Tanenbaum, 'Delayed gratification: Why it took everybody so long to acknowledge that bacteria cause ulcers', *Journal of Young Investigators*, 2005. https://www.jyi.org/2005-february/2005/2/9/delayed-gratification-why-it-took-everybody-so-long-to-acknowledge-that-bacteria-cause-ulcers.
- Discovery of the role of sugar in heart disease: see D Lundell, 'Heart surgeon speaks out on what really causes heart disease', *Sign of the Times* blog, 2012. http://www.sott.net/article/242516-Heart-surgeon-speaks-out-on-what-really-causes-heart-disease.

The blood-letting consensus was so strong, it still continues today in some places. See E Engelhaupt, 'Bloodletting is still happening, despite centuries of harm', *National Geographic*, 2015. http://phenomena.nationalgeographic.com/2015/10/27/bloodletting-is-still-happening-despite-centuries-of-harm/.

163 In addition to the list in Endnote 162, see below for further examples of prior consensuses that were overturned. These were not perceived as interim concepts from which to evolve further (I imagine there must have been many thousands of those). The consensuses were enforced, socially promoted, clung-to and defended against evidence to varying degrees, sometimes with much damage. For clarity, the point here is not: 'because some consensuses are wrong and inappropriately defended, means that all other consensuses are wrong'. Or even that any other particular consensus must by comparison also be wrong. But merely that neither Diethelm and McKee, or Hoofnagle or anyone else, can cite 'a major consensus' as being the ultimate criterion for determining who is speaking to the evidence, and who is not. This criteria may well be right most of the time, but clearly it is not guaranteed to be right.

- Static continents: http://www.smithsonianmag.com/science-nature/when-continental-drift-was-considered-pseudoscience-90353214/.
- The miasma theory: https://en.wikipedia.org/wiki/Miasma_theory.
- The Ptolemaic system: https://en.wikipedia.org/wiki/Ptolemaic_system.
- Eugenics: http://www.newstatesman.com/society/2010/12/disabled-america-immigration.

And, related to the 'saturated fats cause heart disease' and the true role of sugar in Endnote 162, saturated fats were also blamed for causing obesity and diabetes:

- I Leslie, 'The sugar conspiracy', *The Guardian*, 2016. http://www.theguardian.com/society/2016/apr/07/the-sugar-conspiracy-robert-lustig-john-yudkin.

164 See A West. 'Contradiction on emotional bias in the climate domain', *Climate Etc* blog, 2015. http://judithcurry.com/2015/04/24/Contradiction-on-emotional-bias-in-the-climate-domain/.

165 For instance, bias within a report from climate advocacy group Influence Map. See: S Everley, 'Influence Map continues green campaign to dupe the press on "climate denial"'. *Energy in Depth*, 2016. https://www.energyindepth.org/influencemap-green-campaign-press-climate-denial/?154.

166 I Hyman, 'Smart people, dumb decisions', *Psychology Today*, 2014. https://www.psychologytoday.com/us/blog/mental-mishaps/201408/smart-people-dumb-decisions.

167 HIV to AIDS replication is documented at the link below, which is to the US National Institute of Allergy and Infectious Diseases, including the tragic incidents mentioned in the main Chapter 6 text.

- Link: https://web.archive.org/web/20160910072901/http://www.niaid.nih.gov/topics/hivaids/understanding/howhivcausesaids/pages/hivcausesaids.aspx.

168 D&M2009 says regarding proofs of climate change:

> For example, those denying the reality of climate change point to the absence of accurate temperature records from before the invention of the thermometer.

> Others use the intrinsic uncertainty of mathematical models to reject them entirely as a means of understanding a phenomenon.

However, both sides know the formal temperature record only goes back so far, and this is neither an argument for or against calamitous climate change (and indeed it is not typically presented as such). For earlier periods, both sides use temperature proxies. Sceptics question the divergence between model output and observations, and the use of immature models as the basis for world-transforming policy decision, but I doubt that many reject models entirely.

169 I was struck by a comment from Dr Michael Siegel (see TA-link, which includes his profile), as summarised at TA-link. Siegel is very far from occupying the opposite position to Deithelm and McKee regarding secondhand smoke (or ETS – environmental tobacco smoke), and in fact is a tobacco control advocate who apparently argued that ETS kills over 50,000 Americans each year. His testimony contributed towards a 145 billion dollars verdict against tobacco companies. Yet to say the least, he is robust in his criticism of D&M2009, here's an excerpt:

> Diethelm and McKee have endangered the integrity of public health by comparing those who challenge the conclusion that secondhand smoke causes heart disease and lung cancer with those who deny the Holocaust. As a primarily science-based movement, public health is supposed to have room for those who dissent from consensus opinions based on reasonable scientific grounds. To argue that those who fail to conclude that the small relative risk for lung cancer of 1.3 among persons exposed to secondhand smoke is indicative of a causal connection are comparable to Holocaust deniers is to turn public health into a religion, where the doctrines must be accepted on blind faith to avoid being branded as a heretic.

Diethelm and McKee reply to Siegel; see the 13th February eletter at the *European Journal of Public Health*. OJ-link below. A robust debate develops, which is worth following up. Epidemiologist Geoffrey Kabat accuses Diethelm and McKee of a 'sleight of hand'. The editor of *Reason* magazine, Jacob Sullum (his profile is at JS-link below) also jumps in to support Siegel at his own publication, see RM-link.

- TA-link: http://tobaccoanalysis.blogspot.co.uk/2009/02/response-to-article-comparing.html.
- MS-link: https://www.allamericanspeakers.com/celebritytalentbios/Michael+Siegel/407872.
- OJ-link: https://web.archive.org/web/20090621120535/http:/eurpub.oxfordjournals.org/cgi/eletters/19/1/2.
- RM-link: http://reason.com/blog/2009/02/13/if-you-question-the-deadliness.
- JS-link: http://reason.com/people/jacob-sullum/all.

Tobacco researcher Kamal Chaouachi complains (see 17th February eletter at the same OJ-link above) about Diethelm and McKee's 'abuse of strong phrases and words such as "fake experts", "denialism", etc', and also that they are defaming Siegel.

Some of the other writers also claim that Diethelm and McKee have essentially been cherrypicking in the assembly of their cases, both for ETS and for 'denialism'.

What are we to make of such criticisms? Are the promoters of 'denialism' partaking of denialist behaviour as they themselves would define it? Or are all their detractors simply wrong? More to the point, how could we tell? Certainly not, it would seem, by applying D&M2009's own criteria.

170 DNA was not discovered until almost a century later, so definitive proof was not available and indirect evidence had not yet been properly marshalled. Various theories on the mechanisms of evolution put forward during this intervening century turned out to be wrong, including Darwin's own (see D-link below).

- D-link: https://www.britannica.com/science/pangenesis.

171 The authors might argue that denialism (e.g. from evolution deniers in the 1870s) is excusable when the science that they are criticising is immature; i.e. they are not really 'deniers'. But if that's the case, then we are faced with having to determine whether a science really is mature before we can know that actual 'denialist' behaviour is taking place. However, in a contested domain, opinions on the relevant science's level of maturity will also be divided.

172 For instance, a decades-long consensus on links between consumption of fats and heart disease appears to be slowly subsiding, allowing an evidential position to emerge. Previously, consensus mechanisms suppressed contrary evidence. See 'Illustrated History Of Heart Disease 1825–2015' at https://web.archive.org/web/20160426053320/http://www.dietheartpublishing.com/diet-heart-timeline.

173 There are many different conceptions of ideology. I would define it along the following lines:

> A construction of symbolic forms, images and texts that encodes and transmits belief systems, these usually understood to be modern belief systems, and in some interpretations also associated with modern mass communications. As such they are often viewed as filling the vacuum left by declining religious systems. Generically, both ideologies and religions are cultural phenomena, and in some cases consist a bounded cultural entity, supported on mechanisms that came out of gene-meme co-evolution.

174 Eccentricity and idiosyncrasy are, by definition, individualistic behaviours, and so do not benefit from social amplification and (subconscious) social co-ordination. Ideology (see Endnote 173) and faith are based on group dynamics, and so do benefit from such co-ordination and amplification. We'd therefore expect them to produce completely different patterns of behaviour, and D&M2009's suggestion of a common pattern seems unlikely.

Greed is ultimately a personal motivation too, but can have some expression in groups (e.g. a criminal gang or a cartel).

175 Mark Hoofnagles's Denialism blog ('about') is at http://scienceblogs.com/denialism/about/. Apart from the format going wild, this still seems to reflect the

2007 content. An archive link, from before the format changes, is much easier to read. See: https://web.archive.org/web/20180905050017/htp://scienceblogs.com/denialism/about/.

176 For instance, 'inversionism'. Hoofnagle says:

> There is also a variant of conspiracy theory, inversionism, in which some of one's own characteristics and motivations are attributed to others.

See MH-link at Endnote 177.

177 From Mark Hoofnagle's blog (see MH-link below):

> Denialists are not honest brokers in the debate (you'll hear me harp on this a lot)...This tendency towards quote-mining and misrepresentation of science is really the clearest proof of the dishonesty inherent in denialist tactics...Cranks and denialists aren't honest brokers in a debate, they stand outside of it and just shovel horse manure into it to try to sow confusion and doubt about real science.

At Endnote 178 he says he 'knows' they are 'spouting BS', implying deliberate lying.

Hoofnagle also indicates that 'a lot of people get suckered in by denialist arguments', saying that this means such people aren't 'denialists' in themselves, so I presume not liars either. Yet he provides no method to distinguish between these people and those he considers as genuine denialists. In these circumstances, inappropriate labelling is inevitable.

- MH-link: M Hoofnagle, 'About', *Denialism* blog (undated). http://scienceblogs.com/denialism/about/. At the time of publication this still reflects original (2007) content.

178 From Mark Hoofnagle's blog (see MH-link at Endnote 177):

> Cranks are a bit more deserving of pity, a bit closer to delusion and mental illness than the pure denialist, who knows that they are spouting BS to sow confusion.

179 One can't help but speculate that Diethelm and McKee also dropped dishonesty because they sensed this would result in blowback from people who'd feel inappropriately branded.

Some much more bounded/small-scale social phenomena can appear to be driven by dishonesty, for instance the 2008 global banking crisis. Yet research suggests (see SD-link below) that even these cases are more complex; the dishonesty seems to be a proxy for a toxic culture. Those involved are no more dishonest than their peers, and outside of business they remain so.

- SD-link: University of Zurich, 'Business culture in banking industry favors dishonest behavior', *Science Daily*, 2014, https://www.sciencedaily.com/releases/2014/11/141119132524.htm.

180 There are many largely overlapping lists of fallacies. There is some history of them provided by Stanford University.

- http://plato.stanford.edu/entries/fallacies/ (Stanford University).

181 For instance, at Cultural Cognition (CC-link), Dan Kahan is doing some great work on 'identity-protective cognition', a powerful effect that occurs when an evidential position generated by science threatens someone's cultural identity. The major biases this produces are comprehensively explored within several topic domains, and Kahan has developed tools to measure them. These insulate the investigator from domain knowledge and bias through use of appropriate questioning styles, and to good effect.

- CC-link: http://www.culturalcognition.squarespace.com/blog/.

182 M Specter, *Denialism: How irrational thinking hinders scientific progress*. Penguin, 2009.

183 Published about 9 months after D&M2009, Specter's book does not reference D&M2009 or Hoofnagle. The introduction proposes a theory that 'denialists' are primarily driven by fear of technology gone out of control, saying 'nothing scares us quite so much'. This fear is largely a cultural effect; technology threatens existing cultural modes, and one would expect a defensive reaction, which Specter acknowledges can encompass 'an entire segment of society'. This is therefore a much more viable proposition regarding causation. And denialism in the Holocaust or Creationism domains, for instance, is not driven by fear of technology going out of control in any case; this is indeed cultural defence. Anti-Semitism predates technology by millennia.'

The strong critique of Specter's book at *Grist* (see G-link below) is interesting. Specter is admonished for not spending enough time on climate denialism, and apparently 'he lurches toward a kind of denialism of his own', according to author Tom Philpott. That's the major problem with a concept that has such weak, in fact essentially non-existent, underpinning as 'denialism'. Everyone can use it against anyone and everyone else.

- G-link: T Philpott, 'Michael Specter's new book *Denialism* misses its targets', *Grist*, 2009. http://grist.org/article/2009-10-31-michael-specter-denialism-organic-gmo/.

184 Even in our most definitive case, of AIDS/HIV, which, being replicable, should be crystal clear, there is discourse on the 'bad' side that is 'not strictly one of denial', and also 'forms of denial' on the 'good' side. Didier Fassin says (reference below) 'There is a blind spot in denial, which is the denial of those who accuse the others of denial.' As he points out, this does not mean that the situation is symmetrical. However, it does mean that even apparently straightforward cases may contain major complexities, and that the ill-defined concept of denialism, far from being useful, merely serves to amplify emotions and further fuel the contest. Fassin's book is cited in the 'balance' section of the 'denialism' Wiki entry, headed 'Prescriptive and polemic'.

- Link: D Fassin, When Bodies Remember: Experiences and Politics of AIDS in South Africa. University of California Press, 2007.

185 The most productive approach is to develop tools that insulate the researcher as much as possible from domain knowledge when attempting to see what people think, and thus understand what their cultural identities and influences are. This minimises domain-related bias. Kahan has made great progress along these lines; see discussion in Chapter 11. The understanding that denialism is ultimately a function of cultural identity is a hopeful sign that approaches will improve. Massimo Pigliucci, a professor of philosophy, says:

> Participants at the conference agreed that what the large variety of denialisms have in common is a very strong, overwhelming, ideological commitment that helps define the denialist identity in a core manner. This commitment can be religious, ethnical or political in nature, but in all cases it fundamentally shapes the personal identity of the people involved, thus generating a strong emotional attachment, as well as an equally strong emotional backlash against critics.

- SS-link: Massimo Pigliucci, 'The varieties of denialism', *Scientia Salon* blog, 2014. https://scientiasalon.wordpress.com/2014/10/28/the-varieties-of-denialism/.

186 In addition to issues already reported, a specific framing by D&M is very noticable. For instance, their 2010 paper, which largely overlaps D&M2009 (see BMJ-link) consistently points to government agencies or NGOs as the targets of denialist misinformation, but never as the generators of such information. In practice, governments and NGOs may also be part of a false cultural consensus. The paper emphasises 'corporate interests', yet not governmental interests, and likewise D&M2009 cites 'corporate largesse' and 'powerful corporate interests', which are all fine to point out. Yet neither paper ever cites 'governmental interests' or 'NGO interests'; not so fine. And an example of Republican bias is given. Fine. But not one of Democrat bias. One doesn't always have to state a balance explicitly for each given point, this would be tedious. Yet overall balance is highly desirable, and a collection of minor imbalances like this are a route to much more serious problems when there is no theoretical underpinning provided for the effect they're attempting to describe.

- BMJ-link: M McKee, 'How the growth of denialism undermines public health', *BMJ* 2010; 341: c6950.

187 The case of *Slate* contributor Amanda Marcotte (see AM-link below), as reported in the *Boston Review* (BR-link), demonstrates how denialism has become a pernicious meme that is used as a means to slap down the opposing side in any dispute; in this case, in the guise of 'rape denialism'. The concept has mushroomed out of control. Web searches reveal that the term has been deployed as: science denialism, race denialism, Trump denialism, GMO denialism, terror denialism, Renewables denialism, demographics denialism, technological denialism, maths denialism, of course climate denialism, and more. Whether or not the side deploying such labels is favoured by any actual evidence, the accusation of denialism provides no benefit and simply makes debates more polarised and toxic. Generally speaking, little or no evidence for denialism is supplied, not even that which might pass the D&M2009 criteria. However, this doesn't mean the label is deployed cynically;

groupthink or cultural bias means that those making the accusations will typically believe they are true.

- AM-link: https://twitter.com/AmandaMarcotte.
- BR-link J Levine, 'Feminism can handle the truth', *Boston Review*, 2014. http://web.archive.org/web/20211017222456/http://bostonreview.net/blog/judith-levine-uva-rape-denialism-rolling-stone-hoax-feminism.

188 See: M Hoofnagle, 'Is the holocaust denial/climate change denial comparison apt?', *Mark Hoofnagle's Denialism* blog, 2012. http://scienceblogs.com/denialism/2012/05/18/is-the-holocaust-denialclimate/.

189 S Lewandowsky et al., 'Science and the public: Debate, denial, and skepticism', *Journal of Social and Political Psychology* 2016; 4(2): 1–99.

190 M McKee and P Diethelm, 'How the growth of denialism undermines public health', *British Medical Journal* 2010; 341: c6950.

191 N Oreskes and E Conway, *Merchants of Doubt: How a handful of scientists obscured the truth on issues from tobacco smoke to global warming*. Bloomsbury, 2012.

192 Such social theory as is presented for the causation of 'denialism' appears wafer-thin. The work implies that right-leaning scientists are more inclined to use scientific arguments for political purposes, yet this presented viewpoint might be a cherry-pick, at least according to the review by Dagfinn Reiersøl (see DO-link). While a presented social theory doesn't have to be founded in current literature, beyond the above there appears to be no alternative logic-chain, and regarding the main conflicted topic (climate-change) the book essentially settles on a deliberate and nefarious conspiracy which clouds knowledge of, and thwarts action upon, this issue. See CE-link below for a Climate Etc. post on the book (and movie based on same), which contains the following quote from sociologist Reiner Grundman:

> 'It is disappointing to see professional historians reduce the complexity to a black and white affair where it goes without saying what the preferred colour is. The social science literature relevant to the understanding of policymaking in the face of uncertainty is largely absent. This raises the question of what epistemological status it can claim. Its authors have been critical of the scientific credentials of the contrarians, quoting the lack of peer review or selective use of information. But it is what the title and subtitle suggest: less a scholarly work than a passionate attack on a group of scientists turned lobbyists. I wonder if it does not do a disservice to the cause it is advocating.'

- DO-link: D Reiersøl, 'Debunking Oreskes Part 4: Disinformation or debate?' *Evil Questions* blog, 2014. https://www.evilquestions.com/2014/10/05/debunking-oreskes-part-4-disinformation-or-debate/.
- CE-link: J Curry, 'Bankruptcy of the "merchants of doubt" meme', *Climate Etc* blog, 2015. https://judithcurry.com/2015/03/15/bankruptcy-of-the-merchants-of-doubt-meme/.

193 Many people have accused Lewandowsky of hypocrisy, notably in comments left on the blog of one of his co-authors (Nick Brown, see NB-link below). I by no means endorse all the critiquing comments, and some of the author team's defence may have merit. However, the general flavour of the criticisms, namely that the (main two) authors by no means practice what they preach, appears to be valid. My own comments point out major methodological problems with two of the papers they cite: D&M2009 (see Chapter 6) and Cook et al. 2013 (see JD-link).

- NB-link: N Brown, 'It's a small world', *Nick Brown's Blog*, 2016. http://steamtraen.blogspot.co.uk/2016/08/its-small-world.html.
- JD-link: J Duarte, 'Cooking stove use, housing associations, white males, and the 97%', *José Duarte* blog, 2014. http://www.joseduarte.com/blog/cooking-stove-use-housing-associations-white-males-and-the-97.

194 Namely:

- Issues with his methods: S McIntyre, 'Lewandowsky's fake correlation', *Climate Audit* 2012. http://climateaudit.org/2012/09/18/lewandowskys-fake-correlation/. (CA1-link).
- Issues with principles, S McIntyre, 'The "Ethics Application" for Lewandowsky's fury', *Climate Audit*, 2014. http://climateaudit.org/2014/04/07/the-ethics-application-for-lewandowskys-fury/ And: J Duarte, 'More scams', *José Duarte* blog, 2014. http://www.joseduarte.com/blog/more-fraud.

195 See:

- S Lewandowsky et al., 'Scientific uncertainty and climate change: Part I. Uncertainty and unabated emissions', *Climatic Change* 2014; 124: 21–37.
- S Lewandowsky et al., 'Scientific uncertainty and climate change: Part II. Uncertainty and mitigation'. *Climatic Change* 2014; 124: 39–52.

These papers are also the backbone of a 2014 article in *The Guardian* by environment writer Dana Nuccitelli, aided by John Cook: The climate change uncertainty monster – more uncertainty means more urgency to tackle global warming (see G-link below).

- G-link: D Nuccitelli, 'The climate change uncertainty monster – more uncertainty means more urgency to tackle global warming", *The Guardian*, 2014. http://www.theguardian.com/environment/climate-consensus-97-per-cent/2014/apr/04/climate-change-uncertainty-stronger-tackling-case.

The heart of Lewandowsky's argument is (emphasis mine):

> …in the case of the climate system, it is very clear that greater uncertainty will make things even worse. This means that *we can never say that there is too much uncertainty for us to act*. If you appeal to uncertainty to make a policy decision the legitimate conclusion is to increase the urgency of mitigation.'

196 Lewandowsky uses the term 'leakage' to describe the process by which popular memes in the public domain 'leak' into the scientific discourse, and hence in-

fluence conclusions. He claims a particular case here: https://www.youtube.com/watch?v=fYzEWOHWTLk.

197 This appears to be an inappropriate application of the uncertainty principle, as explained in: J Curry, 'Worst case scenario versus fat tail', *Climate Etc* blog, 2014. http://judithcurry.com/2014/04/18/worst-case-scenario-versus-fat-tail/.

198 For instance:

> First, legitimate public concern about a lack of transparency and questionable research practices must be met by ensuring that research lives up to modern standards…
>
> Second, we believe that daylight is the best protection against politically-motivated maneuverings to undermine science…
>
> Finally, skeptical members of the public must be given the opportunity to engage in scientific debate.

199 This can happen in regard to long-debated issues, and newer ones too. For instance, pollsters show that US public opinion is polarised on evolution (versus creationism; a long-standing debate) and fracking (a relatively new issue), meaning there is large-scale resistance. For details see:

- Creationism and evolution: https://www.pewresearch.org/science/2015/07/01/chapter-4-evolution-and-perceptions-of-scientific-consensus/.
- Fracking: http://www.gallup.com/poll/182075/americans-split-support-fracking-oil-natural-gas.aspx.

200 P Weintraub, 'The doctor who drank infectious broth, gave himself an ulcer, and solved a medical mystery', *Discover* magazine, 2010. https://www.discovermagazine.com/health/the-doctor-who-drank-infectious-broth-gave-himself-an-ulcer-and-solved-a-medical-mystery.

201 Depending on how it's measured, about 8 to 10 years; see the timeline on Helicobacter (wiki-link below), which cites a series of references. During this period, the field was gradually catching up with the idea of a bacterial cause, although Marshall argues that scepticism was an issue in the early years; *New York Times* medical correspondent Dr. Lawrence K. Altman said in 2002, 'I've never seen the medical community more defensive or more critical of a story'.

- https://en.wikipedia.org/wiki/Timeline_of_peptic_ulcer_disease_and_Helicobacter_pylori.

202 From *The Stanford Encyclopedia of Philosophy*:

> 'The 5th century sophists develop forms of debate which are ancestors of skeptical argumentation.'
>
> 'The sophists explore the idea that, if things are different for different cultures, there may be no fact of the matter of how those things really are.'

- Link: https://plato.stanford.edu/entries/skepticism-ancient/.

203 *The Stanford Encyclopedia of Philosophy* (SEP-link) says:

> ...scholars may have paid too little attention to skepticism's ancestry in poetry ...Pyrrho seems to have referred to Homer as a proponent of ideas he approves of, ideas about change, the status of human rationality and language, and more.

Zerba provides examples, regarding 'disagreement about a question that has not been decided, cannot be decided, or awaits decision' (Z1-link), a situation which leads to 'suspension of judgement' (Z2-link).

- SEP-link: https://plato.stanford.edu/entries/skepticism-ancient/.
- Z1-link: M Zerba, *Doubt and Skepticism in Antiquity and the Renaissance*, Cambridge University Press, 2012.
- Z2-link: M Zerba, 'What Penelope knew: doubt and scepticism in the "Odyssey"', *The Classical Quarterly*, 2009; 59(2):295–316.

204 For example, from JP Allen, *Middle Egyptian: An Introduction to the Language and Culture of Hieroglyphs*, Cambridge University Press, 2014, p. 400:

> Make holiday, don't weary of [life]!!
>
> Look, there is no one allowed to take their things with them,
>
> and there is no one who goes away comes back again.

And: J Baines and P Lacovara, 'Burial and the dead in ancient Egyptian society: Respect, formalism, neglect'. *Journal of Social Archaeology* 2002; 2(1): 5–36:

> The skeptical texts, which may be part of a tradition far older than the identifiable evidence, reveal tensions in a complex society's relation to its past and its dead members...These discordant attitudes cast doubt upon the purpose of the structures [tombs], which nonetheless continued to be built; such discordance is not confined to Egypt.

205 A translation by M Lichtheim (from *Ancient Egyptian Literature: A book of readings*, Vol. 2. The New Kingdom, 1976) can be found at https://www.ucl.ac.uk/museums-static/digitalegypt/literature/authorspchb.html.A more compact and intuitive translation of the first two paragraphs can be found at Wikipedia: https://en.wikipedia.org/wiki/The_Immortality_of_Writers.

206 Homer's tales were populist and integrated common tropes, oft-repeated over generations. *The Harper's Songs* were used in common funeral rites. *The Immortality of Writers* was an instructional work for schools.

207 Comedy is a useful mode of expressing scepticism and also inducing it in others. Link 1 notes examples from modern times, including the Simpsons and South Park; there are also examples from the past and ancient times (Links 2 and 3). The modern comedic example on YouTube (Link 4) aims to induce scepticism at the environmental and pension policies of the ruling coalition in Germany. Extreme doubt that these policies are viable is expressed in comic terms.

- Link 1: B Radford, 'Critical and thinking: the Ian Harris interview', *Center for Inquiry*, 2015. http://www.centerforinquiry.net/blogs/entry/critical_and_thinking_the_ian_harris_interview/.
- Link 2: MR Martin, *Between Theater and Philosophy: Skepticism in the major city comedies of Ben Jonson and Thomas Middleton*, University of Delaware Press, 2001.
- Link 3: RT Long, 'Ancient Greece's legacy for liberty: the comedy of politics', *Libertarianism.org*, 2016. https://www.libertarianism.org/columns/ancient-greeces-legacy-liberty-comedy-politics.
- Link 4: Sadly, the original link with English subtitles has been taken down (https://youtu.be/-e2U2cYcPro). The original in German, from the *Heute* show on TV channel ZDF (7th March, 2014) is here: https://youtu.be/GXX17-An9Hw?t=1030, link goes to segment start at 17:11, watch to 23:20.

208 As revealed in Episode 1 of the BBC TV series *Egyptian Journeys with Dan Cruickshank*. See http://www.bbc.co.uk/programmes/b0078yvw.

209 Firm detection may be impossible before writing existed. Using images is problematic, because associated writing may be needed for context, and images will only survive anyway if ordinary people produced them in sufficient numbers. Production on this scale tends to go hand in hand with the spread of writing (for instance there is much pictorial graffiti from ancient societies, but it tends to be interspersed with textual graffiti). If priests or shamans (or other members of the elite) made images, these would probably be orthodox in nature, rather than sceptical.

210 There has been debate about scepticism in pre-classical and pre-historic societies since the early 20th century. Since the 1930s, observation of modern oral societies has confirmed the existence of scepticism, alongside faith. Yet it is argued that this scepticism is more specific than in the modern conception, for instance highlighting the fraudulent nature of particular witch-doctors, but not doubting witch-doctoring in general, as there is no framework in which this magical context doesn't exist for the relevant tribe. EE Evans-Pritchard says of the Azande tribe:

> Absence of formal and coercive doctrines permit Azande to state that many, even most, witch-doctors are frauds. No opposition being offered to such statements they leave the main belief in the prophetic and therapeutic powers of witch-doctors unimpaired. Indeed, skepticism is included in the pattern of belief in witch-doctors. Faith and skepticism are alike traditional.

- Quoted in: RA Manners and D Kaplan, *Anthropological Theory*, Taylor & Francis, 2017.

> Others argue that questions about particular practitioners, especially as these can be frequent, will hardly fail to lead to doubts about the practice itself. However, such doubts are less likely to accumulate in a society without writing, and are perhaps only shared privately. The debate appears to have overlapped with the controversial 'literacy thesis', which makes the topic harder to research, and

I have not delved deeply.

See also Endnote 211.

211 Sarah Iles Johnston says (in *Ancient Religions*, Harvard University Press, 2009, p. 130):

> Similar factors [regarding writing enabling the spanning of generations] lie behind the development of explicitly skeptical traditions. Every society includes men and women with unorthodox ideas, people who adopt a radically dissenting attitude to generally accepted views on religion, politics, and social order. But in an entirely oral culture, skepticism tends to die with the individual skeptic. Once the skeptic commits his or her views to the permanency of writing, however, the possibility opens up a whole tradition of skepticism, an alternative counterculture parallel to society's orthodox norms.

212 According to Britannica (Brit-link below), a belief in life after death is attested to by burials around 30 to 50 thousand years old. However, it notes that religious conceptions may not be bound to material objects that we can find in the archaeological record, so they could be far older.

- Brit-link: https://www.britannica.com/topic/prehistoric-religion.

This is easily old enough for cultural evolution and gene-culture co-evolution to take place. The classic example of the latter is adult lactose tolerance, which evolved from about 9000 years ago alongside dairy farming.

- N-link: A Beja-Pereira et al., 'Gene-culture coevolution between cattle milk protein genes and human lactase genes'. *Nature Genetics* 2003; 35: 311–313.

213 https://en.oxforddictionaries.com/definition/scepticism, last sampled December 2021.

214 See CF Bond and M Robinson, 'The evolution of deception', *Journal of Nonverbal Behavior* 1988; 12: 295–307:

> Deception has evolved under natural selection, as has the capacity to detect deceit. In this article, we describe the adaptive significance of deception in plants, fireflies, octopi, chimpanzees, and Homo sapiens. We review behavior genetic research to find that heredity affects human deceptiveness and theorize that genetically-transmitted anatomical features prefigure human success at deceit.

215 See David Livingstone Smith, 'The evolution of the unconscious', *Psychoanalytische Perspectieven* 2002; 20(4): 525–548.

216 For whom via culture and gene / culture interaction, the pace of evolution increases while in addition presenting more modes available for selection.

217 L McNally and AL Jackson, 'Cooperation creates selection for tactical deception'. *Proceedings of the Royal Society B: Biological Sciences* 2013; 280(1762): 20130699.

218 The concept of withheld judgment is also a major feature of early Greek philosophical scepticism. From the Stanford Encyclopedia of Philosophy (https://plato.stanford.edu/entries/skepticism-ancient/):

> Hellenistic discussions envisage three attitudes that cognizers take to impressions (how things seem to them): assent, rejection, and suspension of judgment (epochê).
>
> Suspension is a core element of skepticism: the skeptic suspends judgment.

219 Shermer lists a set of clues that reveal lying. He invokes the 'selfish gene' model as part of the causal explanation for the arms race of deceit and detection, but it is co-operation that has engendered an environment where lying may prosper as a side effect (see Endnote 217). In a wholly competitive environment lacking a co-operative system, lying would be meaningless; there is no trust to betray. So it is cultural group selection, fostering co-operation, which is the dominant relevant driver, rather than a mainly 'selfish gene' driven process.

- Link: M Shermer, 'Lies we tell ourselves: how deception leads to self-deception', *Scientific American*, 2012. http://www.michaelshermer.com/2012/02/lies-we-tell-ourselves/.

220 For example, see L ten Brinke et al., 'Darwin the detective: Observable facial muscle contractions reveal emotional high-stakes lies', *Evolution and Human Behavior* 2011; 33(4): 411–416.

221 The Continued Influence Effect (CIE) is an effect whereby misinformation (e.g. propaganda or other information biased by cultural or worldview outlook), continues to influence people even after it has been retracted or corrected, and even in the face of warnings before the information is communicated that such a thing as the CIE exists. However, a sceptical stance considerably reduces the CIE.

The CIE is described in Ecker et al. (E2010), Lewandowsky is a co-author, which also notes the following regarding its reduction via scepticism (various other sources including two papers with Lewandowsky as the lead author, in square brackets, are referenced by E2010 in support of the finding; my emphasis):

> The second factor that seems to reduce the CIE is suspicion toward the source of the misinformation. In the WMD studies discussed earlier, belief in the existence of WMDs in Iraq was correlated with support for the war and was especially pronounced in those people who obtained news from sources that supported the invasion (e.g., Fox News; Kull et al., 2003). Lewandowsky et al. (2005)[L2005] uncovered a more direct link between suspicion and the ability to update misinformation related to the Iraq War. They operationalized suspicion as the extent to which respondents doubted the official WMD-related reasons for the invasion. Lewandowsky et al. (2005) found that, when this measure was used as a predictor variable, it explained nearly a third of the variance in people's belief in misinformation. Moreover, once suspicion was entered as a predictor, previously striking mean differences between respondents in the U.S. and two other countries (Ger-

> many and Australia) disappeared and were, instead, found to reflect differing degrees of suspicion between those countries. *Lewandowsky, Stritzke, Oberauer, and Morales (2009) [L2009] extended the notion of suspicion by suggesting that it may be related to a more stable personality trait of skepticism—skeptics will generally tend to question the motives behind the dissemination of information.*'

This finding casts scepticism as a positive and healthy trait. This is also echoed in Lewandowsky et al. (L2012):

> Skepticism: A key to accuracy. We have reviewed how worldview and prior beliefs can exert a distorting influence on information processing. However, some attitudes can also safeguard against misinformation effects. In particular, skepticism can reduce susceptibility to misinformation effects if it prompts people to question the origins of information that may later turn out to be false.

- L2005: S Lewandowsky et al., 'Memory for fact, fiction and misinformation: The Iraq War 2003', *Psychological Science* 2005; 16(3): 190–5.
- L2009: S Lewandowsky et al., 'Misinformation and the "War on Terror": when memory turns fiction into fact' in: WGK Stritzke et al. (eds), *Terrorism and Torture: An Interdisciplinary Perspective* Cambridge University Press, 2009.
- L2012: S Lewandowsky et al., 'Misinformation and its correction: continued influence and successful debiasing' *Psychological Science in the Public Interest* 2012; 13(3), 106–131.
- E2010: UKH Ecker et al., 'Explicit warnings reduce but do not eliminate the continued influence of misinformation', *Memory & Cognition* 2010; 38(8): 1087–1100.

222 It's possible that via evolutionary processes, innate scepticism is maintained at an optimum level within populations. With too much, it is hard to see how collective deceptions would stick. At some point, cultures would become unstable and so groups would not benefit from their significant net advantages. Too little, and successive collective deceptions might sweep too swiftly and deeply through populations, perhaps resulting in a permanent and disruptive switching between extremes, such as modern fascism and communism. Similar optimum evolutionary balances between opposing constraints occur in biology, for example balanced polymorphism.

223 All cultural beliefs are new at one time and so must make initial headway against innate scepticism. This can happen relatively swiftly (in generational terms) if a new culture fills the vacuum left by an old and crumbling belief system that can no longer compete. Or the new culture features highly emotive stories which, via selection, have hit upon the latest means of countering scepticism and so are irresistible to large numbers of people. But innate scepticism will eventually catch up, while cultures themselves evolve constantly, permanently attempting to outflank innate scepticism in this group version of the deceit/detection arms race.

224 See: MS Gazzaniga, *Who's in Charge*, Ecco, 2012, especially Chapter 5, 'The social mind'. While I don't necessarily agree with everything in this book, it makes a convincing case for social thinking.

225 D Kahan et al. *Motivated Numeracy and Enlightened Self-Government*, Working paper 116, The Cultural Cognition Project. http://www.datascienceassn.org/sites/default/files/Science Comprehension Thesis.pdf.

226 It's not an area I've explored, but the theory that strong cultural beliefs disable certain cognitive abilities has some support from MRI scans. In particular, messages from charismatic cultural authority figures may achieve this end in a similar way to hypnosis (see NS-Link and LH-Link). To pick up effects via current MRI scanning seems to require testing of rather committed individuals, but the test scenario falls a long way short of the popular conception of actively imposed hypnosis by extreme cults.

- NS-link: A Coghlan, 'Brain shuts off in response to healer's prayer', *New Scientist*, 2010. https://www.newscientist.com/article/mg20627574.200-brain-shuts-off-in-response-to-healers-prayer.
- LH-link: CH Legare and PA Herrmann, 'Cognitive consequences and constraints on reasoning about ritual', *Religion, Brain & Behavior* 2013; 3(1): 63-65.

227 For instance, Martin Luther adopted a sceptical stance against culturally decadent features of the church of Rome, which he expressed in his *Ninety-five Theses*. He challenged (among other issues) the selling of papal indulgences, and the authority and wealth of the Pope. His stance triggered a much wider Reformation movement (not all of which he approved of), indicating that the older culture had probably been unpopular for quite some time prior. Luther's movement lead ultimately to the breakaway Christian culture of Protestantism.

228 For instance, eugenics, or writ small (i.e. groupthink, perhaps aided by financial interests) the consensus on saturated fats.

229 The authority of science has in some ways replaced that of religions, and is leveraged in all sorts of inappropriate ways. For instance, exaggerated scientific claims are ubiquitously made by the advertising industry, or by bodies, such as the Union of Concerned Scientists, that hijack the authority of science for advocacy. Even scientific institutions, such as the Royal Society, sit uncomfortably with a purely evidential approach, because of their authority and status and their tendency to produce deference. On the other hand, the urge to obtain status encourages innovation and perseverance in reaching scientific goals. However, overall, those benefits are outweighed by the downsides.

230 A public groundswell of innate scepticism about scientific narrative with social impact may motivate a pursuit of the truth via scientific scepticism among those outside the domain, but who are either science-orientated or indeed scientifically adept. If the innate scepticism that motivated the research was inapt – there is no culture in operation – a dead end will be reached, although meaningful science may be done along the way. For apt innate scepticism, however, the research could lead

to a serious and genuine challenge of orthodoxy in an inappropriately entrenched discipline.

231 The classic case is the colonial-era rejection of medical knowledge by native peoples when this was delivered as part of a Christian missionary package. However, the phenomenom is still around. For instance, if a political party 'adopts' and promotes a science issue then there will be cultural consequences: their focus may shift away from the objective science towards political ends, and their political opponents will become more likely to reject the science. An interesting recent case is the politically based resistance to the HPV vaccine (K-link), which was in stark contrast to the straightforward acceptance of the HBV vaccine, which is ultimately of the same medical type and purpose.

The difference was a result of the contrasting ways they were promoted. In the HPV case, the manufacturer, in a hurry to gain a captive market for its product, 'orchestrated a poorly disguised political marketing campaign, one that included adoption of vaccine mandates in state legislatures'. This put the HPV issue into the political arena and made acceptability as much a matter of political identity as of objective assessment of the medical issues. As a result, the first that many parents heard of HPV was through the biasing lens of political conflict.

In the HBV case, the route to market was through standard medical procedure for new treatments, so parents first heard about it from public health authorities and local doctors, trusted sources. So their political identity did not come into it.

- K-link: D Kahan, 'What antagonistic memes look like: the case of the HPV vaccine', *Cultural Cognition* blog, 2016. http://www.culturalcognition.squarespace.com/blog/2016/7/21/what-antagonistic-memes-look-like-the-case-of-the-hpv-vaccin.html.

232 See Judith Curry's take (CE-link below) on the bias in Oreskes & Conway (OC-link). Because the ultimate underpinning of all three tests is the idea that consensus must be right, bias towards consensuses is inevitable, irrespective of whether they are cultural in nature or result from genuine wide agreement on the science.

- CE-link: J Curry, 'Bankruptcy of the "merchants of doubt" meme', *Climate Etc* blog, 2015. https://judithcurry.com/2015/03/15/bankruptcy-of-the-merchants-of-doubt-meme/.
- OC-link: N Oreskes and E Conway, *Merchants of Doubt: How a handful of scientists obscured the truth on issues from tobacco smoke to global warming.* Bloomsbury, 2012.

233 'Facts, beliefs, and identity: The seeds of science skepticism', *Society for Personality and Social Psychology* blog, 2017. https://www.psypost.org/2017/01/facts-beliefs-identity-seeds-science-skepticism-47039.

234 As fostered by small-scale authoritarian groupthink to large-scale cultural entity (e.g. a religion), and as opposed to a scientific consensus, which may be right or wrong but isn't collective deceit.

235 There have been calls for yet more 'climate-change education' in order to dissolve a widespread public scepticism of catastrophic climate-change orthodoxy. Many targeted education programs have been tried, and there is controversy (see link below) about more general climate-change education that has been introduced to school curricula. However, there is little if any sign that public scepticism is waning. This is likely due to the nature of the 'education' and hence its impact on innate scepticism.

- Link: A Montford, *Climate Control, Brainwashing in Schools.* Report 14, The Global Warming Policy Foundation, 2014. https://www.thegwpf.org/publications/climate-control/.

Most 'climate-change education' appears not to teach what the science knows or does not know within various sub-disciplines such as atmospheric physics or oceanography or biological responses. Admittedly this would be complex and difficult to communicate. Instead, it appears to assume a certainty of imminent global catastrophe and attempt to get everyone on board with that message, then discuss how people and organisations could lessen the catastrophic impacts (for example, see the RT-link below).

- RT-link: Rising Tide (UK), 'Popular education pack': https://risingtide.org.uk/files/RTpopedpackUK_PRINT_3.pdf.

236 It seems that articles on scepticism often imply an ancient Greek origin, with little or no mention of the older, instinctive sceptical traits. Philosophical scepticism and scientific scepticism indeed owe much to Greek formalisation, but this process may itself have been prompted by innate scepticism, in part via poetry (see Endnote 203).

237 M Shermer, 'What is skepticism, anyway?', *Huffington Post*, 2013. http://www.huffingtonpost.com/michael-shermer/what-is-skepticism-anyway_b_2581917.html.

238 See further discussion at: J Curry, 'What is skepticism, anyway?', *Climate Etc*, 2014. https://judithcurry.com/2014/06/05/what-is-skepticism-anyway/.

239 Unless the non-conventional idea (let's assume it is a wrong one) acquires a cultural dimension, and especially if then, via emotional selection, it starts to dominate, and becomes a cultural consensus. However, ultimately that would also occur because institutional scientific scepticism didn't maintain objectivity.

240 D Kahan, 'What you "believe" about climate change doesn't reflect what you know; it expresses *who you are*', *The Cultural Cognition Project*, 2014. http://www.culturalcognition.squarespace.com/blog/2014/4/23/what-you-believe-about-climate-change-doesnt-reflect-what-yo.html.

241 YouGov international poll on attitudes to climate change (2019): https://yougov.co.uk/topics/science/articles-reports/2019/09/15/international-poll-most-expect-feel-impact-climate.

242 The 2015 UN 'My World' poll (with around 10 million participants). The poll encompasses very many nations; I use only a subset that has commonality with the other climate surveys. Note: The archive web-copy is very slow indeed. However, it does yield the data. Incredibly, the UN deleted their original interactive site: http://web.archive.org/web/20190802231507/http://data.myworld2015.org/.

243 Too much focus on harm or damage (rather than, say, 'impact') leads to evocation of only half the emotive power of the Catastrophe Narrative, i.e. the 'doom' half and not the 'salvation' half. This leads to an attenuated response, as those more swayed by the salvation half, or by the enhanced power of both halves together, don't react. Framing the question as 'concern' about harm, doesn't have the effect of eliminating potential salvation, and hence attracts the whole spectrum of believers. Likewise, the word 'impact' instead of 'harm' doesn't preclude salvation (impacts can include the changes to avoid doom). The first series in Table 20 in Appendix C, asking about human extinction, precludes salvation and so attracts 'doomsters' only.

244 Leadership positions of the main faiths, on climate change: https://fore.yale.edu/Climate-Emergency/Climate-Change-Statements-from-World-Religions. See also Appendix D.

245 R Haluza-DeLay (2014), 'Religion and climate change: varieties in viewpoints and practices', *Wiley Interdisciplinary Reviews: Climate Change* 2014; 5(2): 261–279.

246 D Kahan, 'What you "believe" about climate change doesn't reflect what you know; it expresses *who you are*', *Cultural Cognition* blog, (2014): http://www.culturalcognition.squarespace.com/blog/2014/4/23/what-you-believe-about-climate-change-doesnt-reflect-what-yo.html.

247 Summary charts in the Excel-Ref show these allied belief (SA) and core belief (FC) trendlines crossing over at the far left-hand side. However, these lines are actually extrapolated by a few percent prior to the least religious nation of Sweden – about 24% religiosity – hence they don't actually cross in practice.

248 About 40% of the publics of even the most religious countries, are seemingly unmoved by the appeal of any unconstrained questions. I think because this bloc in highly religious populations is so hooked on religious belief, they resist any of their faith leaders whose statements move too far away from scripture. Climate change is one example; another the reforms of recent popes, which have been resisted by large numbers of Catholics.

249 World Values Surveys. https://www.worldvaluessurvey.org/WVSOnline.jsp; Wave 5, 2005–2009, Question V111.

250 Due to 'grade inflation' caused by the selection of more emotive narratives over time, the climate-change most-endorsing response 'extremely serious' is now equivalent to a response of 'very serious' in the 2005–2009 window. The latter now represents a more equivocal attitude, and certainly where both response options are offered together it may even be neutral, as shown in Figure 35 in Appendix E. However, for very irreligious nations, either of these endorsing response options

are much lower on the y-axis in modern times than they were in the historic period (Figure 35), because there was far less innate scepticism expressed back then.

A second historic series is included at the 'Extra' sheet of the Excel-Ref, see 'Historic Series 2'. Derived from a BBC international poll, this features even more dispersed data (and unfortunately fewer data points), but once again the average *y*-axis score of the handful of most irreligious nations is very high indeed, as we would expect. Moreover, Turkey and Egypt are shifted the furthest from where one might expect nations to be when using Figure 8 as a guide, indicating cultural relationships that are likely still in flux for these more religious nations.

251 Questions interrogating religion that mix reality-constrained and unconstrained elements (mixed-mode questions – see Chapter 10) should, similar to the 'climate culture' case, generally produce a pattern that is non-linear, but still bounded within an envelope defined by two linear cultural responses.

252 Actually, as noted in Chapter 8, WA1 isn't in quite the same position as WA. It starts in the same place as WA for low religiosities, but has a slightly steeper gradient so ends up higher at the right hand side. Given the strength categories are rather arbitrary anyhow and the trend appeared to be a hybrid, I kept the chart uncluttered by representing both WA and the similar WA1 by a single line. However, this means that within likely margins of error WA1+O2 could equally be a 'Medium-Weak-Aligned + Offset' trend, i.e. MWA1+O2.

253 The orange series in the first chart on the 'PostCovid' sheet of the Excel-Ref is theoretically unconstrained; it comes from the answer option 'a lot of ' to the question: 'Willing to make ____ changes about how you live and work to help reduce the effects of global climate change'. A lot of changes cannot be made without an impact on other activities, which is a reality constraint, even though this is not explicit. The constraint appears to be enough to dissolve any statistically significant trend, but not enough for the data to look like a mixed-mode envelope. It is likely that with more data above 72% religiosity, which this series is unfortunately missing, there would be a statistically significant trend, which would still conform to the question's unconstrained framing, even though it may be modest rather than robust.

254 IMF 2017.

255 Due to very long-term (many generation) development issues, the GDP of nations anti-correlates with religiosity,* so it should theoretically be possible to remove GDP considerations from the depiction of national attitudes to climate change altogether, i.e. even as a secondary variable. The systemic variance about the trends ought to be expressible as a function of the religiosity of each nation relative to its religio-regional peer-group average. However, I haven't attempted to do this.

256 Special Eurobarometer 490: https://www.kantar.com/-/media/project/kantar/global/articles/files/2019/ebs490_report.pdf.

257 From ~Y31/X24 down to ~Y8/X88; $R = 0.59$, $p = 2.6 \times 10^{-3}$.

* And quite robustly, notwithstanding exceptions such as Ireland, Singapore, the oil-rich states, and the US too, although we don't considering the latter until Chapter 11.

258 Typically, when one take averages of bucketed groups, then checks these values for correlation against another variable, there will be a better R^2 than if the averaging wasn't done first. This is because the averaging will narrow the overall data spread. In this case, nations are the buckets.

259 $R=0.77$ and $R^2=0.59$. See James H. Steiger's (Vanderbilt University) teaching slides: http://www.statpower.net/Content/312/Lecture Slides/BasicRegression.pdf.

260 I Ruiz et al., 'Climate change perception: Driving forces and their interactions', *Environmental Science & Policy* 2020; 108: 112–120.

261 B Kvaløy et al. 'The publics' concern for global warming: A cross-national study of 47 countries', *Journal of Peace Research* 2012; 49(1): 11–22.

262 A McCright et al., 'Political ideology and views about climate change in the European Union', *Environmental Politics* 2016; 25(2): 338–358.

263 S Levi, 'Country-level conditions like prosperity, democracy, and regulatory culture predict individual climate change belief', *Communications Earth & Environment* 2021; 2: 51.

264 Papers may measure at the individual level, national level, or both.

265 W Poortinga et al. 'Climate change perceptions and their individual-level determinants: A cross-European analysis', *Global Environmental Change* 2019; 55(22): 25-35.

266 From Figure 8, about 63% of people in nations with 100% religiosity give climate-change most-endorsing responses for the SA series. If all religious individuals in every nation responded identically then, say for Sweden, the least religious nation in my data (debiased religiosity = 22%), the SA trendline at that point should score at least (22 × (63/100)) = ~14% on the *y*-axis. However, not only is the actual figure lower (at ~9.5%), the majority of the response here is probably provided by core believers, who in turn are much more likely to be irreligious anyhow (it's hard to achieve a high commitment to both cultures). In other words, religious people must be acting at least somewhat differently in Sweden (and similarly irreligious countries such as the Czech Republic) than they do in say Thailand or Nigeria or Ghana. Therefore, as noted in the main text, aggregating all their responses will lose information. It is probably not a coincidence that the SA trendline can be very closely imitated by adding the FC trendline (representing core believers) to a pseudo trendline (A_{max}, maximum underlying SA) that runs from $x=20$, $y=0$, up to $x=100$, $y=60$ (see this charted in SA Recon in the Main Trends sheet of the Excel-Ref, column T row 308). This suggests that there is little or no ($y=0$) contribution to SA from religious people in overwhelmingly secular nations; their innate scepticism remains enabled in this setting, and the contribution from such people only surpasses that of core believers when debiased national religiosity is higher than about 40%.

267 S Andi, 'How people access news about climate change', *Digital News Report*, 2020. http://www.digitalnewsreport.org/survey/2020/how-people-access-news-about-climate-change/.

268 A Lo and A Chow 'The relationship between climate change concern and national wealth', *Climatic Change* 2015, 131, 335–348.

269 Lo and Chow also try other predictors, which I think work largely according to their own relationships with GDPpc. As the tables in the paper show, energy usage per capita and the ND-GAIN index of climate adaptation are good proxies for GDPpc, and hence yield overall degraded but still respectable correlations when substituted for GDPpc. Per-capita carbon dioxide emissions is a significantly poorer proxy.

270 I haven't investigated, but as noted in Section 10.3.3, converting their 'perception of danger' variable to an additive Likert-type scale may have caused some perception issues as to the nature of what they're investigating, and could also impact correlation value, though probably not for such an emotive option as 'extremely dangerous'.

271 Hornsea et al. support this position. See MJ Hornsea et al., 'Relationships among conspiratorial beliefs, conservatism and climate scepticism across nations', *Nature Climate Change* 2018; 8: 614–620.

272 My own religiosity scale only has values for whole nations, inclusive of the US but not separately for Democrat and Republican supporters. However, Pew undertook a very comprehensive survey of US religiosity (see P-link below) that includes results by political affiliation, and though it's a little elderly (2014), religiosity doesn't turn on a dime and this will still be accurate enough to use with the US attitude surveys (2018–2020). The nearest two values to my scale are 'absolutely certain' belief in God, and religion being 'very important', which score 55% and 47% respectively for the Democrats and Democrat leaners. These average to 51%.

- P-link: Pew Research Centre. 'Religious Landscape Study, Party Affiliation', across 35,000 participants (2014): https://www.pewforum.org/religious-landscape-study/party-affiliation/.

273 From IPSOS 2021 (see I-link below). US Democrats report their primary identity as: Party ID (38%), religion (27%), with the remainder split between four lesser factors. Given that we're only interested in motivation from the first two, we can make the approximation of splitting the rest of the pie in the same ratio. This gives 58% Dem/Lib motivation to 42% religious motivation.

- I-link: IPSOS, The Fault Lines of America (2021): https://www.ipsos.com/en-us/knowledge/society/The-Fault-Lines-of-America.

274 'How Religion Intersects With Americans' Views on the Environment', from the Pew Research Centre (2022). The relevant chart is the eighth one on the webpage for the Pew study, see P-Link below. The religiously unaffiliated Democrats / Democrat leaners give an 86% response that climate change is a 'very serious' or 'extremely serious' problem. The average response for the 3 main religious brands of Democrat supporters (Evangelical Protestants, mainline Protestants, Catholics), is 83%.

- P-link: https://www.pewresearch.org/religion/2022/11/17/how-religion-intersects-with-americans-views-on-the-environment/.

275 The Economist/YouGov Poll, March 2019: https://d25d2506sfb94s.cloudfront.net/cumulus_uploads/document/a5islr4zij/econTabReport.pdf.

276 Kaiser Family Foundation/Washington Post, Nov 2019: https://www.kff.org/report-section/the-kaiser-family-foundation-washington-post-climate-change-survey-main-findings/.

277 Quinnipiac University National Poll, Dec 2018: https://poll.qu.edu/national/release-detail?ReleaseID=2590.

278 The Hill/HarrisX Poll, *Most important issues facing US*, Dec 2019: https://thehill.com/hilltv/rising/474327-voters-name-health-care-as-top-issue-going-into-2020.

279 Climate Nexus poll, *Pr1922 for US*, Aug 2019: https://climatenexus.org/wp-content/uploads/pdf/National-Poll-Toplines-Crosstabs-PR1922.pdf.

280 The Centre for American Progress/GBAO poll, *America adrift*, Jan 2019: https://www.americanprogress.org/issues/security/reports/2019/05/05/469218/america-adrift/.

281 CBS News Poll, *Climate change an issue for most voters in 2020*: https://www.cbsnews.com/news/cbs-news-poll-climate-change-will-be-an-issue-for-most-voters-in-2020/. By YouGov, Sept 2019. See table Personal Issues – Climate-change: https://www.e-mc2.gr/sites/default/files/2019-11/cbs_climate_Monday-top-tabs.pdf.

282 Pew Research, *US Public Views on Climate and Energy*, November 2019: https://www.pewresearch.org/science/2019/11/25/u-s-public-views-on-climate-and-energy/.

283 'Planning Engineer', 'Myths and realities of renewable energy'. *Climate Etc*, 2014. https://judithcurry.com/2014/10/22/myths-and-realities-of-renewable-energy/.

284 Schussler wrote under the pseudonym 'Planning Engineer', revealing his identify only after his retirement.

285 V Jayaraj, 'The myth of glorious renewables', *Watts Up With That?* blog. https://wattsupwiththat.com/2020/10/26/the-myth-of-glorious-renewables/.

286 'Age of Un-reason: How fear and ignorance drives wind and solar worship cult'. *Stop These Things* blog, 2020. https://stopthesethings.com/2020/10/31/age-of-un-reason-how-fear-ignorance-drives-wind-solar-worship-cult/.

287 D Archibald. 'Australia's energy plan', *Jo Nova* blog, 2020. http://joannenova.com.au/2020/05/great-civilizations-are-built-on-good-fuel-not-on-hydrogen/.

288 MZ Jacobson et al., 'Low-cost solution to the grid reliability problem with 100% penetration of intermittent wind, water, and solar for all purposes'. *PNAS* 2105; 112(49): 15060–15065.

289 UN 'My World' data from 2015. http://web.archive.org/web/20190802231507/ http://data.myworld2015.org/. *Very* slow archive; the UN deleted their original site!

290 Deployed wind power by country from a variety of sources, usefully collected and tabulated by Wiki: https://en.wikipedia.org/wiki/Wind_power_by_country. Sampled in October 2020.

291 Deployed solar power by country from a variety of sources, usefully collected and tabulated by Wiki: https://en.wikipedia.org/wiki/Solar_power_by_country. Sampled in October 2020.

292 Michael Shellenberger profile: https://www.linkedin.com/in/michael-shellenberger-019631a8.

293 Michael Shellenberger TED talk 'Why renewables can't save the planet': https://www.ted.com/talks/michael_shellenberger_why_renewables_can_t_save_the_planet/transcript.

294 R value for the correlation between national religiosity and sunshine hours is 0.56. Note that annual sunshine duration per country is derived from the average values for between two cities (smaller countries) and five cities (larger countries), but ten for Russia. See chart F5 in the Excel-Ref for data tables plus original sources.

295 See, for example, 'The Planet's Most Destructive: The Climate Culprit 100'. https://medium.com/@climateculprits/the-planets-most-destructive-the-climate-culprit-100-79c59eba59d7.

296 C Melore, 'Baby boomers are the new climate change villains, study claims', Studyfinds.org, 26 March 2022. https://studyfinds.org/baby-boomers-climate-change/.

297 Joe Duggans's 'Is this how you feel' site: https://www.isthishowyoufeel.com/this-is-how-scientists-feel.

298 This letter from 2015 is also in the CN-Archive, Group 6 example p).

299 This letter from 2014 is also in the CN-Archive, Group 7 example hb).

300 This is well documented. Two examples I found to be particularly insightful are portrayed in the DVD 'The Nazis: A warning from history' (see link below). The first is revealed by the astonishment of US researchers who worked through the papers of a captured Gestapo regional HQ. They'd expected a huge staff and a population suppressed from the top. What they found was an incredibly small staff and a population that effectively ruled themselves through fear. Earnest letters of denouncement poured into the Gestapo HQ; with most of the job done for them agents merely netted up the worst 'offenders'.

The second is the self-motivation of local doctors regarding the killing of disabled children. Armed with sanction from Hitler obtained using the single letter from a father asking to euthanise his disabled baby, chief of the chancellery Philipp Bouhler instigated a pseudo-legal system requiring three doctors to fill in a form agreeing to the euthanizing of disabled babies. According to the DVD, over several years the system extended and evolved (lower thresholds, increased scope from babies to children) largely through actions of the doctors themselves. There was no further instruction from Hitler, and the doctors eventually dispensed with Bouhler's forms too. They simply decided themselves who to kill and put 'measles' or some such on the death certificate. They thought that they were cleansing the race and they wanted to please officials such as Bouhler, who wanted what he thought would please Hitler.

Within visionary cultural systems that dominate or override the law, schemes with radically different morality can self-establish at frightening speed.

- Link: The Nazis: A Warning from History. http://www.amazon.com/Nazis-A-Warning-History-The/dp/B00097DY66.

301 How long it takes for an institution to fall prey to cultural influence will depend on where the culture originated relative to the social institution. If a new culture first flourishes in some corner of a particular institution, say education, it is likely to spread relatively swiftly through the rest of the educational system. For instance, while having precursors in fringe politics and (mainly legal) academia, Critical Race Theory first appeared to achieve its current cultural form, along with a critical mass of emotive support, as it propagated among US educational theorists and teacher-training colleges, which was followed by the cultural capture of teacher's unions. Hence the nascent culture was largely 'inside' education from its early blossoming, and so was very well-placed to quickly inundate the whole institution, for which see 'The Critical Classroom', Chapter 11, 'The Credential Cartel—How Colleges of Education Spread Critical Race Theory': https://thf_media.s3.amazonaws.com/2022/2022_TheCriticalClassroom_FINAL_WEB.pdf.

302 For example, 120 top UK lawyers have recently refused to prosecute peaceful climate change protestors, or work on behalf of fossil fuel projects. See M Scott, 'Eco-cultist lawyers are undermining the rule of law'. *The Spectator*, 25 March 2023. https://www.spectator.co.uk/article/eco-cultist-lawyers-are-undermining-the-rule-of-law/.

303 The most famous historic case of this kind is the Scopes monkey trial, where the science of evolution was effectively on trial. However the cultural entity involved here (religion) was acting defensively as science encroached on its territory (the authority to say, and teach, where humans came from). For the two examples in the main text, the culture was the proactive party, challenging scientific reality. For the Scopes trial see: https://www.britannica.com/event/Scopes-Trial.

304 A Vaughan, 'World court should rule on climate science to quash sceptics, says Philippe Sands', *The Guardian*, 2015. https://www.theguardian.com/environment/2015/sep/18/world-court-should-rule-on-climate-science-quash-sceptics-philippe-sands.

305 P Sands, 'Climate Change and the Rule of Law: Adjudicating the future in international law', Lecture, 2015. http://web.archive.org/web/20190116004415/http://www.kcl.ac.uk/law/newsevents/climate-courts/assets/CLIMATE-CHANGE-INT-COURTS-17-Sept.pdf.

306 At time of writing, the bill is on hold after a challenge from the higher authority of the UK Government. See discussion at https://www.spectator.co.uk/article/nicola-sturgeons-bungled-gender-crusade-has-undermined-trans-rights/.

307 For instance, the parent who is (in February 2023) suing a school for mandating that her young child attend a Pride Parade. Pushed mainly by extreme trans rights culture (which has hi-jacked Pride) rather than gay rights interests, a great deal of highly questionable teaching practice that has spread throughout US and UK school systems, has only recently reached mainstream public attention, in part due to this culture attempting to change the law. Some of the resulting conflict will be enacted in the courts. See The Spectator: https://www.spectator.co.uk/article/playgrounds-are-no-place-for-pride-parades/.

308 General information on the Eugenics movement: https://www.sciencedirect.com/topics/computer-science/eugenics-movement.

309 See T Plaenkers, *Landscapes of the Chinese Soul: The Enduring Presence of the Cultural Revolution*, Routledge, 2018, Ch. 3. For a personal take on the Red Guard, see: https://www.amazon.com/Born-Red-Chronicle-Cultural-Revolution/dp/0804713693.

310 Brief overview of the Hitler Youth, from Sky History: https://www.history.com/news/how-the-hitler-youth-turned-a-generation-of-kids-into-nazis.

311 Most media coverage of the cost of renewables is based on unsubstantiated claims by industry, government or NGOs. However, data on the UK is highly transparent: financial accounts for all of the UK's offshore windfarms are in the public domain, as are those of dozens of onshore ones. A series of studies has exploited this hard data to show that onshore costs are high and rising, while offshore ones are even higher, but perhaps falling slightly. See:

- J Aldersey-Williams et al., 'Better estimates of LCOE from audited accounts – A new methodology with examples from United Kingdom offshore wind and CCGT', Energy Policy 2019; 128: 25–35.
- A Montford, *Offshore Wind: Cost predictions and cost outcomes*, Briefing 52, The Global Warming Policy Foundation. https://www.thegwpf.org/content/uploads/2021/02/Offshore-Wind-LCOE.pdf.
- A Montford, *The Rising Cost of Onshore Wind*, Briefing 59, The Global Warming Policy Foundation. https://www.thegwpf.org/content/uploads/2022/01/Onshore-Wind-LCOE.pdf.
- G Hughes, *Wind Power Economics, Rhetoric & Reality: Volume I, Wind Power Costs in the United Kingdom*. Renewable Energy Foundation, 2020.

312 *Planet of the Humans*. https://planetofthehumans.com/.

313 'Sri Lanka brings the importance of sustainable nitrogen management to the climate change discourse': http://web.archive.org/web/20211112173904/https://mfa.gov.lk/sl-nitrogen4netzero/.

314 The original News Nation World link for this, here: https://nationworldnews.com/spiral-food-crisis-hits-sri-lanka-as-shock-waves-from-fertilizer-ban/, seems to have mysteriously disappeared from its website. Articles before and after the date of this one (June 16th, 2022) are still present. Here is an archived version from the WayBackMachine: http://web.archive.org/web/20220816104836/https://nationworldnews.com/spiral-food-crisis-hits-sri-lanka-as-shock-waves-from-fertilizer-ban/.

315 M Livermore. *Burnt Offering: The biomess of biomass*. Report 37, The Global Warming Policy Foundation, 2019.

316 'Juliana, et al. v. United States of America, et al. is a climate-related lawsuit filed in 2015 by 21 youth plaintiffs against the United States and several executive branch officials': https://en.wikipedia.org/wiki/Juliana_v._United_States.

317 Via the BBC, January 2022. 'Climate change: Children going to court to force government action': https://www.bbc.co.uk/newsround/59829183.

> A group of children in Portugal are using a human rights law to force European politicians to tackle climate change – but they're not just taking action against their country but a total of 33 European countries.

318 Dawn King's play *The Trials* shows what the plaintiffs would like to happen. As an article in The Guardian explains:

> The play imagines a world a few decades into the future where a group of people are on trial, *Nuremberg-style*, for their culpability in the climate crisis. How many flights did they take? Did they eat meat? Sure, they recycled, but so what? *The penalties for exceeding personal carbon allowances are severe*; the jurors are played by teenagers who have inherited the mess. The defendants are clearly stand-ins for the rest of us, who have fiddled while Rome (and many other places) burned.'

- See: A Dickson, 'Twelve angry children: young jurors call adults to account for climate crisis in *The Trials*', *The Guardian*, 2022. https://www.theguardian.com/stage/2022/aug/04/climate-crisis-the-trials-dawn-king-play.

319 From a poll by McCain foods:

> Nearly half of parents have been eco-shamed by their kids — inspired by teen climate campaigner Greta Thunberg. And 57 per cent were angry at how adults responded to green issues, while 43 per cent of kids would go on a climate change protest.

Another article shows that issues are conflated too; waste plastic is perceived to some extent as a climate-change issue:

- C Jones, 'Greta's shame blame: Nearly half of parents have been eco-shamed by their kids — inspired by teen climate campaigner Greta Thunberg', *The Sun*, 2020. https://www.thesun.co.uk/uncategorized/11179440/parents-eco-shamed-children-greta-thunberg/.

320 Research has shown that children are pushing parents on the subject of global warming. See:

- S Guyoncourt, 'Children push parents to be more conscious about global warming', *The Independent*, 2022: https://inews.co.uk/news/climate-change-children-push-parents-global-warming-conscious-research-1382018.

321 This has been confirmed by formal study:

> Thunberg is not alone. Other young people can be equally convincing, according to a paper published May 6 in *Nature Climate Change*. The team of social scientists and ecologists from North Carolina State University who authored the report found that children can increase their parents' level of concern about climate change because, unlike adults, their views on the issue do not generally reflect any entrenched political ideology.

- L Denworth, 'Children change their parents' minds about climate change', *Scientific American* 2019. https://www.scientificamerican.com/article/children-change-their-parents-minds-about-climate-change/.

Denworth suggests the children are so persuasive because their views don't reflect ideology. This is exactly backwards; they are culturally (ideologically) convinced.

322 J Freedland, 'The school climate change strikes are inspiring – but they should shame us', *The Guardian* 2019. https://www.theguardian.com/commentisfree/2019/feb/15/young-climate-change-greta-thunberg-children.

323 F O'Toole, 'Shame on us for forcing children to wake us up to climate change', *Irish Times* 2019. https://www.irishtimes.com/opinion/fintan-o-toole-shame-on-us-for-forcing-children-to-wake-us-up-to-climate-change-1.3814034.

324 T Soutphommasane, 'Children are right to youth shame our leaders', *Sydney Morning Herald* 2019. https://www.smh.com.au/national/children-are-right-to-youth-shame-our-leaders-20190920-p52t8z.html.

325 A Lubrano, 'Climate change is not only worrying kids; it's making them angry', *The Toronto Star* 2020. https://www.thestar.com/life/2020/03/03/climate-change-is-not-only-worrying-kids-its-making-them-angry.html.

326 T Stuart, 'A new generation of activists is taking the lead on climate change', *Rolling Stone* 2019. https://www.rollingstone.com/politics/politics-features/climate-change-school-strike-825719/.

327 UN. 'Youth in action', https://www.un.org/en/climatechange/youth-in-action.

328 A Sanson and M Bellemo, 'Children and youth in the climate crisis', *BJPsych Bulletin* 2021; 45(4): 205–209.

329 E Salter, 'Children aren't the future: where have all the young climate activists gone?', *The Guardian*, 2022: https://www.theguardian.com/commentisfree/2022/jun/29/children-arent-the-future-where-have-all-the-young-climate-activists-gone.

330 A Montford and J Shade, *Climate Control, Brainwashing in Schools.* Report 14, The Global Warming Policy Foundation. https://www.thegwpf.org/content/uploads/2022/02/Climate-Education.pdf

331 FJ Baumgartner, *Longing for the End: A History of Millennialism in Western Civilization.* Palgrave MacMillan, 2001.

332 Babism, *Encylopaedia Iranica*: https://iranicaonline.org/articles/babism-index.

333 Summary list of millennarian movements within colonial societies (sampled September 2022): https://en.wikipedia.org/wiki/Millenarianism_in_colonial_societies.

334 JDY Peel and CC Stewart, *Popular Islam South of the Sahara.* Manchester University Press, 1986. p. 391.

335 The article at link NW-link below describes victimhood in so-called 'anti-racist' culture, and the article at TC-link does the same for extreme trans rights culture. For some authors, victimhood is a phenomenon in its own right (see for example CM-link below), whereas for this book, it is a characteristic of most cultural entities, especially millennarian ones.

- NW-link: AB Coleman. 'Progressives have fallen in love with black victimhood'. *Newsweek*, 20 April 2022. https://www.newsweek.com/progressives-have-fallen-love-black-victimhood-opinion-1699459.
- TC-link: H Joyce, 'Hard lessons in life'. *The Critic*, April 2023. https://thecritic.co.uk/issues/april-2023/hard-lessons-in-life/.
- CM-link: B Campbell and J Manning, *The Rise of Victimhood Culture: Microaggressions, Safe Spaces, and the New Culture Wars.* Palgrave Macmillan 2018.

336 See, for example, S Lazare, '"Colonizing the Atmosphere": How Rich, western nations drive the climate crisis'. *In These Times* blog, 2020: https://inthesetimes.com/article/climate-change-wealthy-western-nations-global-north-south-fires-west.

337 L Zimmerman, 'Our climate discourse is gradually normalizing an atrocity', *Current Affairs,* 2022. https://www.currentaffairs.org/2022/02/our-climate-discourse-is-gradually-normalizing-an-atrocity.

338 O Milman, 'Climate experts call for "dangerous" Michael Moore film to be taken down'. *The Guardian,* 2020: https://www.theguardian.com/environment/2020/apr/28/climate-dangerous-documentary-planet-of-the-humans-michael-moore-taken-down.

339 'Climate activists to White House: Ban fossil fuel extraction', MSNBC: http://www.msnbc.com/msnbc/climate-activists-white-house-ban-fossil-fuel-extraction.

340 M Bachelard, 'Some say cows are killing the earth. So do we need to ban beef?', *Sydney Morning Herald,* 2011. http://www.smh.com.au/environment/climate-change/some-say-cows-are-killing-the-earth-so-do-we-need-to-ban-beef-20110924-1kr2a.html.

341 L Hickman, 'James Lovelock: Humans are too stupid to prevent climate change', *The Guardian*, 2010: http://www.theguardian.com/science/2010/mar/29/james-lovelock-climate-change.

342 S Hayward, 'Climate change democracy deniers', *Powerline* blog 2015: http://www.powerlineblog.com/archives/2015/10/climate-change-democracy-deniers.php.

343 *Climate Etc* article, 'Adjudicating the future: silencing climate dissent via the courts': http://judithcurry.com/2015/10/11/adjudicating-the-future-silencing-climate-dissent-via-the-courts/.

344 Start at this blog-post on *The Law is my Oyster*: https://the-law-is-my-oyster.com/2015/11/21/in-violation-of-aarhus/. There's much more on this blog regarding the trampling of law by big green energy. Note, the blog appears to have gone dark; see it on the Wayback Machine here: http://web.archive.org/web/20211024190012/https://the-law-is-my-oyster.com/2015/11/21/in-violation-of-aarhus/.

345 *Climate Etc* article, 'Adjudicating the future: silencing climate dissent via the courts', featuring text from Donna LaFramboise and Robin Guenier: http://judithcurry.com/2015/10/11/adjudicating-the-future-silencing-climate-dissent-via-the-courts/.

346 E Pilkington, 'Put oil firm chiefs on trial, says leading climate change scientist' (the scientist being James Hansen), *The Guardian*, 2008. http://www.theguardian.com/environment/2008/jun/23/fossilfuels.climatechange.

347 L Zimmerman, 'Our climate discourse is gradually normalizing an atrocity', *Current Affairs* magazine, 2022: https://www.currentaffairs.org/2022/02/our-climate-discourse-is-gradually-normalizing-an-atrocity:

> Very occasionally, some writers have in fact registered the violence at the heart of such anodyne terms as 'global warming' and 'climate change'. Almost twenty years ago, journalist Ross Gelbspan wrote in *Boiling Point* that fossil fuel executives were 'criminals against humanity'. More recently, Kate Aronoff has argued that, because 'fossil-fuel executives are mass murderers', we 'should put them on trial for crimes against humanity'. Natasha Lennard has surveyed efforts to define 'ecocide' as an 'international crime, on a par with war crimes and genocide, prosecutable by the International Criminal Court'.

The Aronoff article referred to is: K Aronoff, 'It's time to try fossil-fuel executives for crimes against humanity', *Jacobin*, 2019. https://jacobin.com/2019/02/fossil-fuels-climate-change-crimes-against-humanity.

348 *Climate Etc* article, 'RICO!': http://judithcurry.com/2015/09/17/rico/.

349 J Curry, 'Carbon mandate: an account of collusion, cutting corners and costing Americans billions'. *Climate Etc* blog, 2015. http://judithcurry.com/2015/09/25/carbon-mandate-an-account-of-collusion-cutting-corners-and-costing-americans-billions/.

350 J Curry, 'Conflicts of interest in climate science'. *Climate Etc* blog, 2015. http://judithcurry.com/2015/02/25/Conflicts-of-interest-in-climate-science/.

351 Religiosity self-assessment figures from Gallup (G-link), as tabulated by Wiki (W-link).

- G-link: https://news.gallup.com/poll/142727/religiosity-highest-world-poorest-nations.aspx.
- W-link: https://en.wikipedia.org/wiki/Importance_of_religion_by_country.

352 Irreligiosity self-assessment figures from mainstream pollsters (see M-links) as tabulated by Wiki (W-link).

M-links:

- Win/Gallup, 'Religion prevails in the world' https://web.archive.org/web/20171114113506/http://www.wingia.com/web/files/news/370/file/370.pdf.
- Win/Gallup, 'Global Index of Religiosity and Atheism' https://web.archive.org/web/20131021065544/http://www.wingia.com/web/files/news/14/file/14.pdf.
- DentsuCommunicationInstitute,http://web.archive.org/web/20070223012306/https://www2.ttcn.ne.jp/~honkawa/9460.html.
- Win Gallup 'Losing our Religion? Two-thirds of people still claim to be religious'. https://www.gallup-international.bg/en/33531/losing-our-religion-two-thirds-of-people-still-claim-to-be-religious/.
- P Zuckerman, 'Atheism: Contemporary numbers and patterns', in M Martin (ed.), *The Cambridge Companion to Atheism*'. Cambridge University Press, 2007.
- W-link: https://en.wikipedia.org/wiki/List_of_countries_by_irreligion.

353 *Climate Etc* article, 'Can religiosity predict cultural climate beliefs'; https://judithcurry.com/2020/04/17/can-religiosity-predict-cultural-climate-beliefs/. Note; the measurements presented here are now much enhanced and better explained in Chapters 8–10 of this book. However, that Vietnam is non-conforming is not demonstrated in the book, but is within this post.

354 Pew Research Centre. 'The World's Muslims: Unity and diversity', (2012): https://www.pewforum.org/2012/08/09/the-worlds-muslims-unity-and-diversity-2-religious-commitment/.

355 M Cutting and M Walsh, 'Religiosity Scales: What are we measuring in whom?', *Archive for the Psychology of Religion* 2008; 30(1), 137–154.

356 https://web.archive.org/web/20171114113506/http:/www.wingia.com/web/files/news/370/file/370.pdf.

357 See sheet 'Relig DB and Renewables' in the Excel-Ref.

358 https://fore.yale.edu/Climate-Emergency/Climate-Change-Statements-from-World-Religions.

359 T Winright et al., 'Statement of Ethicists Without Borders: "Christian ethics, climate emergency and nonviolent direct action"'. *Daily Theology*, 2019. https://

dailytheology.org/2019/06/09/statement-of-ethicists-without-borders-christian-ethics-climate-emergency-and-nonviolent-direct-action/.

360 See CN Archive, Group 5.

361 'Archbishop Justin Welby on the Climate Emergency'. https://www.youtube.com/watch?v=AgsJE2UnGk4.

362 A Dutney et al., 'An open letter from Australian religious leaders', reproduced in: T Ormerod and N Ormerod, 'Religions speak with one voice on climate policy ', ABC News, 2013. https://www.abc.net.au/religion/religions-speak-with-one-voice-on-climate-policy/10099774.

363 From IPSOS. 'Two thirds of citizens around the world agree climate change is as serious a crisis as Coronavirus' (2020): https://www.ipsos.com/en/two-thirds-citizens-around-world-agree-climate-change-serious-crisis-coronavirus.

364 M Hall et al., 'Believing in climate change, but not behaving sustainably: Evidence from a one-year longitudinal study', *Journal of Environmental Psychology* 2018; 56: 55–62.

365 S Wilson, 'The efficacy of youth protests in Indonesia', from *Young Australians in International Affairs*, 2019: https://www.youngausint.org.au/post/the-efficacy-of-youth-protests-in-indonesia.

366 V Pandey, 'Youth protests sweep India', *The Deccan Chronicle*, 2019: https://www.deccanchronicle.com/nation/current-affairs/201219/youth-protests-sweep-india.html.

367 'The history of student protests in India', from *The Wire* (2020): https://thewire.in/politics/the-history-of-student-protests-in-india.

368 'India's young are the real heroes of the year', from *The Wire* (2019): https://thewire.in/rights/young-india-students-caa-protests.

369 Web address for original Extinction Rebellion 'local groups' map: https://rebellion.earth/act-now/local-groups/.

370 The huge 2015 UN 'My World' poll, now from archive as the UN deleted the original(!): http://web.archive.org/web/20190802231507/http://data.myworld2015.org/. And the 2019 YouGov international survey: https://yougov.co.uk/topics/science/articles-reports/2019/09/15/international-poll-most-expect-feel-impact-climate. The former has a huge list of nations, but unfortunately not Taiwan and Hong Kong from the latter.

371 Web address for reorganised Extinction Rebellion global membership map, from which the groups per nation can be determined: https://rebellion.global/.

372 The wind turbine list used for Section 12.3 is in the table 'Win Cap' on sheet 'Relig DB and Renewables', in the Excel-Ref. I realised later that picking nations from this list may introduce a little bias. The seven nations picked are random apart from their required religiosity, yet all 40 nations in this list deploy wind turbines, so have a minimum associated position on climate-change. But deliberately picking

outside a list of this type might hit group threshold problems, or at least more noisy data as the XR group numbers per nation get low. I stuck with my choice.

373 Children's Strikes Weekly for Climate website: https://www.fridaysforfuture.org/statistics/map.

374 B Kvaløy et al. 'The publics' concern for global warming: A cross-national study of 47 countries', *Journal of Peace Research* 2012, 49(1), 11–22.

375 B Tranter and K Booth, 'Scepticism in a changing climate: A cross-national study', *Global Environmental Change* 2015; 33: 154–164.

376 A McCright et al., 'Political ideology and views about climate change in the European Union', *Environmental Politics*, 25(2): 338–358.

377 A Low and A Chow, 'The relationship between climate change concern and national wealth', *Climatic Change* 2015; 131: 335–348.

378 MJ Hornsey et al., 'Meta-analyses of the determinants and outcomes of belief in climate change', *Nature Climate Change* 2016; 6: 622–626.

379 S Levi, 'Country-level conditions like prosperity, democracy, and regulatory culture predict individual climate change belief', *Nature Communications Earth & Environment* 2021; 2: 51.

380 O Kácha et al., 'Four Europes: Climate change beliefs and attitudes predict behavior and policy preferences using a latent class analysis on 23 countries', Preprint, Psyarchiv.: https://psyarxiv.com/rfz98/download.

Printed in Great Britain
by Amazon

25113318R00248